T0122450

Studies in Computational Intelligence

Volume 541

Series editor

Janusz Kacprzyk, Polish Academy of Sciences, Warsaw, Poland
e-mail: kacprzyk@ibspan.waw.pl

For further volumes:
http://www.springer.com/series/7092

About this Series

The series "Studies in Computational Intelligence" (SCI) publishes new developments and advances in the various areas of computational intelligence—quickly and with a high quality. The intent is to cover the theory, applications, and design methods of computational intelligence, as embedded in the fields of engineering, computer science, physics and life sciences, as well as the methodologies behind them. The series contains monographs, lecture notes and edited volumes in computational intelligence spanning the areas of neural networks, connectionist systems, genetic algorithms, evolutionary computation, artificial intelligence, cellular automata, self-organizing systems, soft computing, fuzzy systems, and hybrid intelligent systems. Of particular value to both the contributors and the readership are the short publication timeframe and the world-wide distribution, which enable both wide and rapid dissemination of research output.

Robert Bembenik · Łukasz Skonieczny
Henryk Rybiński · Marzena Kryszkiewicz
Marek Niezgódka
Editors

Intelligent Tools for Building a Scientific Information Platform: From Research to Implementation

Editors
Robert Bembenik
Łukasz Skonieczny
Henryk Rybiński
Marzena Kryszkiewicz
Faculty of Electronics and Information Technology
Institute of Computer Science
Warsaw University of Technology
Warsaw
Poland

Marek Niezgódka
Interdisciplinary Centre for Mathematical and Computational Modelling (ICM)
University of Warsaw
Warsaw
Poland

ISSN 1860-949X ISSN 1860-9503 (electronic)
ISBN 978-3-319-38239-5 ISBN 978-3-319-04714-0 (eBook)
DOI 10.1007/978-3-319-04714-0
Springer Cham Heidelberg New York Dordrecht London

Printed on acid-free paper

Springer is part of Springer Science+Business Media (www.springer.com)

Preface

This book is inspired by the SYNAT 2013 Workshop held in Warsaw, which was a forum for exchange of experience in the process of building a scientific information platform. It is a consecutive book of the series related to the SYNAT project, and it summarizes the workshop's results as well as captures the current work progress in the project. The previous volumes entitled "Intelligent Tools for Building a Scientific Information Platform" and "Intelligent Tools for Building a Scientific Information Platform: Advanced Architectures and Solutions," have been also published in Springer's Studies in Computational Intelligence.

SYNAT is a program that was initiated in 2010 and has been scheduled for a period of 3 years. It has been focused on meeting the challenges of global digital information revolution, especially in the context of scientific information. The final results of the program encompass two main areas:

- research on consistent strategies for the development of domain-specific systems of repositories that constitute a basis for future-oriented activities in the areas recognized as critical for the national knowledge infrastructure, inter alia, artificial intelligence, knowledge discovery and data mining, information retrieval, and natural language processing,
- development of an integrated ICT platform for operating the entire complex of knowledge resources (i) equipped with a comprehensive set of functionalities, among those a range of novel tools supporting enhanced forms of scholarly communication, (ii) allowing for two-way collaborations on various levels, breaking the barriers between creation and use, (iii) facilitating its interoperability with leading international knowledge infrastructures.

The papers included in this volume cover topics of diverse character, which is reflected in the arrangement of this book. The book consists of the following parts: Challenges, Research, Research Environments, and Implemented Systems. We will now outline the contents of the chapters.

Part I, "Challenges," deals with challenges facing information science and presents trends likely to shape it in the years to come as well as gives an outline of an informational model of open innovation.

- Bruno Jacobfeuerborn and Mieczysław Muraszkiewicz ("Some Challenges and Trends in Information Science") argue that contemporary information science is

a vivid discipline whose development is inspired and driven by many other disciplines such as theory of information, mathematics, computer science, psychology, sociology and social communications, librarianship, museology and archival sciences, linguistics, law, and cognitive sciences. In the chapter the authors briefly take a look at a collection of assorted phenomena, methodologies, and technologies that will have a significant impact on the scope and the ways information science will unfold in the coming years. The chapter provides the authors' understanding of information science as a dynamic discipline that extends its boundaries as new methodologies and technologies come along, as a result of scientific discoveries, engineering achievements, and emergence of new business models. It presents and elaborates on the challenges that constitute a framework within which new trends in information science have started appearing and most likely will shape the face of information science and its applications in the years to come.
- Bruno Jacobfeuerborn ("An Informational Model of Open Innovation") proposes an outline of an informational model of open innovation. This model can be a foundation for establishing an innovation facility by a production or service provider entity. The pillars of the model are: (i) access to and extensive use of up-to-date information, knowledge, and the best practices; (ii) the use of efficient and user friendly knowledge management systems based on advanced semantic tools; (iii) extensive use of ICT business intelligence tools, including Web 2.0 facilities and Web services, and social networks; (iv) a customer experience management system and customers' participation in identifying new products and/or value-added services (prosumerism); (v) close collaboration with academia and relevant nongovernmental organizations; and (vi) access to a collaborative work platform anytime and from anywhere, implemented as a cloud computing facility. It seems that the proposals included in the chapter can contribute to the research within new areas and trends in management, in particular with respect to modeling innovation processes and their evaluation.

Part II, "Research", is devoted to theoretical studies in the areas of artificial intelligence, machine learning, text processing, and knowledge engineering addressing the problems of implementing intelligent tools for building a scientific information platform.

- Marzena Kryszkiewicz and Bartłomiej Jańczak ("Basic Triangle Inequality Approach versus Metric VP-Tree and Projection in Determining Euclidean and Cosine Neighbors") discuss three approaches to efficient determination of nearest neighbors, namely using the triangle inequality when vectors are ordered with respect to their distances to one reference vector, using a metric VP-tree, and using a projection onto a dimension. The techniques are well suited to any distance metrics such as the Euclidean distance, but they cannot be directly used for searching nearest neighbors with respect to the cosine similarity. The authors provide an experimental comparison of the three techniques for determining nearest neighbors with regard to the Euclidean distance and the cosine similarity.

- Karol Draszawka, Julian Szymański and Henryk Krawczyk ("Towards Increasing Density of Relations in Category Graphs") proposes methods for identifying new associations between Wikipedia categories. The first method is based on Bag-of-Words (BOW) representation of Wikipedia articles. Using similarity of the articles belonging to different categories allows to calculate the information about categories' similarity. The second method is based on average scores given to categories while categorizing documents by dedicated score-based classifier.
- Hung Son Nguyen, Michał Meina, and Wojciech Świeboda ("Weight Learning in TRSM-based Information Retrieval") present a novel approach to keyword search in Information Retrieval based on Tolerance Rough Set Model (TRSM). Bag-of-Word representation of each document is extended by additional words that are enclosed into inverted index along with appropriate weights. The extension words are derived from different techniques (e.g., semantic information, word distribution, etc.) that are encapsulated in the model by a tolerance relation. Weight for structural extension are then assigned by unsupervised algorithm.
- Krzysztof Goczyła, Aleksander Waloszek, and Wojciech Waloszek ("An Analysis of Contextual Aspects of Conceptualization: A Case Study and Prospects") present a new approach to development of modularized knowledge bases. The authors argue that modularization should start from the very beginning of modeling, i.e., from the conceptualization stage. To make this feasible, they propose to exploit a context-oriented, semantic approach to modularization. This approach is based on the Structural Interpretation Model (SIM) presented earlier elsewhere. The first part of the chapter presents a contextualized version of the SYNAT ontology developed using the SIM methodology as well as a set of tools needed to enable a knowledge engineer to create, edit, store, and perform reasoning over contextualized ontologies in a flexible and natural way. The second part of the chapter gives a deeper insight into some aspects of using these tools, as well as into ideas underlying their construction. The work on contextualization of knowledge bases led the authors to further theoretical investigation of the hierarchical structure of a knowledge base system. Indeed, in a system of heterogeneous knowledge sources, each source (a knowledge base) can be seen in its own separate context, as being a part of a higher level contextual structure (a metastructure) with its own set of context parameters. The theoretical background of this conception is presented in the third part of the chapter.

Part III, "Research Environments," presents environments aimed at research purposes created within SYNAT. These environments include: Music Discovery and Recommendation System, Web Resource Acquisition System for Building Scientific Information Database, tools for log analysis (LogMiner, FEETS, ODM), Content Analysis System (CoAnSys), System for Fast Text Search and Document Comparison, Chrum—a tool for convenient generation of apache Oozie Workflows and PrOnto—A Local Search Engine for Digital Libraries.

- Bożena Kostek, Piotr Hofmann, Andrzej Kaczmarek, and Paweł Spaleniak ("Creating a Reliable Music Discovery and Recommendation System") discuss problems related to creating a reliable music discovery system. The SYNAT database that contains audio files is used for the purpose of experiments. The files are divided into 22 classes corresponding to music genres with different cardinality. Of utmost importance for a reliable music recommendation system are the assignment of audio files to their appropriate genres and optimum parameterization for music-genre recognition. Hence, the starting point is audio file filtering, which can only be done automatically, but to a limited extent, when based on low-level signal processing features. Therefore, a variety of parameterization techniques are briefly reviewed in the context of their suitability to music retrieval from a large music database. In addition, some significant problems related to choosing an excerpt of audio file for an acoustic analysis and parameterization are pointed out. Then, experiments showing results of searching for songs that bear the greatest resemblance to the song in a given query are presented. In this way a music recommendation system may be created that enables to retrieve songs that are similar to each other in terms of their low-level feature description and genre inclusion. The experiments performed also provide the basis for more general observations and conclusions.
- Tomasz Adamczyk and Piotr Andruszkiewicz ("Web Resource Acquisition System for Building Scientific Information Database") describe the architecture and findings made as an effect of integration of a complex resource acquisition system with a frontend system. Both underlying and frontend systems are mentioned briefly with reference to their root publications. The main accent has been put on data architecture, information processing, user interaction, obtained results, and possible future adaptation of the system.
- Janusz Sosnowski, Piotr Gawkowski, Krzysztof Cabaj, and Marcin Kubacki ("Analyzing Logs of the University Data Repository") claim that identification of execution anomalies is very important for the maintenance and performance refinement of computer systems. For this purpose they use system logs. These logs contain vast amounts of data, hence there is a great demand for techniques targeted at log analysis. The chapter presents authors' experience with monitoring event and performance logs related to data repository operation. Having collected representative data from the monitored systems the authors have developed original algorithms of log analysis and problem predictions that are based on various data mining approaches. These algorithms have been included in the implemented tools: LogMiner, FEETS, and ODM. Practical significance of the developed approaches has been illustrated with some examples of exploring data repository logs. To improve the accuracy of problem diagnostics the authors have developed supplementary log database which can be filled in by system administrators and users.
- Piotr Jan Dendek, Artur Czeczko, Mateusz Fedoryszak, Adam Kawa, Piotr Wendykier, and Łukasz Bolikowski ("Content Analysis of Scientific Articles in Apache Hadoop Ecosystem") describe algorithms currently implemented in CoAnSys (Content Analysis System), which is a research framework for mining

scientific publications using Apache Hadoop. The algorithms include classification, categorization, and citation matching of scientific publications. The size of the input data classifies these algorithms in the range of big data problems, which can be efficiently solved on Hadoop clusters.

- Maciej Wielgosz, Marcin Janiszewski, Marcin Pietroń, Pawel Russek, Ernest Jamro, and Kazimierz Wiatr ("Implementation of a System for Fast Text Search and Document Comparison") present an architecture of a system for fast text search and documents comparison with a main focus on N-gram-based algorithm and its parallel implementation. The algorithm which is one of several computational procedures implemented in the system is used to generate a fingerprint of analyzed documents as a set of hashes which represent the file. The work examines the performance of the system, both in terms of a file comparison quality and a fingerprint generation. Several tests were conducted of N-gram-based algorithm for Intel Xeon E5645, 2.40 GHz which show approximately 8x speedup of multi over single core implementation.
- Piotr Jan Dendek, Artur Czeczko, Mateusz Fedoryszak, Adam Kawa, Piotr Wendykier, and Łukasz Bolikowski ("Chrum: the Tool for Convenient Generation of Apache Oozie Workflows") argue that conducting a research in an efficient, repetitive, evaluable, but also convenient (interms of development) way has always been a challenge. To satisfy these requirements in a long term and simultaneously minimize costs of the software engineering process, one has to follow a certain set of guidelines. The article describes such guidelines based on the research environment called Content Analysis System (CoAnSys) created in the Center for Open Science (CeON). In addition to the best practices for working in the Apache Hadoop environment, the tool for convenient generation of Apache Oozie workflows is presented.
- Janusz Granat, Edward Klimasara, Anna Mościcka, Sylwia Paczuska, and Andrzej Piotr Wierzbicki ("PrOnto: A Local Search Engine for Digital Libraries") describe system PrOnto version 2.0 and results of work on this system in the SYNAT project. After the introduction, the chapter presents briefly the functionality of PrOnto that is a system of personalized search for information and knowledge in large text repositories. Further, the chapter presents elements of the personalized ontological profile of the user, the problem of finding similar concepts in many such profiles, and the issue of finding interesting documents in large text repositories, together with tests of the system and conclusions.

Part IV, "Implemented Systems," showcases production systems that were developed during the course of the SYNAT project.

- Andrzej Czyżewski, Adam Kupryjanow, and Janusz Cichowski ("Further Developments of the Online Sound Restoration System for Digital Library Applications") describe new signal processing algorithms that were introduced to the online service for audio restoration available at the web address: http://www.youarchive.net. Missing or distorted audio samples are estimated using a specific implementation of the Jannsen interpolation method. The algorithm is based on the autoregressive model (AR) combined with the iterative complementation of signal samples. The chapter is concluded with a presentation of experimental results of application of described algorithmic extensions to the online sound restoration service.

- Adam Dudczak, Michał Dudziński, Cezary Mazurek, and Piotr Smoczyk ("Overview of Virtual Transcription Laboratory Usage Scenarios and Architecture") outline findings from the final stage of development of the Virtual Transcription Laboratory (http://wlt.synat.pcss.pl, VTL) prototype. VTL is a crowdsourcing platform developed to support creation of the searchable representation of historic textual documents from Polish digital libraries. This chapter describes identified usage scenarios and shows how they were implemented in the data model and architecture of the prototype.
- Jakub Koperwas, Łukasz Skonieczny, Marek Kozłowski, Henryk Rybiński, and Wacław Struk ("University Knowledge Base: Two Years of Experience") summarize 2-years development and exploitation of the repository platform built at Warsaw University of Technology for the purpose of gathering University research knowledge. The implementation of the platform in the form of the advanced information system is discussed. New functionalities of the knowledge base are presented.
- Cezary Mazurek, Marcin Mielnicki, Aleksandra Nowak, Krzysztof Sielski, Maciej Stroiński, Marcin Werla, and Jan Węglarz ("CLEPSYDRA Data Aggregation and Enrichment Framework: Design, Implementation and Deployment in the PIONIER Network Digital Libraries Federation") describe the architecture for aggregation, processing, and provisioning of data from heterogeneous scientific information services. The implementation of this architecture was named CLEPSYDRA and was published as an open source project. This chapter contains an overview of the CLEPSYDRA system design and implementation. It also presents the test deployment of CLEPSYDRA for the purpose of the PIONIER Network Digital Libraries Federation, focusing on aspects such as agent-based data aggregation, data normalization, and data enrichment. Finally, the chapter includes several scenarios for the future use of the system in the national and international contexts.

This book could not have been completed without the involvement of many people. We would like to express our high appreciation to all the contributors and to thank the reviewers. We are grateful to Janusz Kacprzyk for encouraging us to prepare this book.

Warsaw, November 2013

Robert Bembenik
Łukasz Skonieczny
Henryk Rybiński
Marzena Kryszkiewicz
Marek Niezgódka

Contents

Part I
Challenges

Some Challenges and Trends in Information Science

Bruno Jacobfeuerborn and Mieczyslaw Muraszkiewicz

Abstract Contemporary information science is a vivid discipline whose development is inspired and driven by many other disciplines such as theory of information, mathematics, computer science, psychology, sociology and social communications, librarianship, museology and archival sciences, linguistics, law, and cognitive sciences. In this chapter we briefly take a look at a collection of assorted phenomena, methodologies and technologies that will have a significant impact on the scope and the ways information science will unfold in the coming years. The chapter will provide our understanding of information science as a dynamic discipline that extends its boundaries as new methodologies and technologies come along, as a result of scientific discoveries, engineering achievements and emergence of new business models. It will present and elaborate on the challenges that constitute a framework within which new trends in information science have started appearing and most likely will shape the face of information science and its applications in the years to come.

Keywords Information science · Information and communications technologies (ICT) · Challenges · Trends

1 Introduction

The world of today is determined by the world of tomorrow. However paradoxical such a statement may sound we are convinced that one of the distinguishing features of the contemporary people's mind-set is to see and organise the tasks and

B. Jacobfeuerborn (✉)
Deutsche Telekom AG, Bonn, Germany
e-mail: Bruno.Jacobfeuerborn@t-mobile.de

M. Muraszkiewicz (✉)
Institute of Computer Science, Warsaw University of Technology, Warszawa, Poland
e-mail: M.Muraszkiewicz@ii.pw.edu.pl; M.Muraszkiewicz@elka.pw.edu.pl

R. Bembenik et al. (eds.), *Intelligent Tools for Building a Scientific Information Platform: From Research to Implementation*, Studies in Computational Intelligence 541,
DOI: 10.1007/978-3-319-04714-0_1,

actions we undertake through the lens of hopes, plans, and expectations we consciously or unconsciously address to the future. Hence the slogan saying that *The future is now* is not only a nimble wording but also a legitimate key and guidance to understand and mould our present settings, initiatives and endeavours. In this chapter we briefly take a look at a collection of assorted phenomena, methodologies and technologies that, as we think, will have a particularly significant impact on the scope and the ways information science will unfold in the years to come. In the most general terms information science will have to address such great highlights and challenges as networking, intelligence/smartness, openness/transparency, property rights, uncertainty, innovativeness, entrepreneurship, and subjectivity, whose importance will tremendously increase when it comes to defining the bottom line and practices of information science in times to come.

At first glance this chapter could be classified as an attempt to forecast what the future of information science will look like. This is however not the case because of two reasons. Firstly, we realise the warning by Yogi Berra that *It's tough to make predictions, especially about the future* and we do not want to take a risk to mislead our readers, and secondly, the trends we shall describe have already their roots in the present and our only contribution is to argue (that is a safe bet) that they will become influential and significant in the process of information science advancements and unleashing its future epic developments.

In what follows the chapter will provide our understanding of information science as a dynamic discipline that extends its boundaries as new technologies for acquiring, storing, processing, transferring, retrieving, evaluating and making available information to its users and customers come along, mainly as a result of scientific discoveries, engineering achievements and emergence of new business models. Then we elaborate on the highlights and challenges that constitute a framework within which new trends in information science have already started taking place. The chapter will be completed by a brief list of arbitrarily assorted trends that will most likely shape the face of information science and its applications in the prospective years.

2 Information Science

Information science is an old discipline whose origins have a lot to do with the process of institutionalisation of science, which goes back to the Enlightenment, in particular to the British Royal Society and its first scientific journal, namely *Philosophical Transactions*, and to the Library Company of Philadelphia, the first public library in the U.S., set up by Benjamin Franklin. Since then the development of information science has closely accompanied the development of formal, natural and social sciences. However, in a sense, information science existed and operated in the shadow of these sciences, as a kind of meta-level observatory point and a facility to classify scientific publications or a supporting tool to document their outcomes and help researchers identify and reach these outcomes (e.g. Dewey decimal classification system). The advent of computers, high capacity data storages, and wide-area networks in the second part of the previous century, and a tremendous progress of information and communications technologies, including a spectacular power of the internet that now permeates and infiltrates almost all the domains of human activates have provided momentum to and dramatically changed the role and status of information science. Today, information science is a dynamic, substantive, well-established and recognised discipline with its university faculties, conferences, and journals. There are no doubts that information science as of today has definitely left away from the shadows of other disciplines and enjoys its autonomous, impartial and self-governing position in the realm of sciences.

Owing to its interdisciplinary nature information science is a special discipline that taps into methodologies, best practices and results of many other disciplines and theories among which are the theory of information, philosophy, logics and mathematics, computer science, psychology, sociology and social communications, library and archival sciences, museology, linguistics, law, cognitive sciences, and commerce and management, to mention only the most often quoted disciplines. On the one hand information science pursues efforts to better understand, define and represent its fundamental notions of information and knowledge that are at its core and underpin any research information scientists carry out. On the other hand, in practical terms, it deals among others with collecting, classifying, clustering, storing, retrieving, evaluating, aggregating and disseminating information on sciences or disciplines it takes into account by means of a wide range of media starting with a word of mouth, to catalogues, to printed and/or electronic newsletters and bulletins, to personalised information updates, to running specialised portals, and to bibliometric analysis and reports. This is how information science supports researchers, scientific communities, and also practitioners operating in administration, business, education, health and other domains of life. Information science has been developing its own methodology, subject to constant progress and change, which is focused on methods of knowledge acquisition, representation, and classification, on textual and multimedia objects retrieval, on semantic analysis of text, and on users' needs analysis, to mention just a few topics out of many. What characterises information science and makes it a fascinating area of research is its horizontal approach that traverses across an array of sciences borrowing inspirations, methodologies and tools from them. Thus information science while being focused on its own foundations, methodologies and specific problems can also tangibly contribute to the disciplines and domains it describes and/or supports.

By definition information is a central notion of information science and its understanding determines any further discussion on the status and possible evolution of this discipline. Conventionally, there are two generic approaches to look at and define information. The first approach is deliberately devoid of any effort to capture the meaning (semantics) of information and is only focused on a purely technical aspect of representing information as a sequence of bits that has to be efficiently stored in a memory and transmitted through a communication channel as quickly as possible with the least possible error rate. This approach was developed and studied by Claude Shannon and Warren Weaver [1]. Their concern was focused on how to ensure the reliability, efficiency and accuracy of the information transmission between the issuer and the receiver rather than on the meaning and value of information. They of course realised the limits of this approach in the situations being beyond the engineering problems of information storing and transmitting, and therefore came out with a Mathematical Theory of Communication to quantitatively estimate the value of information. In this theory, which is based on statistics and probability, they assumed that the value of information is inversely proportional to predictability of the message the information carries out. Insofar as the first approach is scientifically complete, well known and widely applied by computer and telecommunications engineers, the second one, i.e. the attempt to capture and valuate the meaning of information is still an on-going exercise, subject to various proposals and trials, including such aspects as availability, timeliness and completeness of information, the degree of relevance, precision and reliability of information, manageability of information, what is at stake as an outcome of acquiring and applying the information, the price of information, uniqueness or the next best substitute of the requested information, or ability to download information. An interesting summary of such aspects can be found in [2]. Incidentally, for a classic paper proposing an information value theory the reader is referred to [3].

Given the objective of this chapter our proposition to define information is based on [4]. Information is a quadruple of the following interlinked components:

I = < **IO**, **IC**, **II**, **IV** >

where: **IO** is the *information object* that is the thing that actually carries information (e.g. paper, a magnetic or optical disk); **IC** is the *information content* that is the intended meaning, the communication, the message that the issuer of the information wanted to record on the information object; **II** is the *information issuer*. This notion includes three sub-elements that are as follows: (i) the *information author* which can be a man or a machine that creates information content; (ii) the *information publisher,* which is the agent (person, device, organisation) that publishes the information; it may happen that the information author and information publisher is the same agent; (iii) the *information disseminator,* which is the agent (person, device, organisation) that takes care about the information dissemination in a given way and environment; it may well happen that the information publisher and disseminator and even the author may be the same agent; **IV** is the *information value* that is relative to its receiver who values the information content depending on his/her/it subjective criteria. It is important to note that the value of information can be assessed in cognitive, affective or behavioural terms. Cognitive value is related to knowledge, affective value is related to emotions, and behavioural value is related to skills that allow one to execute given tasks. As mentioned above measuring information value is difficult. Some proposal towards this end is given in [5].

Having defined information, now we present how we understand the notion of knowledge, still bearing in mind the purpose of this chapter. Following the approach proposed in [4] we formally define knowledge in two steps. Firstly, we say that we take into account a finite set composed of information to which we add a set of logical axioms (tautologies) and a set of allowed inference rules (e.g. deduction, induction, inference by analogy), and secondly, each information and adopted inference rules are subject to validation by an authoritative agency (e.g. communities of scientists, law makers). Note that this definition is open to an arbitrary decision on what is considered knowledge. It is entirely up to the authoritative agency to classify a given set of information along with accompanying inference rules as knowledge or not-knowledge. Perhaps at first sight this way of defining knowledge may appear striking, odd and even ill but after a moment of a reflection one has to admit that this is exactly how humans construct knowledge basing on their experience, research and laws of reasoning that, if organised and structured according to specific and detailed rigorous methodologies and procedures, we dub science.

3 Challenges and Trends

We do not intend to open any discussion on whether the Hegelian notion of *Zeitgeist* is a real thing or merely a handy metaphor to explain visible and invisible currents of thoughts, mems, lifestyle and fashions, preferences, habits, and even prejudices, which populate intellectual and emotional universes of human beings. For one thing, we welcome this metaphor for it allows us to establish the highlights and challenges that institute a framework within which we can identify new vital trends in information science, in particular these tendencies that will enhance its frontiers. What are then these highlights and challenges? No doubt there are many. We pick up those that we believe will entail a deep "reengineering" of information science. They are as follows: (i) networking; (ii) intelligence/smartness; (iii) a cluster consisting of openness, transparency, collaborativeness and shareness; (iv) a cluster comprising uncertainty and risk; (iv) intellectual property; (v) innovativeness; (vi) entrepreneurship; (vii) a cluster encompassing subjectivity and affectivity. Let us now comment on these items.

Networking
The inexorable march of the internet and subsequently of mobile technologies that have conquered almost all the facets of human activities has caused a paradigmatic change in terms of how we perceive, conceptualise and describe the world that surrounds us. Until recently we looked at the world through the systemic lens for the major notion to depict complex structures was the concept of *system*. This has changed and now the tool of choice to portray, explain and comprehend what we see and/or what we plan to design and implement is the notion of *network*. More on this significant alteration and a transformative shift caused by networking is said in [6]. Here let us only note that retackling well-known problems by means of the networking conceptual apparatus, which have originally their descriptions and solutions within the systemic approach, can bring innovative answers and

resolutions. For instance, the task of organising large innovation exercises that have conventionally been carried out by corporate research labs and/or governmental agencies and academia, now can be democratised by such phenomena as crowdsourcing or open innovation models. To conclude: it seem that the edifice of information science, as it is perceived by many of its researchers, scholars and practitioner, is still systemic, which we consider a serious obstacle limiting and hampering its potential and development. A broader application of the networking approach, especially towards establishing various research and development, and business ecosystems can indubitably enhance the horizon of information science and thereby make it more relevant to and useful for scientists, scholars, and other beneficiaries of its actual and prospective offer.

Intelligence
Arthur C. Clarke is the author of the saying: *Any sufficiently advanced technology is indistinguishable from magic*. No doubt this witty observation applies also to information and communication technologies, especially when they take advantage of artificial intelligence. Nowadays, we observe a fashion, or perhaps just a fad, in which before the name of artefacts is added the word "smart", for instance a smartphone, smart car, or smart building. People expect that machines and services will be more and more intelligent, that our environment will be infused with the so-called ambient intelligence provided by information and communication technologies along with various sensors and actuators, which are observant, sensitive and responsive to the presence and needs of humans. One of the latest topics discussed by the artificial intelligence community revolves around the so-called *intelligence explosion* [7, 8]. The main assumption of this concept is that a truly intelligent machine will be able to devise and manufacture even a more intelligent machine, i.e. an offspring will be more advanced than the parents, and the process of reproduction can be recursively continued. As a result sophisticated machines surpassing human intelligence will emerge and will be in everything, cohabiting with human beings. The issue is however that the social and moral values of that robust artificial intelligence will not necessarily be the same as the ones of humans. Already now information science has to take into account this fact and strengthen its focus on artificial intelligence methodologies, tools, and solutions taking into consideration ethical aspects of the intelligence explosion. As a memento comment to this recommendation we quote Irving J. Good, a distinguished researcher who worked at Bletchley Park on cryptography issues with Alan Turing: *Let an ultraintelligent machine be defined as a machine that can far surpass all the intellectual activities of any man however clever. Since the design of machines is one of these intellectual activities, an ultraintelligent machine could design even better machines; there would then unquestionably be an 'intelligence explosion' and the intelligence of man would be left far behind. Thus the first ultraintelligent machine is the last invention that man need ever make* [7]. We do not feel capable enough to elaborate any further on this statement; however, it seems that an ultraintelligent machine, how I. J. Good dubbed it, is already within a reach of contemporary computer science and engineering and that its appearance

and co-existence with humans may have an enormous impact on humans' *Weltanschauung* and daily life. We think that the above note on the intelligence explosion should be complemented by a comment on the attempts to work out a methodology to understand and describe the so-called *collective intelligence* and semantics of large entities and processes that occur in the network settings. In this respect let us point to the work by Pierre Levy and its IEML programme: *The IEML research program promotes a radical innovation in the notation and processing of semantics. IEML (Information Economy MetaLanguage) is a regular language that provides new methods for semantic interoperability, semantic navigation, collective categorization and self-referential collective intelligence* [9]. We believe that artificial intelligence and collective intelligence are two threads of paramount importance to be tackled by future information science, especially by studying their possible co-existence and interweaving.

Openness

The spirit of democracy has been haunting the world. Even those who do not practice democracy refer to it, mention it, and often label their actions, deeds and activities by means of this term. They do so since democracy is an attractive setting and facility that refers to basic human desires. Democracy assumes, among other features, openness and transparency. The users of information systems and information technologies in both democratic and non-democratic countries increasingly emphasise the need for openness of services, tools, and procedures. As practicing openness inevitably builds trust it consequently boosts collaborative work and readiness to share resources and assets. Noteworthy, attitudes that include openness are getting more and more common among researchers, teachers, students, free-lanced developers, and civil servants. A good example that illustrates and supports this claim is the movement of *open access* that conquers universities and research communities and forces scientific publishers to revise their policies [10]. Another example is the movement of *open data* that argues that certain types of data, mainly those that are generated by applications working at the service of public institutions, meaning financed by taxpayers, should be made available for free to everyone who wants to make use of them. This concept goes hand in hand with the doctrine of *open government* that assumes and promotes the idea that citizens of democratic countries have the right to access all the documents (that are not protected by other laws) and proceedings of the public administration entities in order to allow them to control the administration [11, 12]. While advocating for openness one has to clearly assert that openness should not exclude or reduce privacy. Practicing openness and protecting privacy are not contradictory in terms of one's policy. Even a glimpse at the present agenda of information science proves that openness is merely a tiny part of its methodology, what undoubtedly constraints its scope and coverage, unless openness will be much more heeded and regarded.

Uncertainty and Risk

Ulrich Beck in its seminal book [13] introduced the notion of "risk society" in order to reflect uncertainty and the changing nature of the conditions societies work and live in. Risk, the readiness to accept and deal with uncertainty or to avoid it, is for him a central category to depict the "mechanics" of contemporary society, its dynamics, relationships, conflicts, and challenges. Contemporary information science whose main methodological focus is on deterministic situations and processes has not fully adopted risk to its core agenda. Although some encouraging results to tame risk by means of the probability theory and statistics have already been achieved by information science, especially in decision-making theories, the lack of comprehensive and innovative approach to risk is certainly a serious drawback. A new and inspirational approach to uncertainty, randomness, and risk can be found in [14, 15].

Intellectual Property

Intellectual property, as opposed to the property of physical objects, is a form of property of intangible entities that have been produced by human minds through creative acts of imagination and/or intellectual efforts. The role of intellectual property and the role of intellectual property law and rights have significantly increased in the era of digital representation, storage, transmission, and dissemination of symbolic goods such as literary texts, music, patents, computer programmes, trademarks, etc. [16]. The recent controversies regarding the ACTA law (Anti-Counterfeiting Trade Agreement) clearly proved that this sphere is far from being regulated in a satisfactory manner and that reaching a consensus will be a tough task to accomplish. Another example of how difficult is the issue of intellectual property is provided by attempts to define a software patent, i.e. a patent whose subject is a computer programme; the European Patent Office does not accept computer programmes as it considers them not patentable by their very nature unless they cause "further technical effect". Note that an essential component of all symbolic goods are information and knowledge. Nowadays we can witness a struggle between those who want to migrate the existing (conventional) intellectual property law, perhaps with minor modifications, to the cyberspace, and those who opt for a deep rethinking and reengineering of the law in order to reflect a different "physics" of the cyberspace and different behavioural patterns of its inhabitants. Information science cannot neglect this issue and has to clearly express its stance for the aspect of intellectual property is one of the defining elements of its epistemic and pragmatic status. The readers interested in the issues of intellectual property in the digital age, including a librarian's point of view are referred to [17, 18], and those who want to learn how intellectual property rights can be engaged in privatising various forms of expression may get acquainted with [19].

Innovativeness

In the course of their evolution humans proved strongly committed to innovate and innovativeness. Discoveries, inventions and innovations have been the flywheel of technical, economic and societal progress. Although the real breakthroughs and revolutions have mainly been owed to discoveries and inventions, what sometimes dwarfed the role of innovations, its importance and impact cannot be neglected. This said we want to firmly emphasise that we distinguish between the three mentioned above notions. In [20] we characterised the difference between them as follows: *Discovery is an outcome of a fundamental scientific research even if it emerges as a result of serendipity (e.g. gravitation law, penicillin). Invention is about creating an artefact that has not existed so far (e.g. light-bulb, transistor). Innovation however is a result of purposeful and painstaking endeavour driven by a pre-defined objective, often undertaken in an organizational framework (ex. iPhone by Apple Inc.). Discoveries and inventions are hardly subject to management; they are like looking for a needle in a haystack. Innovativeness however can be planned and managed.* Today, organising innovativeness processes is a crucial task not only for business and corporate world, but above all for communities and nations since given the scarcity of raw resources and natural assets innovation becomes the major instrument to maintain and pursue economic and societal progress. For companies it is also one of the most effective tools to provide and uphold competitive cutting–edge, and for non-for-profit organisation to achieve and sustain a comparative advantage. On the other hand, in order to succeed modern innovativeness needs a strong information support that has to be provided in a well-organised scientific manner. These are serious reasons why information science has to include innovation and innovativeness into its pivotal agenda. Incidentally, an interesting approach how to boost innovativeness is the model of *open innovation* [21] in which innovation, as opposed to a classic model of walled garden, is organised as an open-shop environment, where internal ideas and proposals generated in a company are combined with the external ones of different provenances. In this model private companies collaborate with academia, non-governmental organisations and individuals and jointly pave the path to success.

Entrepreneurship

Interestingly enough Peter Drucker, the father of modern management and a founder of scientific approach to management, bounded innovation with entrepreneurship and wrote: *Innovation is the specific instrument of entrepreneurship* [22]. We do share this opinion that becomes especially valid in the context of a new type of entrepreneurship. In [23] we depicted the entrepreneurs of a new kind as internet savvy, highly media and information literate, knowledge-intensive, operating worldwide, self-reliant and self-driven, cooperative, risk taking, socially responsible and aware of environmental issues. Such entrepreneurs take a full advantage of the stationary and mobile internet and new business models, and in their daily makings and operations as well as in strategic planning they badly need informational and knowledge support. The present information science curricula does not offer a comprehensive approach specific to the profile and needs of this new class of entrepreneurs, including nomadic small and medium companies and individuals, whose members are counted in hundreds of millions and quickly grow in developed and developing countries. In [23] we made a few proposals regarding the information science curricula to take into account the needs of entrepreneurs of a new type, which among others include the proposals regarding innovation, crowdsourcing, open access and open data, social networking, and cloud computing.

Subjectivity and Affectivity
Information science, as other sciences, is legitimately founded on the paradigm of rationality and has a strong deterministic flavour. Rationality is mainly based on empirical experience, logic and causation, scrutiny, and Popperian falsification. Although it permits and includes statistical reasoning, it definitely excludes subjectivity and affectivity. But people are not “rational animals” as Aristotle dubbed our species. On the contrary, as we have learned from thorough artful and inventive research by Amos Tversky and Daniel Kahneman [24] humans when it comes to judgements and decision-making are hardly rational creatures, especially in the presence of uncertainty and emotional pressures. This is why it is so difficult to work out a consensus regarding a general definition of information value that would take into account semantics of information. Information science, should it want to address actual patterns and makings of various processes in which humans are involved, has to accept the fact that human reasoning is biased by emotions and hidden mental processes. This said, we argue that information science needs to find ways and means to tackle the issue of subjectivity and affectivity.

Above we established a draft framework of highlights and challenges information science has to face with in order to maintain and even increase momentum. These are strong mental, intellectual, and societal currents of global and civilizational dimensions, which have a robust transformative and amplification power. Now we should mention that within this framework there are some interesting new trends that we consider especially promising and representative to the present period of ideas fermentation and a transition from a classical way of treating and practicing information science to a new model of this discipline. Here are some of the trends: new media and social networking (e.g. Youtube, Facebook, Twitter); a bourgeoning interest in new/alternative forms of publication, evaluation and dissemination of the results of scientific works (e.g. open access solutions, digital libraries and repositories, virtual research environments, epistemic web); a mushrooming of non-institutional sources of information and education (e.g. wikis, research blogs, webinars, agorae of minds such as TED conferences, www.ted.com, Khan Academy, www.khanacademy.org, Coursera, www.coursera.org); advanced methods of acquiring knowledge (e.g. crowdsourcing, big data, text and data mining, mobile receivers such as drones or Google glasses); new self-adaptive voice and gesture user interfaces (e.g. the Kinect interface); new forms of collaboration (e.g. gamification), and—to close this incomplete list by grand ventures: coupling information science with biology and genetics (bioinformatics), and with neurosciences.

4 Final Remarks

A sharp-eyed reader has certainly noticed that networking was placed as the first item on our list of challenges. We did so since it is the most noticeable and prevalent phenomenon, an emblematic feature of today’s world. Networks are everywhere either visible or invisible; they permeate our life and activities. We have to confess however that we had the following dilemma: Is it networking or intelligence that should open the list? The reason we hesitated is well expressed by Eliezer S. Yudkovsky's opinion: *I would seriously argue that we are heading for the critical point of all human history. Modifying or improving the human brain, or building strong AI, is huge enough on its own. When you consider the intelligence explosion effect, the next few decades could determine the future of intelligent life. So this is probably the single most important issue in the world* [25]. Indeed, for a longer term the issue of intelligence, or to be more exact of artificial intelligence, is dramatically critical for the mankind; yet for the short term, networking seems to have the most significant and implicational impact on the overall organisation of human work and life. Incidentally, a real fuss or enthusiasm (depending on one's attitude) will start when both facilities, namely dense, widespread and ubiquitous networking and strong artificial intelligence synergistically join forces and give rise to a number of innovative solutions.

Let us conclude the chapter with a somewhat lofty note that information science finds itself now on a tipping point; it faces a few already identified challenges and probably even more challenges in times to come as new technologies, behavioural patterns and lifestyles will usher in science and quotidian life. This chapter was conceived as an entry point to a further discussion on identifying and addressing challenges and opportunities and on how today's information science should address them. Undoubtedly, the present decisions in this respect will determine the future scope and coverage of information science, and thereby its usefulness for researchers, scholars, teachers, students, and ordinary citizens.

Acknowledgments The authors wish to thank Professor Henryk Rybinski of Warsaw University of Technology and Professor Barbara Sosinska-Kalata of University of Warsaw for stimulating and inspiring discussions on the trends in information science and scientific information.

Also thanks are addressed to the reviewer who in order to provide a broader context of our discourse aptly suggested supplementing the Section of References with the following readings: (i) for the discussion on information and knowledge in Sect. 2—with publications [26–30]; (ii) for the discussion on intelligence in Sect. 3—with publication [31].

The National Centre for Research and Development (NCBiR) supported the work reported in this chapter' under Grant No. SP/I/1/77065/10 devoted to the Strategic Scientific Research and Experimental Development Program: "Interdisciplinary System for Interactive Scientific and Scientific-Technical Information".

References

1. Shannon, C.E., Weaver, W.: The mathematical theory of communication. University of Illinois Press, Urbana (1949)
2. Wang, R.Y., Strong, D.M.: Beyond accuracy: what data quality means to data consumers. J. Manage. Inf. Syst. **12**(4), 5–33 (1996)
3. Howard, R.A.: Information value theory. IEEE Trans. Syst. Sci. Cybern. (**SSC-2**), 22–26 (1966)

4. Jacobfeuerborn, B.: Reflections on data, information and knowledge. Studia Informatica, vol. 34, No. 2A(111), pp. 7–21, Silesian University of Technology Press, New York (2013)
5. Gilboa, I., Lehrer E.: The value of information - An axiomatic approach. J. Math. Econ. **20**, 443–459 (1991)
6. Muraszkiewicz, M.: Mobile network society. Dialog Universalism **14**(1–2), 113–124 (2004)
7. Good, I.J.: Speculations concerning the first ultraintelligent machine. In: Alt, F.L., Rubinoff, M. (eds.), Advances in Computers, vol. 6, pp. 31–88. Academic Press, New York, (1965)
8. Muehlhauser, L.: Facing the Intelligence Explosion. Machine Intelligence Research Institute (2013)
9. Levy, P.: From social computing to reflexive collective intelligence: the IEML research program. University of Ottawa (2009). http://www.ieml.org/IMG/pdf/2009-Levy-IEML.pdf. Accessed 2 June 2013
10. Suber, P.: Open access. The MIT Press, Cambridge (2012)
11. Lathrop, D., Ruma L.: Open government: collaboration, transparency, and participation in practice. O'Reilly Media, Sebastopol (2010)
12. Tauberer, J.: Open government data (2012). http://opengovdata.io/. Accessed 2 June 2013
13. Beck, U.: Risk society: towards a new modernity. SAGE Publications Ltd, London (1992)
14. Taleb, N.M: Fooled by randomness: The hidden role of chance in life and in the markets. Thomson/Texere, New York (2004)
15. Taleb, N.M: Antifragile: things that gain from disorder. Random house, New York (2012)
16. Jacobfeuerborn, B.: A netist scenario and symbolic goods consumerism, Miscellanea Informatologica Varsoviensia, SBP, 41–46 (2012)
17. Elkin-Koren, N., Salzberger, E.: The Law and economics of intellectual property in the digital age: the limits of analysis. Routledge, New York (2012)
18. Wherry, T.L.: Intellectual property: everything the digital-age librarian needs to know. American Library Assn Editions, Chicago (2009)
19. McLeod, K., Lessig, L.: Freedom of expression: resistance and repression in the age of intellectual property. University of Minnesota Press, Minneapolis (2007)
20. Jacobfeuerborn, B. Muraszkiewicz, M.: ICT Based information facilities to boost and sustain innovativeness. Studia Informatica, vol. 33, no 2B(106), pp. 485–496. Silesian University of Technology Press, Poland (2012)
21. Chesbrough, H.: Open innovation: the new imperative for creating and profiting from technology. Harvard Business School Press, Boston (2003)
22. Drucker, P.F.: Innovation and entrepreneurship. Harper and Row, New York (1985)
23. Jacobfeuerborn, B.: Information science agenda for supporting entrepreneurs of a new type. Miscellanea Informatologica Varsoviensia, SBP, vol. VI, 147–165 (2013)
24. Kahneman, D.: Thinking fast and slow. Farrar, Straus and Giroux, New York (2013)
25. Yudkovsky, E.S.: 5-Minute singularity intro. http://yudkowsky.net/singularity/intro. Accesses 2 June 2013
26. Bar-Hillel, Y., Carnap, R.: Semantic information. Br. J. Philos. Sci. **4**(14), 147–157 (1953)
27. Dretske, F.: Knowledge and the flow of information. Chicago University Press, Chicago (1981)
28. Barwise, K.J., Seligman, J.: Information flow: the logic of distributed systems. Cambridge University Press, Cambridge (1997)
29. Adriaans, P., van Benthem, J. (eds.): Philosophy of information. Handbook of Philosophy of Science, vol. 8. Elsevier, Amsterdam (2008)
30. Floridi, L.: The Philosophy of information. Oxford University Press, New York (2011)
31. T. Poggio, T., Smale, S.: The mathematics of learning: dealing with data. Notices AMS **50**(5), 537–544 (2003)

An Informational Model of Open Innovation

Bruno Jacobfeuerborn

Abstract The chapter proposes an outline of an informational model of open innovation. This model can be a foundation for establishing an innovation facility by a production or service provider entity. The pillars of the model are: (i) access to and extensive use of up-to-date information, knowledge, and best practices; (ii) the use of efficient and user friendly knowledge management systems based on advanced semantic tools; (iii) extensive use of ICT business intelligence tools, including web 2.0 facilities and web services, and social networks; (iv) a customer experience management system and customers' participation in identifying new products and/or value-added services (prosumerism); (v) close collaboration with academia and relevant non-governmental organizations; (vi) access to a collaborative work platform any-time and from anywhere, implemented as a cloud computing facility. It seems that the proposals included in this chapter can contribute to the research within new areas and trends in management, in particular with respect to modeling innovation processes and their evaluation.

Keywords Informational model · Innovation · Open innovation · ICT

1 Introduction

Today, for a company or institution to achieve competitiveness or comparative advantage in a crammed and widely competitive world without being innovative in terms of its internal processes and offered products and/or services is nearly impossible. Therefore, innovation has become a core element of survival and development strategies. It combines technology, management, business, and what is equally important, psychology; it is usually born in a multifaceted environment

B. Jacobfeuerborn (✉)
Deutsche Telekom AG, Bonn, Germany
e-mail: Bruno.Jacobfeuerborn@t-mobile.de; jbf.bruno@gmail.com

R. Bembenik et al. (eds.), *Intelligent Tools for Building a Scientific Information Platform: From Research to Implementation*, Studies in Computational Intelligence 541,
DOI: 10.1007/978-3-319-04714-0_2,

where the minds of technologists, managers, researchers, and marketing specialists meet and collaborate towards a common objective.

Despite a myriad of chapters, books, conferences, declarations, collections of best practices, and ordinary conversations innovation still remains an untamed concept that tends to spin out of control of practitioners and researchers. There is no satisfactory theory of innovation that has proved its general validity and practical applicability. There are but successful and unsuccessful cases of innovativeness that are hardly prone to generalizations and abstractions. This is because innovation and innovativeness are more an art than a craft, more a matter of ingenuity than rigorous procedures. Indeed, on the groundwork of any innovation lie such intangible elements as knowledge, creativity, and intellectual (sometimes also physical) prowess. However, in spite of its subjective ingredients innovativeness is a process that can be managed and accounted; it is not merely an act of creativity or a result of serendipity. Theodore Levitt, editor of the Harvard Business Review, was credited to say: "Creativity is thinking up new things. Innovation is doing new things." Incidentally, this is why although a lot of people are creative, yet they are not innovative, for the road from thinking to doing is certainly not straightforward. Usually, an innovation endeavor is framed by organized efforts and purposeful activities driven by a predefined goal. Moreover, more and more often innovation does not happen in walled gardens. Should a company or institution want to innovate in a systematic way, it needs to collaborate, exchange and share intellectual efforts with the external world, it has to open to and share the innovation cycle with other stakeholders.

The chapter proposes an outline of an informational model of open innovation. This model can be a foundation for establishing an innovation facility by a production or service provider entity. The pillars of the model are: (i) access to and extensive use of up-to-date information, knowledge, and best practices; (ii) the use of efficient and user friendly knowledge management systems based on advanced semantic tools; (iii) extensive use of ICT business intelligence tools, including web 2.0 facilities and web services, and social networks; (iv) a customer experience management system and customers' participation in identifying new products and/or value-added services (prosumerism); (v) close collaboration with academia and relevant non-governmental organizations; (vi) access to a collaborative work platform any-time and from anywhere, implemented as a cloud computing facility. In the course of the discourse we compare two approaches to innovation, i.e. closed and open innovation methodologies. It seems that the proposals included in this chapter can contribute to the research within new areas and trends in management, in particular with respect to modeling innovation processes and their evaluation.

2 Open Innovation

Classically, innovation is considered a bolster of comparative advantage, and therefore, work on innovative products or services is carefully concealed from actual and potential competitors. This is in a nutshell the philosophy of the Closed

Innovation paradigm within which a company maintains full control on an innovation project from its start, to research, to concept development and prototyping, to production, and to marketing and distribution. However, in an increasingly complex word and a growing sophistication of new products and services self-reliance and a walled-garden approach to innovation become a suboptimal strategy. These days less and less companies can rely on their own resources only in order to carry out innovation projects. Innovation requires a well-managed and focused mobilization of creative human resources, funds, and organizational and managerial measures. It is on the overall a resources and time-consuming process loaded with various types of risks. An operational risk is due to company's constraints and includes, among others, failure in meeting the specification of the product/service, failure in matching a final deadline delivery, or exceeding the allocated budget. The company is also exposed to a financial risk when the return on investment is less than planned. In addition, the company has to take into account a market and commercial risk when competitors have launched a competing product/service at the same time, or when prospective costumers abstain from purchasing the product/service; In [1] one can find striking innovation statistics that read: "(i) over 90 % of innovations fail before they reach the market; (ii) over 90 % of those innovations that do reach the market will also fail; (iii) over 90 % of innovations are delivered late, over-budget or to a lower quality than was originally planned." A solution to reduce the risks and courageously face the above mentioned statistics is to open the innovation cycle, and thereby to strengthen the capacity of the innovation team and to distribute the risks among the stakeholders. This was in the main the rationale behind the Open Innovation approach proposed in [2]. The Open Innovation scheme elaborated by Professor Henry Chesbrough combines indigenous and external concepts and ideas as well as company's own and external research facilities and resources, and paths and access points to market for advancing the development and establishment of new technologies, products and services. The comparison of both Closed and Open Innovation approaches is given in Table 1 [2].

By accepting the open approach to innovation one has instantly wider access to experienced top-notched experts and developers, to venture capital, to university research, to the networking potential of non-governmental organizations (NGO), and to the market access points. Agents seeking competitive advantages cannot neglect such gains.

Within the Open Innovation paradigm many business models are possible. In [3] we presented an open approach to innovation that was established by the Faculty of Electronics and Technology of Warsaw University of Technology, Polska Telefonia Cyfrowa (then rebranded to T-Mobile Polska), the Foundation of Mobile Open Society through wireless Technology operating in Central Europe, and Polish eMobility Technological Platform that associates over 40 mobile technology companies. The site where innovative projects have been executed by students and young researchers is the BRAMA Laboratory located at the Faculty, http://brama.elka.pw.edu.pl/home. The Lab is a success story of bridging industry, business, academia and NGOs since its start in the year of 2006 BRAMA has been a host of some 120 projects.

Table 1 Comparison of closed and open innovation

Closed innovation principles	Open innovation principles
The smart people in the field work for us	Not all the smart people in the field work for us. We need to work with smart people inside and outside the company
To profit from R&D, we must discover it, develop it, and ship it ourselves	External R&D can create significant value: internal R&D is needed to claim some portion of that value
If we discover it ourselves, we will get it to the market first	We don't have to originate the research to profit from it
The company that gets an innovation to the market first will win	Building a better business model is better than getting to the market first
If we create the most and the best ideas in the industry, we will win	If we make the best use of internal and external ideas, we will win
We should control our IP, so that our competitors don't profit from our ideas	We should profit from others' use of our IP, and we should buy others' IP whenever it advances our business model

3 An Informational Model of Open Innovation

Architecting models is an enterprise loaded with subjectivity, mainly expressed by a set of adopted assumptions regarding the modeled reality and the choice of attributes characterizing the modeled situation and entities.[1] The model we propose in this chapter is also constrained by our arbitrary assumptions and qualifies for what Professors S. Hawking and L. Mlodinow dubbed the model-dependent-realism principle, which reads *There may be different ways in which one could model the same physical situation, with each employing different fundamental elements and concepts* [4]. The readers interested in innovation models are referred to surveys [5, 6], and the books [7, 8]. Here, to establish our model we assume three major premises:

(a) The innovation model is focused on informational aspects of innovativeness.
(b) The concept of open innovation is embedded in the model.
(c) The model substantially and extensively exploits state-of-the-art ICT facilities.

In so far as premise (c) is self-explanatory in the digital age we live in, premises (a) and (b) require a justification. The reason why we confine the boarders of our model to informational aspects is twofold. First, we are convinced that out of the three canonical constituents of innovation mentioned at the outset of this chapter, i.e. knowledge, creativity, and prowess, knowledge is the most perceptible and measurable; and secondly, most of the innovation models one can find in literature

[1] Perhaps, it would be more appropriate to use the wording 'knowledge model of open innovation' instead of 'informational model of open innovation'; on the other hand side, it seems that the term 'informational' better resorts to common mindsets.

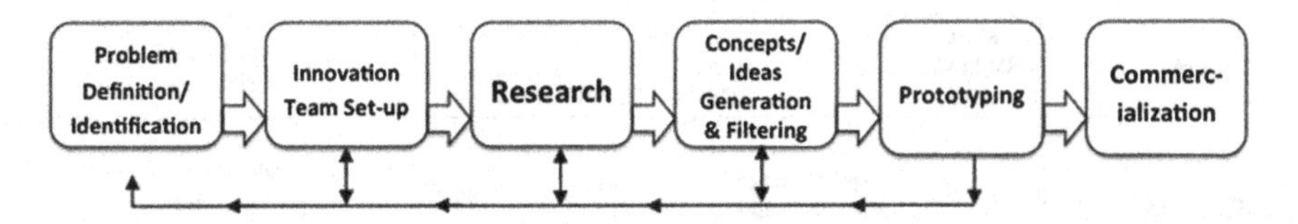

Fig. 1 A General scheme of innovation cycle

deal with processual aspects of innovation considering information and knowledge as given for granted. As to justifying premise (b) we are convinced that the paradigm of openness increasingly gains importance and recognition as a methodology to solve complex problems that require an engagement of many agents. This methodology is present in science, culture, lifestyle, and to a growing extent in economy. Its spectacular and successful incarnations are the FLOSS movement (Free/Libre/Open Source Software) with the most eminent and known instance of the Linux operating system, the Open Access model of scientific publishing that is practiced by leading universities and even to a certain degree by private publishers, and the famous case of Wikipedia in the area of producing high-quality content. Open Innovation attempts to tap into best practices of these experiences while trying to respect the constraints dictated by competition and market requirements.

Before we enter upon any further details of the informational model of innovation let us sketch the general scheme of innovation cycle. Its main stages are depicted in Fig. 1. After having identified and defined the subject to innovation efforts and setting up or hiring the execution team a research on the subject matter is commenced with the aim to collect as much as possible information, including IPR (Intellectual Property Rights), on the topic at hand. The collected information is an essential input to the next stage during which ideas and concepts are generated by the members of the execution team, and then filtered against various feasibility constraints determined by such factors as the state-of the-art of technology, business conditions, or resources available to the company. After a filtering scrutiny one or more of the generated ideas move to the phase of prototyping, and then after evaluation go to the hands of marketing specialist and are prepared for commercialization. Noteworthy, the innovation process includes loops because iterations, i.e. returns to previous stages, might be necessary as a result of assessments and decisions made during the work.

This fact is depicted in Fig. 1 by means of a feedback loop represented by a thin line with arrows indicating the direction of information flow. For the sake of comprehensiveness, as a supplement to our sketchy scheme of innovation, we quote a more elaborating model picked up from [8], which is displayed in Fig. 2. Note this model also features certain openness whose identification we left to the reader.

In [3] we proposed the concept of Innovation Factory where we drafted the core of a generic model of innovation, which prioritized the role of information and knowledge in organizing and executing an innovation cycle in high-tech

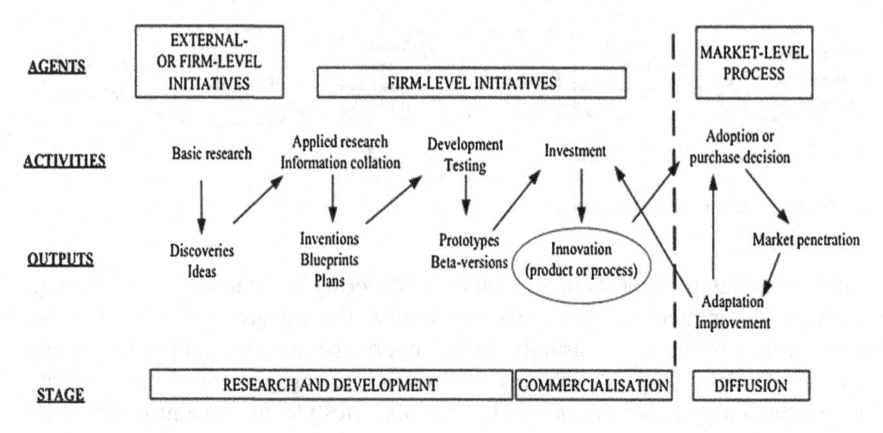

Fig. 2 Innovation process by Greenhalgh and Rogers [8]

Fig. 3 Knowledge framework of innovation

companies. Now, we add to that a Knowledge Framework that is schematically depicted in Fig. 3.

The sources of knowledge that supply the Innovation Factory are very much diversified and multifold. Company's 'Internal Expertise and Information Resources' are the first assets to tap into while carrying out an innovation project.

Part of internal information resources is 'Customer Relationship Management System' (CRM) that is already a classic and widely used tool, especially by larger companies, for working out and implementing company's customer-centered philosophy and marketing strategies. In [9] the following concise definition of CRM can be found: "CRM involves all of the corporate functions (marketing, manufacturing, customer services, field sales, and field service) required to contact customers directly or indirectly". A survey of CRM methodologies and a proposal of a CRM architecture focused on informational aspects can be found in [10]. In the year of 2003 Professor B. Schmitt came out with a more advanced approach to managing customers and maintaining links with them. He dubbed his approach 'Customer Relationship Management' (CEM) and defined it as "the process of strategically managing a customer's entire experience with a product or a company" [11]. In [12] the CEM concept was further elaborated as "a process-oriented satisfaction concept aimed at establishing wide and rich interactive relations with customers. It takes into account not only the product and functionality, but also lifestyle, aesthetics, and social aspects such as prestige, networking, etc. In order for CEM to work it has to be embedded in the company culture and be a unification platform of the company's business processes." Both CRM and CEM are tremendous sources of comprehensive information on company's customers, and therefore we include them in the Knowledge Framework of our innovation model. The same is true for knowledge that can be collected from the archives of the company 'Help Desk and Hot Line' that has a direct and constantly updated contact with actual and prospective customers who through their questions and problems provide incessant feedback information on the products and services of the company.

In our approach we emphasize the need for extensive use of information and knowledge, and that knowledge management systems are essential for feeding and managing the Knowledge Framework. This applies to both explicit and tacit knowledge and non-codified knowledge that are available both in the company and outside the company. Therefore, as far external sources of knowledge are concerned such classic methods as 'Business Intelligence' and 'Market Research' and experts' opinions and advice provide inputs to the Knowledge Framework. To collect information and knowledge from the outside world a company or an organization makes use of ICT business intelligence tools, including web 2.0/3.0 facilities and web services. In particular, useful are the tools based on artificial agents (small programs) whose task is to penetrate the web in order to search new information and update the information stored in the company's archive as well as the application of cloud computing platforms for efficient organization of collaboration among the stakeholders [13].

From [14] one can learn how knowledge can be extracted by means of advanced sematic tools and data and text mining techniques from social networks that are plenty throughout the web. 'Social Networking' systems such as Facebook, Grupon, MySpace, Allegro, Nasza Klasa or Twitter are included as vital and viable information sources to supply the Knowledge Framework, but also contributions via 'Crowdsourcing', the phenomenon identified and presented in [15] that taps into a collective intelligence of customers, suppliers, and a general public in order to collect creative, unbiased, and unexpected solutions is added as a knowledge and expertise source to the Innovation Factory. Official 'Statistics' regularly issued by national statistics bureaus and produced by various professional research and analytical agencies as well as NGOs may turn out really helpful in devising new products and services, especially when it comes to demographic data and data characterizing the standards and conditions of living.

Needless to argue about the importance of 'Patents' and 'Intellectual Property Rights (IPR)' in the process of working out innovative solutions. The analysis of patents can be an inspiring exercise guiding and boosting the process of generating ideas and concepts. Note that IPR is particularly a sensitive topic since any omissions in this respect can cause various undesirable implications such as court summons to face charges of infringing patents, financial penalties, and if worse comes to worst, can severely hamper the image of the company.

Our approach is based on the openness assumption. Therefore the Knowledge Framework includes as a source of knowledge 'External Experts and Consultants'. Owing to its mobility and involvement in many different projects throughout the world this folk is a natural carrier of the cutting-edge best practices and experiences, therefore, while applying a non-disclosure agreement hiring external specialists can bring up a valuable contribution to the innovation project. We also need to observe one of the latest tendencies to implicitly or explicitly involve some customers for identifying new products and/or value-added services or to get them involved in prototyping or evaluation, which transforms them into 'Prosumers'. Although the concept of prosumerism refers to the old works by Alvin Toffler done in the 1980s of the previous century, it only recently has gained popularity owing to the web 2.0 and progress in e-commerce and social networking. In our approach protagonists of the Knowledge Framework, which supply the Innovation Factory with knowledge, are also the stakeholders of the open innovation arrangement for according to our innovativeness model innovation is developed in open shops. Figure 4 presents schematically the mapping of sources of knowledge belonging to the Knowledge Framework of Factory of Innovation (see Fig. 3) onto the stages of the innovation process (see Fig. 1). The directed lines linking sources of knowledge to corresponding stages of the innovation process should be interpreted as informational contributions to the respective stages for supporting the activities performed within these stages.

4 Final Remarks

Undoubtedly, innovation has an idiom of its own yet, as already mentioned, in spite of a number of attempts to capture its very nature and dynamics these efforts are like the unavailing pursuit of the Holy Grail. It is well possible that we still are not able to ask the right questions with respect to innovativeness. Of the three basic elements

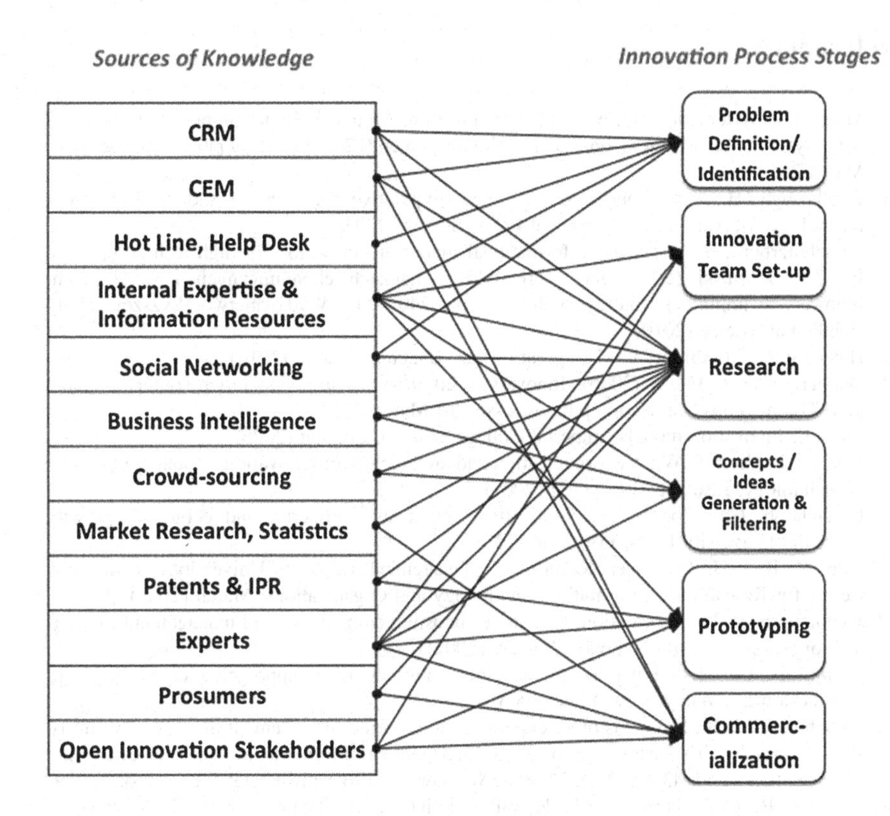

Fig. 4 Mapping of knowledge sources onto innovation process stages

underlying innovation, i.e. knowledge, creativity, and prowess we have a fairly good control of the first one, whereas the two latter ones are difficult to theorize and predict. This is why, for instance, we can hardly make proper use of incentives and understand the role of monetary rewards and their efficiency vis-à-vis other motivators such as satisfaction, team spirit, prestige, and self-improvement. Another example of an unsolved issue is where to host the Innovation Factory: should it be on the premises of the company integrated with its business processes or placed outside the company as an autonomous or even substantive entity. In [16] Professor C. Christensen provides interesting pros and cons arguments to address this issue. Notwithstanding, the more aspects and cases of innovation are recorded and analyzed the more likely is success while implementing business processes aimed at achieving innovative products or services. Let us note that the lack of a general theory of innovation and innovativeness that would be sound and pervasively applicable seems to account for false dilemmas and myths related to innovation itself and innovation endeavors, many of which were depicted in [17]. Whatever the future theory of innovation it has to contain the knowledge component and thoroughly elaborate on it. This chapter is an attempt towards this direction.

References

1. Mann, D.: Systematic Improvement & innovation. Myths & Opportunities (2009), http://www.systematicinnovation.com/download/management%20brief%2009.ppt. Accessed 12 March 2013
2. Chesbrough, H.: Open innovation: the new imperative for creating and profiting from technology. Harvard Business School Press, Boston (2003)
3. Jacobfeuerborn, B.: A generic informational model of innovation for high-tech companies. In: Z.E. Zieliński (ed.) Rola informatyki w naukach ekonomicznych i społecznych. Innowacje i implikacje interdyscyplinarne, pp. 240—248. Wydawnictwo Wyzszej Szkoly Handlowej, Kielce (2010)
4. Hawking, S., Mlodinow, L.: The great design. Bantam, London (2010)
5. Dubberly Design Office (2006). Innovation, http://www.dubberly.com/wp-content/uploads/2008/06/ddo_book_innovation.pdf. Accessed 12 March 2013
6. Tidd, J.: Innovation models. Imperial College London, London (2006)
7. Allen, T.J., Henn, G.W.: The organization and architecture of innovation. Managing the flow of technology. Elsevier, Burlington (2007)
8. Greenhalgh, Ch., Rogers, M.: Innovation, intellectual property, and economic growth. Princeton University Press, Princeton (2010)
9. Gray, P., Byun, J.: Customer Relationship Management, Report of University of California, Center for Research on Information Technology and Organizations, March (2001)
10. Jacobfeuerborn, B.: Information systems for boosting competence and management in high-tech organizations, MOST Press, Warsaw (2005)
11. Schmitt, B.: Customer experience management: a revolutionary approach to connecting with your customers. Wiley, New York (2003)
12. Jacobfeuerborn, B.: From customer experience to customer involvement management. In: B. Jacobfeuerborn (ed.) Customer Experience Management. Informational Approach to Driving User Centricity, pp. 43–46. MOST Press, Warsaw, IINiSB Uniwersytet Warszawski (2009)
13. Sosinsky, B.: Cloud computing bible, Wiley, Indianapolis (2011)
14. Russell, M.A.: Mining the social web: analyzing data from Facebook, Twitter, LinkedIn, and other social media sites. O'Reilly Media, Sebastapol (2011)
15. Surowiecki, J.: The wisdom of crowds. Anchor, New York (2005)
16. Christensen, C.: The innovator's dilemma, Collins Business Essentials New York (2006)
17. Berkun, S.: The Myths of innovation. O'Reilly Media, Sebastapol (2010)

Part II
Research

Basic Triangle Inequality Approach Versus Metric VP-Tree and Projection in Determining Euclidean and Cosine Neighbors

Marzena Kryszkiewicz and Bartłomiej Jańczak

Abstract The Euclidean distance and the cosine similarity are often applied for clustering or classifying objects or simply for determining most similar objects or nearest neighbors. In fact, the determination of nearest neighbors is typically a subtask of both clustering and classification. In this chapter, we discuss three principal approaches to efficient determination of nearest neighbors: namely, using the triangle inequality when vectors are ordered with respect to their distances to one reference vector, using a metric VP-tree and using a projection onto a dimension. Also, we discuss a combined application of a number of reference vectors and/or projections onto dimensions and compare two variants of VP-tree. The techniques are well suited to any distance metrics such as the Euclidean distance, but they cannot be directly used for searching nearest neighbors with respect to the cosine similarity. However, we have shown recently that the problem of determining a cosine similarity neighborhood can be transformed to the problem of determining a Euclidean neighborhood among normalized forms of original vectors. In this chapter, we provide an experimental comparison of the discussed techniques for determining nearest neighbors with regard to the Euclidean distance and the cosine similarity.

Keywords The triangle inequality · Projection onto dimension · VP-tree · The Euclidean distance · The cosine similarity · Nearest neighbors

M. Kryszkiewicz (✉) · B. Jańczak
Institute of Computer Science, Warsaw University of Technology,
Nowowiejska 15/19, 00-665 Warsaw, Poland
e-mail: mkr@ii.pw.edu.pl

B. Jańczak
e-mail: b.janczak@ii.pw.edu.pl

R. Bembenik et al. (eds.), *Intelligent Tools for Building a Scientific Information Platform: From Research to Implementation*, Studies in Computational Intelligence 541,
DOI: 10.1007/978-3-319-04714-0_3,

1 Introduction

The Euclidean distance and the cosine similarity are often applied for clustering or classifying objects or simply for determining most similar objects or nearest neighbors. In particular, the cosine similarity is popular when dealing with texts [9]. It should be noted that the task of searching nearest neighbors is a typical subtask of the tasks of classification and clustering [1, 5–8]. It is challenging if datasets are large and high dimensional. In this chapter, we discuss three principal approaches to efficient determination of nearest neighbors: namely, a basic triangle inequality approach that applies ordering of vectors with respect to their distances to one reference vector [5, 6], using a VP-tree [13], which is a particular metric tree [10, 12, 14], and using a projection onto a dimension [3]. Also, we discuss a combined application of a number of reference vectors and/or projections onto dimensions and compare two variants of VP-tree. All the techniques have been proposed to prune large numbers of objects that certainly are not nearest neighbors of a given vector. Conceptually, the simplest technique is the one based on the projection and most complex is the one using a VP-tree.

These techniques are well suited to any distance metrics such as the Euclidean distance, but cannot be directly used for searching nearest neighbors with respect to the cosine similarity. However, we have shown recently in [4] that the problem of determining a cosine similarity neighborhood can be transformed to the problem of determining a Euclidean neighborhood among normalized forms of original vectors. Thus the techniques can be indirectly applied in the case of the cosine similarity as well. Nevertheless, the question raises if the transformation influences the efficiency of the techniques.

In this chapter, we will provide an experimental evaluation of the discussed techniques for determining nearest neighbors with regard to the Euclidean distance. We will also examine experimentally if the transformation of the problem of the search of cosine similar nearest neighbors to the search of Euclidean nearest neighbors is beneficial from practical point of view and to which degree in the case of each candidate reduction technique. We will use five benchmark datasets in our experiments with a few thousand vectors up to a few hundred thousand vectors and with a few dimensions up to more than a hundred thousand dimensions.

Our chapter has the following layout. In Sect. 2, we recall basic notions and relations between the Euclidean distance and the cosine similarity. In Sect. 3, we recall definitions of neighborhoods and recall how the problem of determining a cosine similarity neighborhood can be transformed to the problem of determining a neighborhood with regard to the Euclidean distance. The usage of the triangle inequality by means of reference vectors for efficient pruning of non-nearest neighbors is recalled in Sect. 4. Section 5 is devoted to presentation of a VP-tree as a means of reducing candidates for nearest neighbors. We also notice there a difference in properties of two variants of a VP-tree: the one based on medians and the other one based on bounds. In Sect. 6, we present pruning of candidates by means of the projection of vectors onto a dimension. Here, we also consider a combined application of reference vectors and/or projections onto dimensions. In Sect. 7, we provide an experimental evaluation of the presented techniques on benchmark datasets. Section 8 summarizes our work.

2 The Euclidean Distance, the Cosine Similarity and Their Relation

In the chapter, we consider vectors of the same dimensionality, say n. A vector u will be also denoted as $[u_1, \ldots, u_n]$, where u_i is the value of the i-th dimension of u, $i = 1 \ldots n$. A vector will be called a *zero vector* if all its dimensions are equal zero. Otherwise, the vector will be called *non-zero*.

Vectors' similarity and dissimilarity can be defined in many ways. An important class of dissimilarity measures are distance metrics, which preserve the triangle inequality.

We say that a measure *dis preserves the triangle inequality* if for any vectors u, v, and r, $dis(u, r) \leq dis(u, v) + dis(v, r)$ or, alternatively $dis(u, v) \geq dis(u, r) - dis(v, r)$.

The most popular distance metric is the *Euclidean distance*. The *Euclidean distance* between vectors u and v is denoted by *Euclidean*(u, v) and is defined as follows:

$$Euclidean(u, v) = \sqrt{\sum\nolimits_{i=1\ldots n} (u_i - v_i)^2}$$

Among most popular similarity measures is the *cosine similarity*. The *cosine similarity* between vectors u and v is denoted by $cosSim(u, v)$ and is defined as the cosine of the angle between them; that is,

$$cosSim(u, v) = \frac{u \cdot v}{|u||v|}, \text{ where:}$$

- $u \cdot v$ is the *standard vector dot product of vectors u and v* and equals $\sum\nolimits_{i=1\ldots n} u_i v_i$;
- $|u|$ is *the length of vector u* and equals $\sqrt{u \cdot u}$.

In fact, the cosine similarity and the Euclidean distance are related by an equation:

Lemma 1 [4]. *Let u, v be non-zero vectors. Then*:

$$cosSim(u, v) = \frac{|u|^2 + |v|^2 - Euclidean^2(u, v)}{2|u||v|}.$$

Clearly, the cosine similarity between any vectors u and v depends solely on the angle between the vectors and does not depend on their lengths, hence the calculation of the $cosSim(u, v)$ may be carried out on their *normalized forms* defined as follows:

A *normalized form of a vector u* is denoted by *NF(u)* and is defined as the ratio of u to its length $|u|$. A vector u is defined as a *normalized vector* if $u = NF(u)$. Obviously, the length of a normalized vector equals 1.

Theorem 1 [4]. *Let u, v be non-zero vectors. Then*:

$$cosSim(u, v) = cosSim(NF(u), NF(v)) = \frac{2 - Euclidean^2(NF(u), NF(v))}{2}.$$

Theorem 1 allows deducing that checking whether the cosine similarity between any two vectors equals or exceeds a threshold ε, where $\varepsilon \in [-1, 1]$, can be carried out as checking if the Euclidean distance between the normalized forms of the vectors does not exceed the modified threshold $\varepsilon' = \sqrt{2 - 2\varepsilon}$:

Corollary 1 [4]. *Let u, v be non-zero vectors,* $\varepsilon \in [-1, 1]$ *and* $\varepsilon' = \sqrt{2 - 2\varepsilon}$. *Then*:

$$cosSim(u, v) \geq \varepsilon \text{ iff } Euclidean(NF(u), NF(v)) \leq \varepsilon'.$$

In the case of very high dimensional vectors with, say, tens or hundreds of thousands of dimensions, there may be a problem with a correct representation of the values of dimensions of normalized vectors, which may result in erroneous calculation of the Euclidean distance between them. One may avoid this problem by applying an α *normalized form of a vector u* that is defined as α *NF(u)*, where $\alpha \neq 0$. Large value of α mitigates the problem of calculating distances between normalized forms of high dimensional vectors.

Let $\alpha \neq 0$. One may easily note that: $Euclidean(NF(u), NF(v)) \leq \sqrt{2 - 2\varepsilon} \Leftrightarrow |\alpha|\ Euclidean(NF(u), NF(v)) \leq |\alpha| \sqrt{2 - 2\varepsilon} \Leftrightarrow Euclidean(\alpha\ NF(u), \alpha\ NF(v)) \leq |\alpha| \sqrt{2 - 2\varepsilon}$. As a result, we may conclude:

Corollary 2 [4]. *Let* $\alpha \neq 0$, *u, v be non-zero vectors,* $\varepsilon \in [-1, 1]$ *and* $\varepsilon' = |\alpha|\sqrt{2 - 2\varepsilon}$. *Then*:

$$cosSim(u, v) \geq \varepsilon \text{ iff } Euclidean(\alpha NF(u), \alpha NF(v)) \leq \varepsilon'.$$

3 ε-Neighborhoods and *k*-Nearest Neighbors

In this section, we first recall the definitions of an ε-Euclidean neighborhood, ε-cosine similarity neighborhood and the method of transforming the latter problem to the former one [4]. Next, we recall how to calculate k-Euclidean nearest neighbors based on an ε-Euclidean neighborhood. Finally, we formally state how to transform the problem of calculating k-cosine similarity nearest neighbors to the problem of calculating k-Euclidean nearest neighbors. In the next sections of the paper, we will assume that the determination of cosine similarity ε-neighborhoods and k-nearest neighbors is carried out as described in the current section; that is, by calculating the Euclidean distance between α normalized vectors instead of calculating the cosine similarity between original not necessarily α normalized vectors.

In the remainder of the paper, we assume that $\varepsilon \geq 0$ for the Euclidean distance and $\varepsilon \in (0, 1]$ for the cosine similarity if not otherwise stated.

3.1 Euclidean and Cosine ε-Neighborhoods

ε-Euclidean neighborhood of a vector p in D is denoted by $\varepsilon NB_{Euclidean}{}^{D}(p)$ and is defined as the set of all vectors in dataset $D \backslash \{p\}$ that are distant in the Euclidean sense from p by no more than ε; that is,

$$\varepsilon NB_{Euclidean}{}^{D}(p) = \{q \in D \backslash \{p\} | Euclidean(p, q) \leq \varepsilon\}.$$

ε-cosine similarity neighborhood of a vector p in D is denoted by $\varepsilon SNB_{cosSim}{}^{D}(p)$ and is defined as the set of all vectors in dataset $D \backslash \{p\}$ that are cosine similar to p by no less than ε; that is,

$$\varepsilon SNB_{cosSim}{}^{D}(p) = \{q \in D \backslash \{p\} | cosSim(p, q) \geq \varepsilon\}$$

Corollary 2 allows transforming the problem of determining a cosine similarity neighborhood of a given vector u within a set of vectors D (Problem P1) to the problem of determining a Euclidean neighborhood of α $NF(u)$ within the vector set D' consisting of α normalized forms of the vectors from D (Problem P2).

Theorem 2 [4]. *Let D be a set of m non-zero vectors* $\{p_{(1)}, \ldots, p_{(m)}\}$, $\alpha \neq 0$, *D′ be the set of m vectors* $\{u_{(1)}, \ldots, u_{(m)}\}$ *such that* $u_{(i)} = \alpha\, NF(p_{(i)})$, $i = 1 \ldots m$, $\varepsilon \in [-1, 1]$ *and* $\varepsilon' = |\alpha|\sqrt{2 - 2\varepsilon}$. *Then*:

$$\varepsilon SNB_{cosSim}{}^{D}\left(p_{(i)}\right) = \left\{p_{(j)} \in D | u_{(j)} \in \varepsilon' \text{-} NB_{Euclidean}{}^{D'}\left(u_{(i)}\right)\right\}.$$

Please note that all vectors considered in the resultant problem P2 have α normalized forms and so have length equal to α.

Example 1. Let us consider determination of an ε-cosine similarity neighborhood of a vector $p_{(i)}$ in dataset $D = \{p_{(1)}, \ldots, p_{(8)}\}$ from Fig. 1 for the cosine similarity threshold $\varepsilon = 0.9856$ (which roughly corresponds to the angle of 9.74°). Let $\alpha = 1$. Then, the problem can be transformed to the problem of determining ε'-Euclidean neighborhood of $u_{(i)} = NF(p_{(i)})$ in the set $D' = \{u_{(1)}, \ldots, u_{(8)}\}$ containing normalized forms of the vectors from D, provided $\varepsilon' = \sqrt{2 - 2\varepsilon} \approx 0.17$. The resultant set D' is presented in Fig. 2.

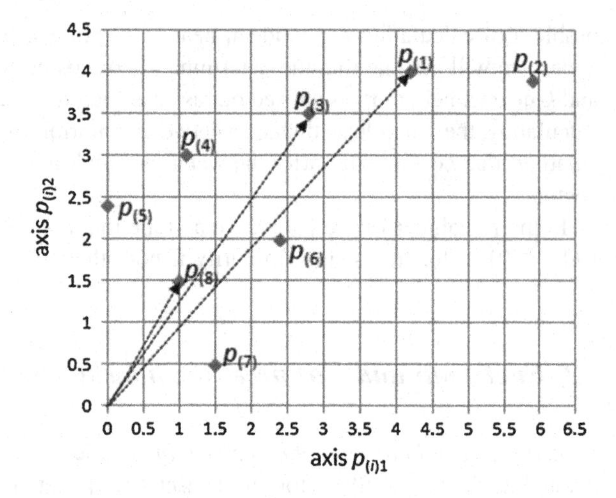

Fig. 1 Sample set D of vectors (after [4])

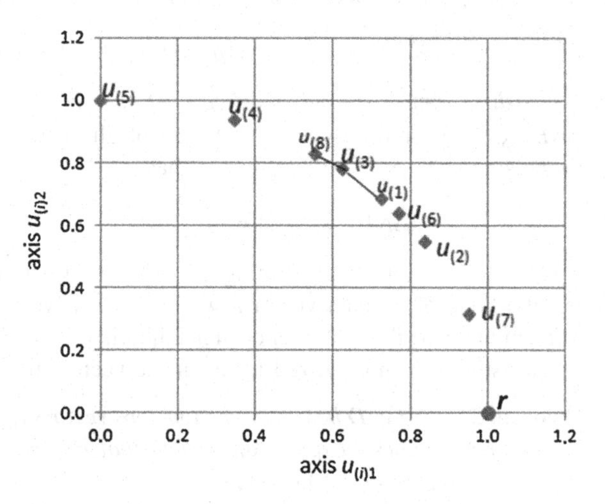

Fig. 2 Set D' containing normalized forms of vectors from D (after [4])

3.2 Euclidean and Cosine k-Nearest Neighbors

Instead of looking for an ε-Euclidean neighborhood (or an ε-cosine similarity neighborhood), one may be interested in determining *k-Euclidean nearest neighbors* (*k-cosine similarity nearest neighbors*, respectively). The task of searching k-Euclidean nearest neighbors can be still considered as searching an ε-Euclidean neighborhood for some ε value (possibly different for different vectors and adjusted dynamically) as follows:

Let K be a set containing any k vectors from $D \backslash \{p\}$ and $\varepsilon = \max\{Euclidean(p, q) | \, q \in K\}$. Then, k-Euclidean nearest neighbors are guaranteed to be found within ε Euclidean distance from vector p; that is, they are contained in

$\varepsilon\text{-}NB_{Euclidean}{}^{D}(p)$. In practice, one may apply some heuristics to determine possibly best value (that is, as little as possible) of ε within which k-nearest neighbors of p are guaranteed to be found and the value of ε can be re-estimated (and thus possibly narrowed) when calculating the distance between p and next vectors from $D \backslash (K \cup \{p\})$ [6].

The above approach to searching k-Euclidean nearest neighbors can be easily adapted to searching k-cosine similarity nearest neighbors based on the following observation.

Proposition 1. *Let D be a set of m non-zero vectors* $\{p_{(1)}, \ldots, p_{(m)}\}$, $\alpha \neq 0$, *D' be the set of m vectors* $\{u_{(1)}, \ldots, u_{(m)}\}$ *such that* $u_{(i)} = \alpha\, NF(p_{(i)})$, $i = 1 \ldots m$. *Then the following statements are equivalent*:

- $\{p_{(j1)}, \ldots, p_{(jk)}\}$ are k-cosine similarity nearest neighbors of $p_{(i)}$ in D;
- $\{u_{(j1)}, \ldots, u_{(jk)}\}$ are k-cosine similarity nearest neighbors of $u_{(i)}$ in D';
- $\{u_{(j1)}, \ldots, u_{(jk)}\}$ are k-Euclidean nearest neighbors of $u_{(i)}$ in D'.

Thus, the problem of determining k-cosine similarity nearest neighbors in a given set of vectors D is transformable to the problem of determining k-Euclidean nearest neighbors in set D' being the set of (α) normalized forms of vectors from D.

4 Basic Approach to Triangle Inequality-Based Search of Euclidean ε-Neighborhoods and Nearest Neighbors

4.1 Determining ε-Euclidean Neighborhoods with the Triangle Inequality

In this section, we recall the method of determining ε-Euclidean neighborhoods as proposed in [5]. We start with Lemma 2, which follows from the triangle inequality.

Lemma 2 [5]. *Let D be a set of vectors. Then, for any vectors u,* $v \in D$ *and any vector r, the following holds*:

$$Euclidean(u, r) - Euclidean(v, r) > \varepsilon \Rightarrow Euclidean(u, v) > \varepsilon \Rightarrow$$
$$v \notin \varepsilon NB_{Euclidean}{}^{D}(u) \wedge u \notin \varepsilon NB_{Euclidean}{}^{D}(v).$$

Now, let us consider vector q such that $Euclidean(q,r) > Euclidean(u,r)$. *If* $Euclidean(u,r) - Euclidean(v,r) > \varepsilon$, then also $Euclidean(q,r) - Euclidean(v,r) > \varepsilon$, and thus, one may conclude that $v \notin \varepsilon\text{-}NB_{Euclidean}{}^{D}(q)$ *and* $q \notin \varepsilon\text{-}NB_{Euclidean}{}^{D}(v)$ without calculating the real distance between q and v. This observation provides the intuition behind Theorem 3.

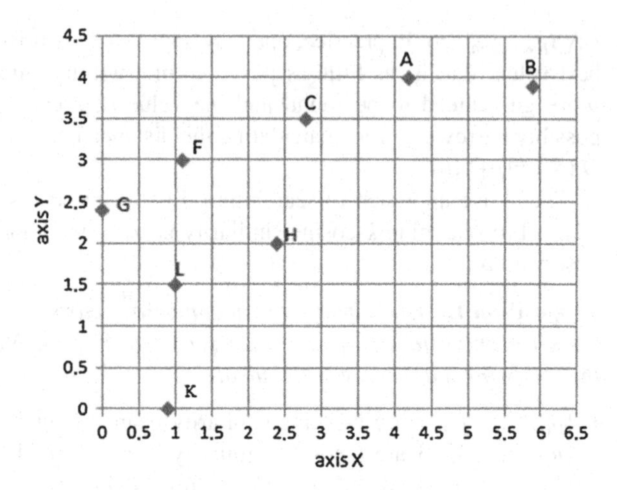

Fig. 3 Set of vectors D [5]

Theorem 3 [5]. *Let r be any vector and D be a set of vectors ordered in a non-decreasing way with regard to their distances to r. Let* $u \in D$, *f be a vector following vector u in D such that Euclidean(f,r) – Euclidean(u,r)* $> \varepsilon$, *and p be a vector preceding vector u in D such that Euclidean(u,r) – Euclidean(p,r)* $> \varepsilon$. *Then*:

(a) f and all vectors following f in D do not belong to $\varepsilon\text{-}NB_{Euclidean}{}^{D}(u)$;
(b) p and all vectors preceding p in D do not belong to $\varepsilon\text{-}NB_{Euclidean}{}^{D}(u)$.

As follows from Theorem 3, it makes sense to order all vectors in a given dataset D with regard to a reference vector r as this enables simple elimination of a potentially large subset of vectors that certainly do not belong to an ε-Euclidean neighborhood of an analyzed vector.

Example 2. (determining Euclidean ε-neighborhood with the triangle inequality). Let r be a vector [0, 0]. Figure 3 shows an example set D of two dimensional vectors. Table 1 illustrates the same set D ordered in a non-decreasing way with respect to the distances of its vectors to vector r. Let us consider the determination of the Euclidean ε-neighborhood of vector F, where $\varepsilon = 0.5$. We note that *Euclidean*(F,*r*) = 3.2, the first vector following F in D such that Euclidean(*f*,*r*) – *Euclidean*(*F*,*r*) $> \varepsilon$ is vector C (*Euclidean*(C,*r*) – *Euclidean*(F,*r*) = 4.5 – 3.2 = 1.3 $> \varepsilon$), and the first vector preceding F in D such that *Euclidean*(F,*r*) – *Euclidean*(*p*,*r*) $> \varepsilon$ is G (*Euclidean*(F,*r*) – *Euclidean*(G,*r*) = 3.2 – 2.4 = 0.8 $> \varepsilon$). By Theorem 3, neither C nor any vectors following C belong to $\varepsilon\text{-}NB_{Euclidean}{}^{D}$(F) as well as neither G nor any vectors preceding G belong to $\varepsilon\text{-}NB_{Euclidean}{}^{D}$(F). As a result, H is the only vector for which it is necessary to calculate its actual distance to F in order to determine $\varepsilon\text{-}NB_{Euclidean}{}^{D}$(F) properly.

In the sequel, a vector to which the distances of all vectors in D have been determined will be called a *reference vector*.

Table 1 Ordered set of vectors D from Fig. 3 with their distances to reference vector $r(0, 0)$

$q \in D$	X	Y	$Euclidean(q,r)$
K	0.9	0.0	0.9
L	1.0	1.5	1.8
G	0.0	2.4	2.4
H	2.4	2.0	3.1
F	**1.1**	**3.0**	**3.2**
C	2.8	3.5	4.5
A	4.2	4.0	5.8
B	5.9	3.9	7.1

In fact, one may use more than one reference vector for estimating the distance among pairs of vectors [5]. Additional reference vectors should be used only when the basic reference vector according to which the vectors in D are ordered is not sufficient to state that a given vector u does not belong to ε-neighborhood of another vector v. The estimation of the distance between u and v by means of an additional reference vector is based on Lemma 2. The actual distance between the two vectors u and v is calculated only when none of reference vectors is sufficient to state that $v \notin \varepsilon NB_{Euclidean}{}^{D}(u)$.

Let us note that the presented method assumes that an object u for which we wish to find its ε-neighborhood among vectors in a set D ordered with respect to vectors' Euclidean distances to a reference vector r also belongs to D. The method, however, can be easily adapted in the case when u does not belong to D. In such a case, it would be useful to calculate the Euclidean distance from u to reference vector r for determining the position in D where u could be inserted without violating the maintained order of vectors. Knowing this position of u, one may determine a respective vector p preceding u in D and a respective vector f following u in D as specified in Theorem 3. This would allow for the candidate reduction which follows from Theorem 3. Clearly, the determination of a respective position of u in D has logarithmic time complexity with regard to the number of vectors in D. In fact, the determination of this position does not require the real insertion of u into D. Let us also note that eventual using of additional reference vectors according to Lemma 2 may require the calculation of the Euclidean distance from u to some or all of these reference vectors.

4.2 Determining k-Euclidean Nearest Neighbors with the Triangle Inequality

In this section, we recall the method of determining k-Euclidean nearest neighbors, as introduced in [4, 6]. The method assumes that all vectors in a given dataset D are ordered with respect to some reference vector r. Then, for each vector u in D, its k-nearest neighbors can be determined in the following steps:

(1) The radius, say ε, within which k-nearest neighbors of u are guaranteed to be found is estimated based on the real distances of k vectors located directly before and/or after u in the ordered set D.[1]
(2) Next, $\varepsilon\text{-}NB_{Euclidean}{}^{D}(u)$ is determined in a way similar to the one described in Sect. 4.1. Clearly, the real distances to u from vectors considered in phase 1, do not need to be calculated again.
(3) k-nearest neighbors of u are determined as a subset of $\varepsilon\text{-}NB_{Euclidean}{}^{D}(u)$ found in step 2.

The above description is a bit simplified. In [6], steps 2 and 3 were not split, and the value of ε was adapted (narrowed) with each new candidate vector having a chance to belong to k nearest neighbors of u. Also, more than one reference vector can be used to potentially reduce the number of calculations of real distances between vectors [6].

5 Using VP-tree for Determining Euclidean ε-Neighborhoods and Nearest Neighbors

A VP-tree (Vantage Points tree) index was originally offered as a tool using the triangle inequality for determining a nearest neighbor of a vector u in a set of vectors D provided the distance between the nearest neighbor and u does not exceed a user-specified threshold value ε [13]. Thus, the problem consisted in searching one nearest neighbor of u in $\varepsilon\text{-}NB_{Euclidean}{}^{D}(u)$. Clearly, it can be easily adapted to searching all vectors in $\varepsilon\text{-}NB_{Euclidean}{}^{D}(u)$ as well as to searching any k nearest neighbors. In the latter case, the estimation of the ε radius within which k nearest neighbors are guaranteed to be found can be carried out as discussed in Sect. 3.2. In the remainder of this section, we focus on describing VP-tree in the context of determining ε-Euclidean neighborhoods.

VP-tree is a binary tree and its structure is as follows: The number of nodes in a VP-tree is equal to the cardinality of D. Each vector v (or its identifier) from D is stored in one node in the VP-tree altogether with the following information:

- *the median* $\mu(v)$ of the distances from v to all vectors stored in the subtree whose root is the node containing v; later on, we will denote this subtree by $S(v)$;
- *a reference to its left child node* being a root of the $LS(v)$ subtree that stores those vectors from $S(v)\backslash\{v\}$ whose distances to v are less than μ;
- *a reference to its right child node* being a root of the $RS(v)$ subtree that stores those vectors from $S(v)\backslash\{v\}$ whose distances to v are at least μ.

[1] Please note that one may estimate the radius ε within which k nearest neighbors of u are guaranteed to be found based on the real distances of any k vectors in set D that are different from u; restricting calculations to vectors located directly before and/or after u in the ordered set D is a heuristic, which is anticipated to lead to smaller values of ε.

Let us now illustrate the way in which ε-$NB_{Euclidean}{}^{D}(u)$ of a given vector u can be determined by means of a VP-tree. First, the distance is calculated between u and the vector stored in the root, say r, of the VP-tree. Let $\mu(r)$ be the value of the median stored in the root. The median is used to determine a'priori if any of the subtrees ($LS(r)$ or $RS(r)$) does not contain the sought neighbors of u. The decision about eventual ignoring any of the subtrees is made based on the following conditions:

(C1) $Euclidean(u,r) - \mu(r) \geq \varepsilon$. Then, for each vector v stored in $LS(r)$, the following holds: $Euclidean(u,r) - Euclidean(v,r) > \varepsilon$ and thus, by Lemma 2, $Euclidean(u,v) > \varepsilon$, so $LS(r)$ does not contain any nearest neighbor of u within the ε radius and should not be hence visited.

(C2) $\mu(r) - Euclidean(u,r) > \varepsilon$. Then, for each vector v stored in $RS(r)$, the following holds: $Euclidean(v,r) - Euclidean(u,r) > \varepsilon$ and thus, by Lemma 2, $Euclidean(u,v) > \varepsilon$, so $RS(r)$ does not contain any nearest neighbor of u within the ε radius and should not be hence visited.

Observation 1. Conditions C1 and C2 are mutually exclusive for $\varepsilon \geq 0$, so at most one subtree of $S(v)$ may be ignored while searching an ε-Euclidean neighborhood based on these conditions.

The presented procedure is repeated recursively for roots of subtrees that were not eliminated by the conditions C1, C2.

The structure and performance of a VP-tree strongly depends on the way vectors are assigned to nodes. As proposed in [13], each vector v in a node of VP-tree should imply the maximal variance among distances from v to all vectors in the subtree $S(v)$. As fulfillment of this condition would result in an unacceptably large time of building the tree, the authors of [13] decided to treat only a random sample of vectors as candidates to be assigned to a node and to determine corresponding medians and variances based on another random sample of vectors. For details of this procedure, please see [13].

As noted in [13], one may use a pair of values called *left_bound*(r) and *right_bound*(r), respectively, instead of the median, to potentially increase the usefulness of the conditions C1, C2. If $S(r)$ is a considered subtree, than *left_bound*(r) is defined as the maximum of the distances from r to all vectors in $LS(r)$, while *right_bound*(r) is defined as the minimum of the distances from r to all vectors in $RS(r)$. Clearly, $left\_bound(r) < \mu(r) \leq right\_bound(r)$ for any vector r. So, $Euclidean(u,r) - \mu(r) \geq \varepsilon$ implies $Euclidean(u,r) - left\_bound(r) > \varepsilon$ and $\mu(r) - Euclidean(u,r) > \varepsilon$ implies $right\_bound(r) - Euclidean(u,r) > \varepsilon$. In fact, the conditions C1 and C2 can be replaced by the following less restrictive conditions C1′ and C2′, respectively:

(C1′) $Euclidean(u,r) - left\_bound(r) > \varepsilon$ (then $LS(r)$ does not contain any nearest neighbor of u within the ε radius and should not be hence visited).

(C2′) $right\_bound(r) - Euclidean(u,r) > \varepsilon$ (then $RS(r)$ does not contain any nearest neighbor of u within the ε radius and should not be hence visited).

Observation 2. Conditions C1′ and C2′ are not guaranteed to be mutually exclusive for $\varepsilon \geq 0$, so it is possible that both subtrees of *S*(*r*) may be ignored while searching an ε-Euclidean neighborhood based on these conditions.

Example 3. Let $\varepsilon = 1$ and *r* be a vector represented by a node in a VP-tree such that *left_bound*(*r*) = 8.5 and *right_bound*(*r*) = 12. Let *u* be a vector such that *Euclidean*(*u*,*r*) = 10. Then, *Euclidean*(*u*,*r*) – *left_bound*(*r*) > ε and *right_bound*(*r*) – *Euclidean*(*u*, *r*) > ε, which means that neither *LS*(*r*) nor *RS*(*r*) contains any nearest neighbor of *r* within the ε radius.

Corollary 3. *Let u be any vector. Then, the determination ε-Euclidean neighborhood of u, where* $\varepsilon \geq 0$*, by means of a VP-tree*:

(a) Requires visiting at least one entire path in the VP-tree leading from its root to a leaf and calculating the Euclidean distances from *u* to all vectors stored in the nodes of this path when using conditions C1 and C2.
(b) May not require visiting any entire path in the VP-tree that ends in a leaf when using conditions C1′ and C2′.

So, if *u* is a vector for which we wish to determine its ε-Euclidean neighborhood, where $\varepsilon \geq 0$, then at least one leaf and all its ancestors in the VP-tree will be visited when applying conditions C1 and C2 (which involve medians), while it may happen that no leaf will be visited when applying conditions C1′ and C2′ (which involve left_bounds and right_bounds).

6 Using Vector Projection onto a Dimension for Determining Euclidean ε-Neighborhoods and Nearest Neighbors

In this section, we recall after [3] how to use vector projection onto a dimension to reduce the number of vectors that should be verified as candidates for ε-Euclidean neighborhoods or *k*-Euclidean nearest neighbors.

One may easily note that for any dimension *l*, $l \in [1, \ldots, n]$, and any two vectors *u* and *v*, $|u_l - v_l| = \sqrt{(u_l - v_l)^2} \leq \sqrt{\sum_{i=1\ldots n}(u_i - v_i)^2} = Euclidean(u, v)$. This implies Property 1.

Property 1. *Let l be an index of a dimension,* $l \in [1, \ldots, n]$*, and u, v be any vectors. Then*:

$$|u_l - v_l| > \varepsilon \Rightarrow Euclidean(u, v) > \varepsilon \Rightarrow$$
$$v \notin \varepsilon NB_{Euclidean}{}^{D}(u) \wedge u \notin \varepsilon NB_{Euclidean}{}^{D}(v).$$

Property 1 provides the intuition behind Theorem 4.

Theorem 4 [3]. *Let l be an index of a dimension, $l \in [1, \ldots, n]$, and D be a set of vectors ordered in a non-decreasing way with regard to the values of their l-th dimension. Let $u \in D$, f be a vector following vector u in D such that $f_l - u_l > \varepsilon$, and p be a vector preceding vector u in D such that $u_l - p_l > \varepsilon$. Then:*

(a) f and all vectors following f in D do not belong to $\varepsilon\text{-}NB_{Euclidean}{}^{D}(u)$;
(b) p and all vectors preceding p in D do not belong to $\varepsilon\text{-}NB_{Euclidean}{}^{D}(u)$.

Please note that Theorem 4 is analogical to Theorem 3, which is based on using the triangle inequality. Both theorems assume that vectors are ordered and use this ordering to ignore some false candidates for ε-Euclidean neighborhoods members. The only difference between the two theorems is that the ordering value for a vector v in Theorem 3 is calculated as the Euclidean distance between v and some fixed reference vector r, while in Theorem 4, the ordering value for v equals the value of some fixed l-th dimension of v.

Clearly, one may also use projections onto additional dimensions to reduce the number of candidates for ε-Euclidean neighborhoods. Projections onto additional dimensions should be used only when the basic dimension according to which the vectors in D are ordered is not sufficient to state that a given vector u does not belong to an ε-neighborhood of another vector v. The estimation of the distance between u and v by means of projection onto any additional dimension is based on Property 1.

In the sequel, we assume that the calculation of k-Euclidean nearest neighbors by means of projection onto a dimension is carried out in an analogical way as their determination by means of the triangle inequality, as described in Sect. 4.2.

Let us also note that one may use both reference vectors and projections onto dimensions for determining Euclidean ε-neighborhoods and k-nearest neighbors, irrespective of whether the set of vectors is ordered with regard to a basic reference vector or with regard to the value of a basic dimension. This approach to reducing candidate vectors was also evaluated by us and its results are provided in Sect. 7.

7 Experiments

In this section, we report the results of the experiments we have carried out on five benchmark datasets (see Sect. 7.1) in order to evaluate the presented techniques of reducing the number of candidates for k nearest neighbors for the Euclidean distance (see Sect. 7.2) as well as cosine similarity measure (see Sect. 7.3).

7.1 Description of Datasets Used in Experiments

In Table 2, we provide a short description of used benchmark datasets.

Table 2 Characteristics of used benchmark datasets

Name	No. of vectors	No. of dimensions	Source
Sequoia	62,556	2	[11]
Birch	100,000	2	[15]
Cup98	96,367	56	http://kdd.ics.uci.edu/databases/kddcup98/kddcup98.html
Covtype	581,012	55	http://ics.uci.edu/pub/machine-learning-databases/covtype/covtype.info
Sports	8,580	126,373	http://glaros.dtc.umn.edu/gkhome/cluto/cluto/download

We have tested how the representation of data influences the efficiency of searching similar vectors. It turned out that sparse datasets (*sports*) should be internally also represented as sparse, dense datasets (*sequoia*, *birch*, *cup98*) should be internally represented as dense or otherwise the calculation time is considerably longer [2]. The *covtype* dataset is neither typically sparse, not typically dense. For this dataset, it turned that the efficiency was more or less the same irrespectively of its internal representation [2].

7.2 Determining k-Euclidean Nearest Neighborhoods

In this section, we compare the performance of calculating k-Euclidean nearest neighbors for 10 % of vectors present in the examined datasets by means of the following algorithms[2]:

- TI$[r_1][r_2]\ldots[r_m]$—this variant requires vectors to be ordered with respect to their Euclidean distances to reference vector r_1. In order to reduce the number of candidates, the triangle inequality is used with respect to at most m reference vectors: the first reference vector r_1 is used according to Theorem 3; remaining specified reference vectors $r_2,\ldots, r_m$ (if any) are used when needed according to Lemma 2;
- Proj$(dim_1)(dim_2)\ldots(dim_m)$—this variant requires vectors to be ordered with respect to their values of dimension dim_1. In order to reduce the number of candidates, projections onto at most m dimensions are used. Projection onto dimension dim_1 is used in accordance with Theorem 4; remaining specified dimensions $dim_2,\ldots, dim_m$ (if any) are used when needed according to Property 1;

[2] In the case when more than one vector is in a same distance from a given vector u, there may be a number of alternative sets containing exactly k nearest neighbors of u. The algorithms we tested return all vectors that are no more distant than a most distant k-th Euclidean nearest neighbor of a given vector u. So, the number of returned neighbors of u may happen to be larger than k.

- TI[r_1][r_2]...[r_m]_Proj(dim_1)(dim_2)...(dim_s)—this variant requires vectors to be ordered with respect to their Euclidean distances to reference vector r_1. In order to reduce the number of candidates, reference vector r_1 is used according to Theorem 3, remaining specified reference vectors $r_2,\ldots, r_m$ (if any) are used when needed according to Lemma 2 and specified dimensions $dim_1,\ldots, dim_s$ are used when needed according to Property 1;
- Proj(dim_1)(dim_2)...(dim_m)_TI[r_1][r_2]...[r_s]—this variant requires vectors to be ordered with respect to their values of dimension dim_1. In order to reduce the number of candidates, projection onto dimension dim_1 is used in accordance with Theorem 4, remaining specified dimensions $dim_2,\ldots, dim_m$ (if any) are used when needed according to Property 1 and specified reference vectors $r_1,\ldots, r_s$ are used when needed according to Lemma 2;
- VP—this variant requires vectors to be stored in a VP-tree and uses medians to determine subtrees that should (not) be visited;
- [VP]—this variant requires vectors to be stored in a VP-tree and uses left_bounds and right_bounds to determine subtrees that should (not) be visited.

In order to underline that more than one reference vector is used by a TI algorithm, we will denote this algorithm interchangeably as TI+. Similarly, if more than one dimension for projecting is used, we will write Proj+ interchangeably with Proj.

All the algorithms which use reference vectors assume that the information about Euclidean distances from each vector in a source dataset to all reference vectors is pre-calculated. We need to mention, nevertheless, that in our experiments each vector u for which its neighbors were looked for played a double role: (i) a role of a vector belonging to the dataset D in which neighbors were searched (in this case, the information pre-calculated for u was treated by an algorithm as available) and (ii) a role of an external vector for which its neighbors were searched in D (in this case, the algorithm treated the information pre-calculated for u as unavailable and calculated it again if needed).

In the sequel, we will apply the following notation related to applied reference vectors and dimensions:

- [max]—a vector that has maximal domain values for all dimensions;
- [min]—a vector that has minimal domain values for all dimensions;
- [max_min]—a vector that has maximal domain values for all odd dimensions and minimal domain values for all even dimensions;
- [rand]—a vector that has randomly selected domain values for all dimensions;
- [rand] $\times$ m—a sequence of m vectors [r_1][r_2]...[r_m], where each vector r_i, where $i = 1\ldots m$, has randomly selected domain values for all dimensions;
- (dmax)—a dimension with most wide domain;
- (drand)—a random dimension;
- (l)—dimension l, where $l \in [1, \ldots, n]$.

Figures 4, 5, 6, 7 and 8 show in a logarithmic scale the average number of the Euclidean distance calculations per an evaluated vector for $k = 1, 2, 5, 10, 20$ and

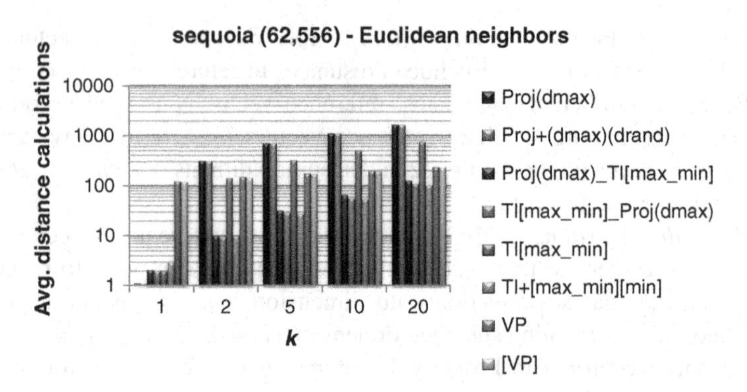

Fig. 4 The avg. no. of the Euclidean distance calculations when searching k-Euclidean nearest neighbors of a vector in 2-dimensional *sequoia* dataset of 62,556 vectors (log. scale)

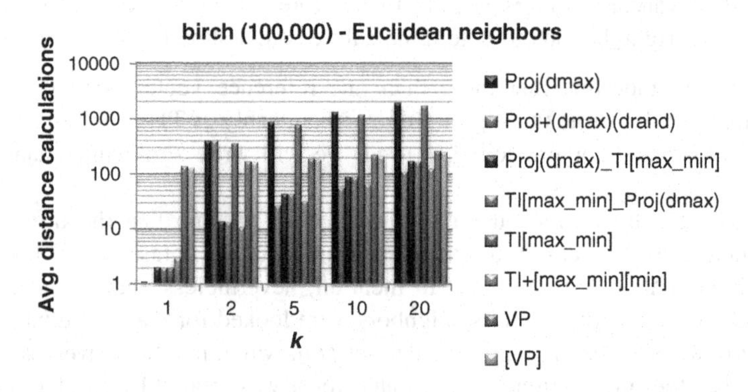

Fig. 5 The avg. no. of the Euclidean distance calculations when searching k-Euclidean nearest neighbors of a vector in 2-dimensional *birch* dataset of 100,000 vectors (log. scale)

respective datasets. In the case of the algorithms that do not use reference vectors, this number is equal to the average number of the candidate vectors in D to which the Euclidean distance from an evaluated vector was calculated. In the current implementation of the algorithms that use reference vectors, the Euclidean distances are calculated from an evaluated vector also to the all reference vectors. These additional Euclidean distance calculations are included in the presented statistics.

For all non-sparse datasets (Figs. 4, 5, 6 and 7), the application of each of the presented optimization techniques reduced the number of the Euclidean distance calculations by orders of magnitude in comparison with the number of vectors in a searched dataset, though to a different degree. For these datasets, it was always possible to determine a respective number of reference vectors that made TI algorithms more efficient than the variants using VP-tree for all tested values of k. In the case of sparse *sports* dataset, the usefulness of the optimization techniques turned out much lower (Figs. 8 and 9 present the experimental results for *sports* in a logarithmic scale and linear scale, respectively).

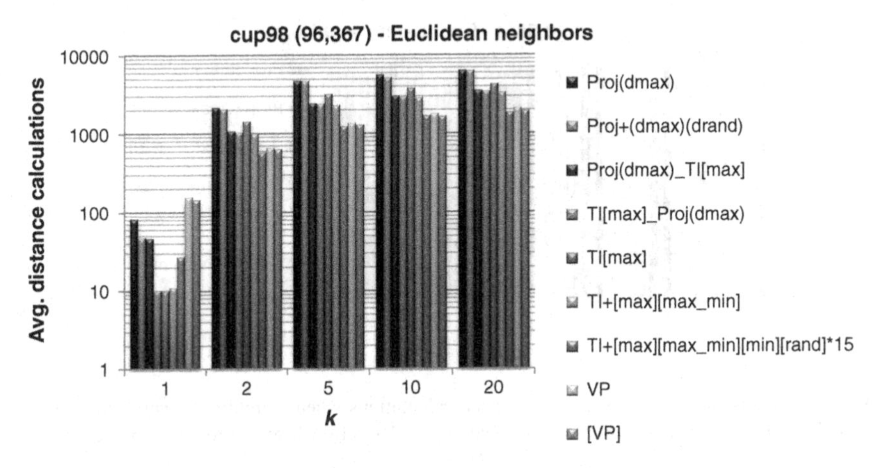

Fig. 6 The avg. no. of the Euclidean distance calculations when searching k-Euclidean nearest neighbors of a vector in 56-dimensional *cup98* dataset of 96,367 vectors (log. scale)

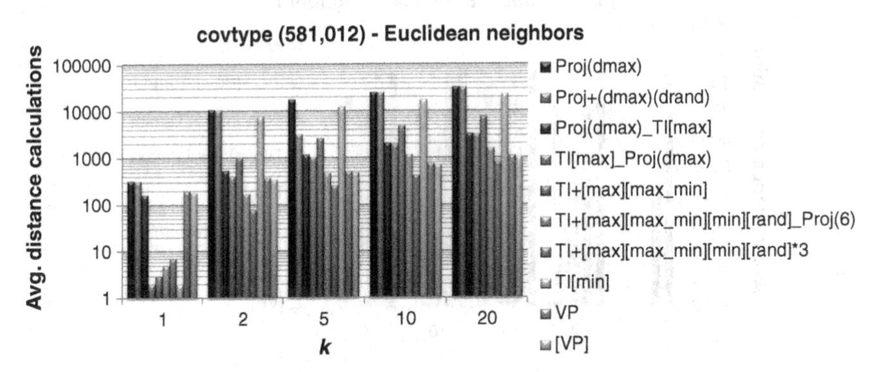

Fig. 7 The avg. no. of the Euclidean distance calculations when searching k-Euclidean nearest neighbors of a vector in 55-dimensional *covtype* dataset of 581,012 vectors (log. scale)

7.3 Determining k-Cosine Similarity Nearest Neighborhoods

In this section, we compare the performance of calculating k-cosine similarity nearest neighbors by equivalent calculation of k-Euclidean nearest neighbors among α normalized forms of original vectors, where $\alpha = 1000$.[3] We tested the same algorithms determining k-Euclidean nearest neighbors as in Sect. 7.2, but the calculations were carried out this time on α normalized forms of original vectors

[3] When applying $\alpha = 1$, the nearest neighbors happen to be incorrectly determined because of errors introduced during normalization of vectors. Hence, we decided to apply larger value of α.

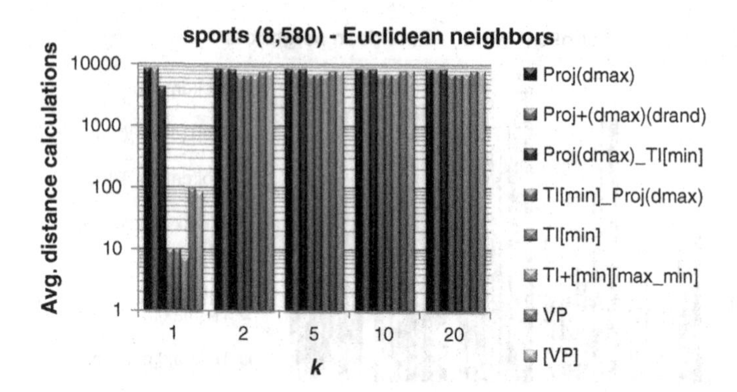

Fig. 8 The avg. no. of the Euclidean distance calculations when searching k-Euclidean nearest neighbors of a vector in 126,373-dimensional *sports* dataset of 8,580 vectors (log. scale)

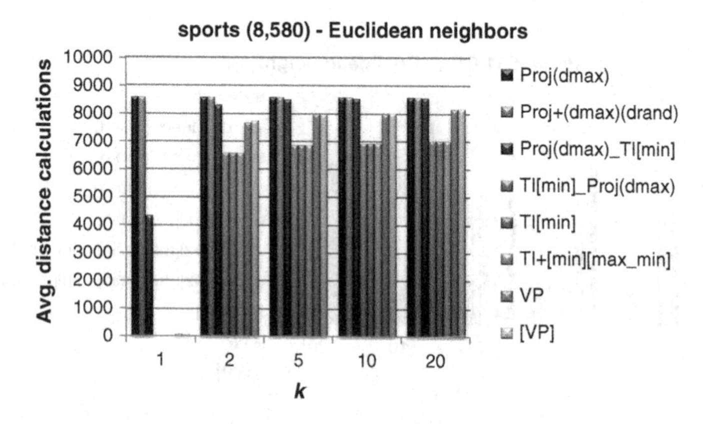

Fig. 9 The avg. no. of the Euclidean distance calculations when searching k-Euclidean nearest neighbors of a vector in 126,373-dimensional *sports* dataset of 8,580 vectors (linear scale)

instead of original vectors themselves and reference vectors were also chosen with respect to α normalized values rather than original ones.

Figures 10, 11, 12, 13 and 14 show in a logarithmic scale the average number of the Euclidean distance calculations per an evaluated vector for $k = 1, 2, 5, 10, 20$ and respective datasets. For all non-sparse datasets (Figs. 10, 11, 12 and 13), the application of each of the presented optimization techniques reduced the number of the Euclidean distance calculations by orders of magnitude in comparison with the number of vectors in a searched dataset, though to a different degree. For three datasets (*sequoia*, *birch*, *covtype*) out of four non-sparse datasets, it was always possible to determine a respective number of reference vectors that made TI algorithms more efficient than the variants using VP-tree. In the case of sparse

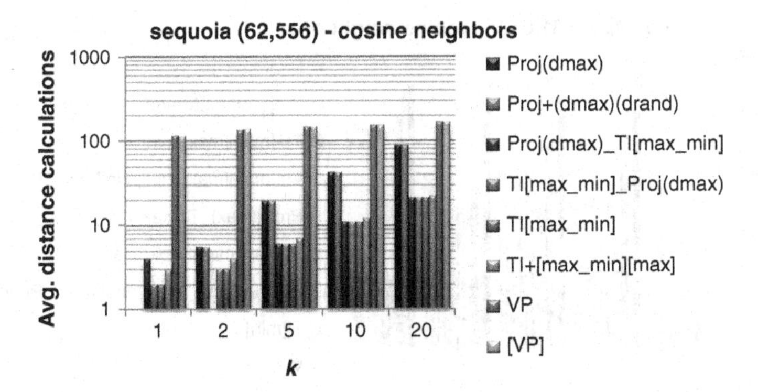

Fig. 10 The avg. no. of the Euclidean distance calculations when searching *k*-cosine nearest neighbors of a vector in 2-dimensional *sequoia* dataset of 62,556 vectors (log. scale)

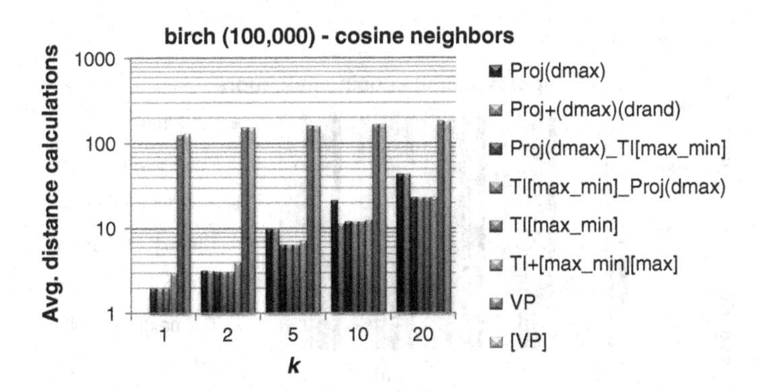

Fig. 11 The avg. no. of the Euclidean distance calculations when searching *k*-cosine nearest neighbors of a vector in 2-dimensional *birch* dataset of 100,000 vectors (log. scale)

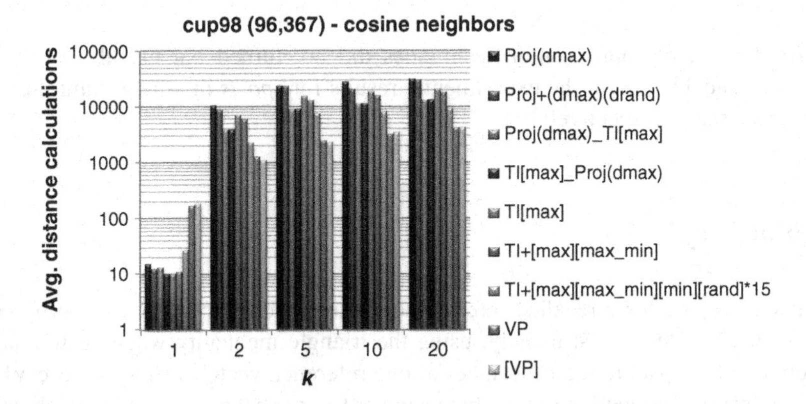

Fig. 12 The avg. no. of the Euclidean distance calculations when searching *k*-cosine nearest neighbors of a vector in 56-dimensional *cup98* dataset of 96,367 vectors (log. scale)

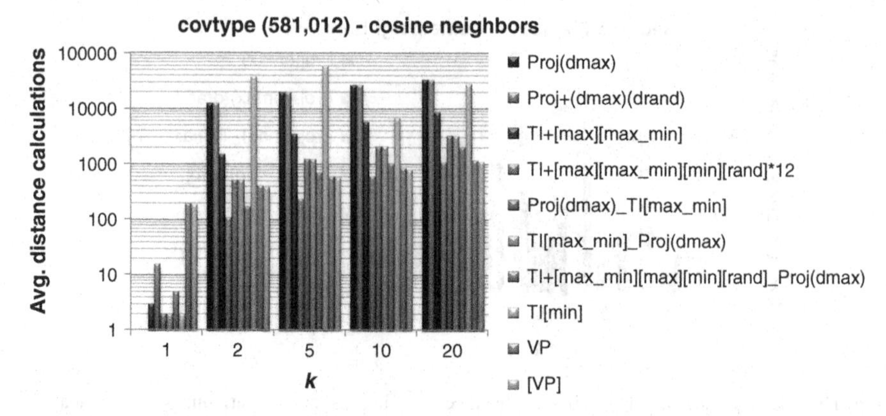

Fig. 13 The avg. no. of the Euclidean distance calculations when searching k-cosine nearest neighbors of a vector in 55-dimensional *covtype* dataset of 581,012 vectors (log. scale)

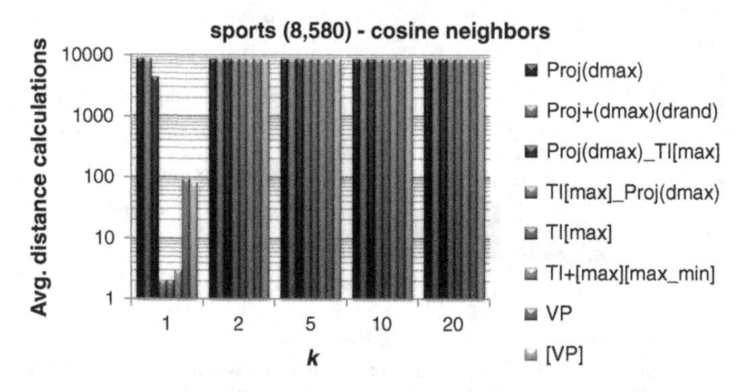

Fig. 14 The avg. no. of the Euclidean distance calculations when searching k-cosine nearest neighbors of a vector in 126,373-dimensional *sports* dataset of 8,580 vectors (log. scale)

sports dataset, the optimization techniques did not turned out useful for $k > 1$ (Figs. 14 and 15 present the experimental results for *sports* in a logarithmic scale and linear scale, respectively).

8 Summary

In this paper, we have recalled three principal approaches to efficient determination of nearest neighbors: namely, using the triangle inequality when vectors are ordered with respect to their distances to one reference vector, using a metric VP-tree and using a projection onto a dimension. Also, we have discussed a combined application of a number of reference vectors and/or projections onto dimensions

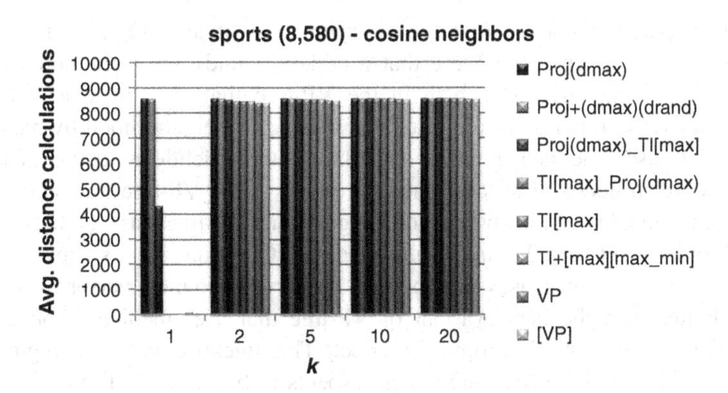

Fig. 15 The avg. no. of the Euclidean distance calculations when searching *k*-cosine nearest neighbors of a vector in 126,373-dimensional *sports* dataset of 8,580 vectors (linear scale)

and have compared two variants of VP-tree. The techniques are well suited to any distance metrics such as the Euclidean distance. They can also be applied directly in the case of the cosine similarity after transforming the problem of finding most cosine similar neighbors into the problem of finding nearest Euclidean neighbors.

In this paper, we provided an experimental comparison of the discussed techniques for determining nearest neighbors with regard to the Euclidean distance and the cosine similarity. All the techniques turned out powerful both for the Euclidean distance and the cosine similarity in the case of used non-sparse benchmark datasets (*sequoia*, *birch*, *cup98*, *covtype*)—the number of calculations of the Euclidean distances turned out by orders of magnitude lower than the number of such calculations when applying brute force approach in which an evaluated vector had to be compared with all vectors in a searched dataset. In the case of the Euclidean distance, it was always possible to obtain a more efficient variant of the algorithm using a few reference vectors (typically one or two, but in the case of the *cup98* dataset-18) than either variant using VP-tree. In the case of the cosine similarity, it was possible to obtain a more efficient variant of the algorithm using a few reference vectors than either variant using VP-tree for all non-sparse datasets except for the *cup98* dataset. Using reference vectors turned out useful for determining Euclidean nearest neighbors also in the case of the sparse *sports* dataset, though to much lower degree than in the case of the non-sparse datasets. Using VP-tree for determining Euclidean nearest neighbors also reduced the number of Euclidean distance calculations for this dataset, but was definitely less efficient than using the algorithm applying even one reference vector. No candidate reduction technique turned out practically useful when searching most cosine similar neighbors in the *sports* dataset.

We note that unlike vantage points that are vectors from a given set of vectors D in which neighbors are searched, reference vectors may be arbitrary. In our experiments, we typically constructed a basic reference vector as a vector with maximal and/or minimal domain values and then ordered vectors in D with respect to their distances to this basic reference vector.

In the paper, we noticed that the determination of an ε-Euclidean distance neighborhood by means of VP-tree that uses left_bounds and right_bounds may not require visiting any entire path in the VP-tree that ends in a leaf. To the contrary, the determination of an ε-Euclidean distance neighborhood by means of VP-tree that uses medians requires calculating the Euclidean distances to all vectors stored in nodes of at least one entire path in the VP-tree that starts in its root and ends in a leaf. This observation suggests that VP-tree using bounds should be not less efficient than VP-tree using medians. Our experiments show that the variant of VP-tree that uses left_bounds and right_bounds performs usually slightly better than the basic variant of VP-tree that uses medians. Sometimes, nevertheless, we observed an opposite effect. This unexpected effect might have been caused among other by randomness aspects in building a VP-tree.

In addition, we would like to note that ordering of vectors with respect to their distances to a basic reference vector can be achieved not necessarily by means of files sorted with respect to distances of source vectors to a first reference vector, but, for instance, by means of B+-trees, which are very easy to create and maintain.

Finally, we would like to note that all our theoretical conclusions related to determining ε-Euclidean neighborhoods and ε-Euclidean nearest neighbors remain valid if we replace the Euclidean distance with any distance metric.

Acknowledgments This work was supported by the National Centre for Research and Development (NCBiR) under Grant No. SP/I/1/77065/10 devoted to the Strategic scientific research and experimental development program: "Interdisciplinary System for Interactive Scientific and Scientific-Technical Information".

References

1. Elkan, C.: Using the triangle inequality to accelerate k-means. In: ICML'03, pp. 147–153. Washington (2003)
2. Jańczak, B.: Density-based clustering and nearest neighborood search by means of the triangle inequality. M.Sc. Thesis, Warsaw University of Technology (2013)
3. Kryszkiewicz, M.: The triangle inequality versus projection onto a dimension in determining cosine similarity neighborhoods of non-negative vectors. In: RSCTC 2012, LNCS (LNAI) 7413, pp. 229–236. Springer, Berlin (2012)
4. Kryszkiewicz, M.: Determining cosine similarity neighborhoods by means of the euclidean distance. In: Rough Sets and Intelligent Systems, Intelligent Systems Reference Library 43, pp. 323–345. Springer, Berlin (2013)
5. Kryszkiewicz M., Lasek P.: TI-DBSCAN: clustering with DBSCAN by means of the triangle inequality. In: RSCTC 2010, LNCS (LNAI) 6086, pp. 60–69. Springer (2010)
6. Kryszkiewicz M., Lasek P.: A neighborhood-based clustering by means of the triangle inequality. In: IDEAL 2010, LNCS 6283, pp. 284–291. Springer (2010)
7. Moore, A.W.: The anchors hierarchy: using the triangle inequality to survive high dimensional data. In: Proceeding of UAI, pp. 397–405. Stanford (2000)
8. Patra, B.K., Hubballi, N., Biswas, S., Nandi, S.: Distance based fast hierarchical clustering method for large datasets. In: RSCTC 2010, pp. 50–59. Springer, Heidelberg (2010)

9. Salton, G., Wong, A., Yang, C.S.: A vector space model for automatic indexing. Commun. ACM **18**(11), 613–620 (1975)
10. Samet, H.: Foundations of Multidimensional and Metric Data Structures. Morgan Kaufmann, San Francisco (2006)
11. Stonebraker, M., Frew, J., Gardels, K., Meredith, J.: The SEQUOIA 2000 storage benchmark. In: Proceeding of ACM SIGMOD, pp. 2–11. Washington (1993)
12. Uhlmann, J.: Satisfying general proximity/similarity queries with metric trees. Inf. Process. Lett. **40**(4), 175–179 (1991)
13. Yanilos, P.N.: Data structures and algorithms of nearest neighbor search in general metric spaces. In: Proceedings of 4th ACM-SIAM Symposium on Descrete Algorithms, pp. 311–321. Philadelphia (1993)
14. Zezula, P., Amato, G., Dohnal, V., Bratko, M.: Similarity Search: The Metric Space Approach. Springer, Heidelberg (2006)
15. Zhang, T., Ramakrishnan, R., Livny, M.: BIRCH: a new data clustering algorithm and its applications. Data Min. Knowl. Disc. **1**(2), 141–182 (1997)

Towards Increasing Density of Relations in Category Graphs

Karol Draszawka, Julian Szymański and Henryk Krawczyk

Abstract In the chapter we propose methods for identifying new associations between Wikipedia categories. The first method is based on Bag-of-Words (BOW) representation of Wikipedia articles. Using similarity of the articles belonging to different categories allows to calculate the information about categories similarity. The second method is based on average scores given to categories while categorizing documents by our dedicated score-based classifier. As a result of application of presented methods we obtain weighed category graphs that allow to extend original relations between Wikipedia categories. We propose the method for selecting the weight value for cutting off less important relations. The given preliminary examination of the quality of obtained new relations supports our procedure.

Keywords Associations mining · Information retrieval · Text documents categorization

1 Introduction

There is a large number of computational methods that allows to process the data using graph-based representation. Some of them requires additional conditions to be fulfilled, e.g.: the vectors representing the relations between nodes can not be

---

K. Draszawka · J. Szymański (✉) · H. Krawczyk
Department of Computer Systems Architecture, Gdańsk University of Technology, Gdańsk, Poland
e-mail: julian.szymanski@eti.pg.gda.pl

K. Draszawka
e-mail: karol.draszawka@eti.pg.gda.pl

H. Krawczyk
e-mail: henryk.krawczyk@eti.pg.gda.pl

R. Bembenik et al. (eds.), *Intelligent Tools for Building a Scientific Information Platform: From Research to Implementation*, Studies in Computational Intelligence 541,
DOI: 10.1007/978-3-319-04714-0_4,

sparse, the graphs should be fully connected etc. The very well known example where such additional conditions are required is calculating Page Rank [1]. While Page Rank is calculated using the power method [2] it requires to make the transition matrix irreducible and stochastic. In the power method ensuring suitable probability distribution is based on weighting schema [3] that adds some additional values to the edges that are considered as important. The matrix irreducibility is achieved by adjustment based on adding to edges with zero values some small numbers. This approach for increasing the graph density based on introducing new edge weights in transition graph is the very simple and as it is random, sometimes leads to unexpected results. As a result we obtain fully connected graph that for large data causes problems with effectiveness of it's processing, so a method that allows to add only significant information to the graph would be more advisable.

Increasing the data density is also method for missing data imputation [4]. It is required while the processed data is incomplete or contains the noise. In machine learning most of the approaches to completing the empty attributes values is based on filling them with averaged values of features related with particular classes. We also had to deal with that problem in our system [5] where we need to separate the textual data using abstract categories that describe the Wikipedia articles. Because the Wikipedia category graph is sparse and incomplete, building the decision tree using it requires additional methods to handle missing relations [6].

As the Wikipedia categories are popular in many NLP applications [7] the problem of their incompleteness has been subject of many studies. The examples where relationships among Wikipedia categories are used include a text topic identification [8], categorization [9] or visualization of large scale cooperative problems e.g. co-authorship [10]. Thus methods for computing similarities among Wikipedia categories are developed [11, 12]. One of the popular approaches is presented in [13] where the number of co-assignments of the same categories in articles is used. The analysis of the connections between articles belonging to the different categories allows to find new relationship between categories. The other method is based on calculating cosine similarity between articles linked by identical keywords [14] and aggregated similarity between pair of categories allows to describe their proximity.

In research presented in this chapter we propose methods for automatic discovery of associations between categories that organize textual repository. As a result a Category Association Graph (CAG) is built and it can be used for extending the original Category Graph.

The chapter is constructed as follows: In Sects. 2 and 3 we present the method based on BOW representation and based on score-based classifier. The construction of referential CAG for evaluation purposes is described in Sect. 4. Then, the evaluation of presented methods are given in Sect. 5. Finally, the last section concludes the chapter.

2 CAG Creation by Examining the Feature Space

Basing on the dataset containing feature vectors of objects and the information about their class labels, a typical way of finding associations between classes is to analyze the distances between classes in the feature space. One can treat two classes as associated with each other, if objects of these classes are close in the feature space.

Firstly, to calculate this similarity, a suitable metric function has to be chosen. In the case of standard Bag-of-Words representation of text documents, the most frequent choice is the cosine metric, but in general Minkowski, Mahalanobis, Jaccard or other metrics can be used [15].

Secondly, a method of determining the similarity between classes based on distances between objects that constitute them should be determined. In the closely related field of agglomerative hierarchical clustering (HAC), such methods are known as linkage functions, because the closest clusters, according to the values returned by these functions, are linked together to form a new, agglomerated cluster. In the problem of CAG creation, there is no such a linking process, because well formed clusters—i.e. classes—are already there. Here, distances returned by linkage functions are calculated not for linkage, but for determining associated classes.

There are many linkage functions [16]. For example, using *single linkage* method, distance between two classes is set to the minimum distance between objects found in these classes. Alternatively, *complete linkage* method settles this value to the distance between the two furthest objects of the classes. Another popular method is *centroid linkage*, that calculates the centroids of each class and then finds the similarity of classes basing on these centroids.

For our purposes, single linkage method is not suitable, because the data is multi-label. This means that the distance between any two classes is simply 0, if only there is a single object labeled with both of them, which is obviously not a desired result. Complete linkage is also too radical and sensitive to outliers. Therefore, we incorporated centroid linkage, as it does not have the disadvantages of previous methods and is easy to implement at the same time.

Thirdly, having similarities between classes, a suitable threshold value has to be determined to binarize them, so that an unweighted CAG can be built. Using the simplest thresholding strategy, score-cut (*S-cut*), if two classes are similar more than the threshold value, then an association edge between these classes is created. Many others, more advanced thresholding strategies can also be exploited here [17–19].

Algorithm 1 briefly describes this method of creation CAG, assuming thecosine metric and centroid linkage function.

```
foreach category i do
    Obtain its centroid c_i by averaging all its articles;
end
foreach category i do
    foreach category j do
        Calculate the cosine similarity between c_i and c_j;
    end
    Associate category i with categories that have the most similar
    centroids – use some kind of thresholding strategy;
end
```

Algorithm 1: The method of centroids' similarity

3 Creating CAG via Score-Based Classifier

As an alternative to the standard method of creating CAGs presented in the previous section, here we introduce a new method of building CAGs. This method uses a text document classifier, developed to categorize articles into correct category or group of categories (multi-label classification). However, our usage here is unconventional—we do not aim to classify the documents, but to find associations between classes.

3.1 Score-Based Classifier

The only requirement to the method is that a classifier must be *score-based*. This means that it contains the first phase in which it explicitly produces scores for all classes and the second phase in which these scores are thresholded, so that the final class prediction is made [18, 19]. Such classifiers are more and more popular for large scale multi-label tasks.

In our experiments, we used the score-based version of *k*-NN classifier presented in [19], where detailed description of the classifier can be found.

3.2 CAG from Average Scores

Our *k*-NN based classifier, using information from local neighbourhood of a query document, returns a vector of scores, one score per class candidate. Such vectors of scores contain valuable information that can be exploited to find associations between categories.

If an object of some class has high score value not only for this class, but also for others, then it is a cue for the multi-label classifier that the object should be labeled more than once. But if many objects from that class have high score value for some other class quite frequently, then this is a cue for a CAG creation method, indicating that the two classes are somehow associated with each other.

This intuition leads us to the method of obtaining CAG, which is described in algorithm 2. Score vectors are calculated for every document of a given category and they are averaged. This way, an averaged score vector can be computed for each category. After that, these averaged scores have to be bipartitioned using some kind of thresholding strategy, in the same manner as with similarities between categories in the standard method.

foreach *document i in the dataset* **do**
    Obtain the vector of class scores using extended k-NN classifier (or other score-based classifier);
end
foreach *category i* **do**
    Average score vectors for all the documents in category i;
    Associate category i with categories with the highest average scores – use some kind of thresholding strategy;
end

Algorithm 2: The method of averaging scores

4 Adjustments to Reference CAG

In their first phases, presented methods of automatic CAG creation make dense, weighed graph. To obtain high quality final graph in the last step of our approach we introduce a method to determine the cut-off threshold that indicates associations which can be treated as a noise. If thresholds' value is low, an automatically created CAG can have a lot of associations, but low quality. On the other hand, too high threshold restricts the number of associations, prevents from introducing not significant associations, but can lead to very sparse CAGs that do not have new interesting relations.

Setting up different threshold values results in obtaining drastically different CAGs. As in our experiments, we do not want to choose arbitrarily the threshold values. For this purpose, we propose the method for selecting it according to predefined criteria. We use a tuning reference for CAG that we call $\mathrm{CAG}_{\mathrm{ref}}$. Using it, the threshold values that indicate coverage factor of CAG according to the $\mathrm{CAG}_{\mathrm{ref}}$ can be set up.

4.1 Construction of CAG_{ref}

For construction of $\mathrm{CAG_{ref}}$ we use existing Wikipedia category structure. Having this quasi-hierarchical organization of classes, the construction of $\mathrm{CAG_{ref}}$ involves applying one of the metrics of similarity between nodes in the graph, such as *cladistic*, *cophenetic* or other distance measures [20]. For example, cladistic distance (d_{clad}) between two nodes is the number of edges in a shortest path joining these two nodes.

We used $\mathrm{CAG_{ref}}$ that has associations between categories that are $d_{\mathrm{clad}} = 1$ away from each other in the Wikipedia organization structure. This means that category is associated only with its direct super categories and direct subcategories. We abbreviate such reference CAG with $\mathrm{CAG_{ref}}(d_{\mathrm{clad}} = 1)$.

4.2 Adjusting Thresholds

Having $\mathrm{CAG_{ref}}$, the F1-score metric can be used for evaluation of discovered relations. The F1-score is a harmonic mean between *precision* (P, the fraction of correctly predicted associations in the set of all predicted associations) and *recall* (R, the fraction of associations in the set of all reference associations that where identified by the algorithm).

The process of finding optimal thresholds consists of scanning through different threshold values and calculating F1-score between resulting CAG and $\mathrm{CAG_{ref}}$. The setting that maximizes F1-score is chosen.

5 Evaluation

The evaluation of the CAGs generated by presented methods was done on the Simple English Wikipedia dataset.[1] The articles were preprocessed in the standard way (stop words removal, stemming process). The dataset consists of all the articles with the bindings to all categories, except for very small categories (i.e. categories having less than 5 articles were discarded).

Also, the so called 'stubs' categories were removed, because they contain very small articles without specific terms, which resulted in being very similar to each other. After such filtering, the test data had 3676 categories and about 55634 articles with about 92700 terms.

In thresholding phase of both methods, we used SS-cut (scaled score-cut) strategy, in which a score vector is first scaled, so that the highest class score in this vector equals 1.

[1] http://dumps.wikimedia.org/ [dumpfile from 01.04.2010].

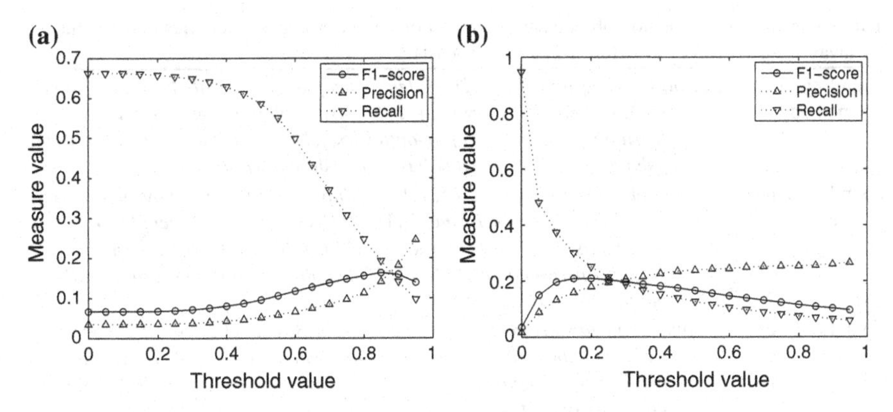

Fig. 1 Adjusting threshold value to optimize similarity to $\mathrm{CAG}_{\mathrm{ref}}(d_{\mathrm{clad}} = 1)$. **a** Method 1 (distance to centroids). **b** Method 2 (score-based k-NN)

5.1 Adjusting CAGs

The results of adjusting thresholds' values to obtain CAG as close as possible to $\mathrm{CAG}_{\mathrm{ref}}(d_{\mathrm{clad}} = 1)$ is presented in Fig. 1. It can be seen, that for method 1 the threshold has to be set relatively high to achieve optimally results. This is because smaller threshold means that classes represented by more distant centroids are associated. Distance to a centroid spread equally in the feature space and thus more and more classes are associated, resulting in high recall of reference associations, but also low precision of resulting bindings. In contrast, the second method incorporate local classifier—k-NN—and because of this locality, optimal thresholds can be lower: not global distance between centroids are to be cut, but averaged locally calculated scores.

For method 1, the maximum F1-score between generated CAG and referential one is 0.1632 ($P = 0.1414\,, R = 0.1929$), for $t = 0.85$. Our method 2 is slightly better in such reconstruction, obtaining $F1 = 0.2094$ ($P = 0.1798$, $R = 0.2508$), for $t = 0.2$.

5.2 Evaluation of New Associations

Our task is to find new, interesting associations that are not the $\mathrm{CAG}_{\mathrm{ref}}$. Moreover, they should be non-obvious, in the sense that they cannot be simply deduced from hierarchical relations between categories—even if such relations are given. Therefore, we investigated such associations returned by presented methods to see whether they are meaningful.

Table 1 presents the lists of new category associations, i.e. not present in the $\mathrm{CAG}_{\mathrm{ref}}$. We showed only the examples of distant categories between which

Table 1 Examples of the non-obvious new associations found between categories that are placed far away in the category structure ($d_{clad} = 5$ at least)

Found by both methods	*Politics-geography* (∞), *2004 deaths-people* (∞), *events by year-history* (∞), *emperors-history* (∞), *1980 births-living people* (∞), *teachers-learning* (∞), *photography-optics* (∞), *Nazi officers-Nazi Germany* (6), *fishing-fish* (10), *lifestyles-basic english 850 words* (7), ...
Found only by method 1	*1806 births-theologians* (∞), *relationships-new religious movements* (∞), *wines-geography of France* (11), *1931 deaths-Danish people* (9), *Magicians-American entertainers* (11), *1799 births-Russian poets* (11), *1820 births-Roman catholics* (∞), *people from north carolina-southern baptist convention* (∞), ...
Found only by method 2	*Supercentenarians*-people (∞), *religion in Brazil-religion* (∞), *theatre-living people* (∞), *presocratic philosophers-history* (5), *1965 deaths-people* (∞), *volcanology-geography* (∞), *science fiction-time* (∞), *navy-transportation* (∞), *human issues-living people* (∞), *Grammy award winners-living people* (∞), ...

Actual d_{clad} for each association is shown in parenthesis

associations were found ($d_{clad} = 5$ at least). To highlight the difference between methods, we showed separately intersection and set differences of results returned by both methods. In fact, 47 % of the new associations returned by method 1 is not found by method 2, and 64 % of those found by method 2 is not indicated by method 1. However, it can be seen that the new associations returned by both methods are meaningful and valuable most of the time.

6 Conclusions

In the chapter we present new methods of extracting useful associations between categories that organize textual repository. We compare achieved characteristics with relations predefined by human created initial categories as well as we manually evaluate new associations.

To evaluate proposed methods we performed tests on Simple English Wikipedia articles. Firstly, we generated a reference CAG from the original Wikipedia category structure. Then, we adjusted threshold settings so that resulting CAGs reflect the reference CAG as good as it can. Here, depending on the type of a CAG_{ref}, different methods are better.

Finally, we analyzed what associations, not present in CAG_{ref}, we obtained. It can be seen, that presented methods return a wide range of various associations, separately from each other. Many of the new associations are correct and useful, especially if they are indicated by at least two of methods independently. Moreover, presented new associations are non-obvious, in the sense that they are located far from each other in Wikipedia category structure.

In the future, it is planned to create a fully valid, dense, gold-standard reference CAG, so that the evaluation could be more precise. It is especially interesting what the impact of using different classifiers on performance of classification errors analysis for associations finding is.

Acknowledgments This work has been supported by the National Centre for Research and Development (NCBiR) under research Grant No. SP/I/1/77065/1 SYNAT: "Establishment of the universal, open, hosting and communication, repository platform for network resources of knowledge to be used by science, education and open knowledge society".

References

1. Page, L., Brin, S., Motwani, R., Winograd, T.: The pagerank citation ranking: bringing order to the web. (1999)
2. Langville, A.N., Meyer, C.D.: Deeper inside pagerank. Internet Math. **1**, 335–380 (2004)
3. Baeza-Yates, R., Davis, E.: Web page ranking using link attributes. In: Proceedings of the 13th International World Wide Web Conference on Alternate Track Papers and Posters, ACM, 328–329, 2004
4. Cleophas, T.J., Zwinderman, A.H.: Missing data imputation. In: Statistical Analysis of Clinical Data on a Pocket Calculator, Part 2, pp. 7–10. Springer (2012)
5. Deptuła, M., Szymański, J., Krawczyk, H.: Interactive information search in text data collections. In: Intelligent Tools for Building a Scientific Information Platform, pp. 25–40, Springer. (2013)
6. Zhang, S., Qin, Z., Ling, C.X., Sheng, S.: Missing is useful: missing values in cost-sensitive decision trees. IEEE Trans. Knowl. Data Eng. **17**, 1689–1693 (2005)
7. Zesch, T., Gurevych, I.: Analysis of the wikipedia category graph for nlp applications. In: Proceedings of the TextGraphs-2 Workshop (NAACL-HLT 2007), pp. 1–8, 2007
8. Schonhofen, P.: Identifying document topics using the wikipedia category network. In: Web Intelligence, 2006. WI 2006. IEEE/WIC/ACM International Conference on, IEEE. pp. 456–462 (2006)
9. Hu, X., Zhang, X., Lu, C., Park, E.K., Zhou, X.: Exploiting wikipedia as external knowledge for document clustering. In: Proceedings of the 15th ACM SIGKDD International Conference on Knowledge Discovery and Data Mining, ACM, 389–396, 2009
10. Biuk-Aghai, R.P., Pang, C.I., Si, Y.W.: Visualizing large-scale human collaboration in wikipedia. Future Gener. Comput. Syst. **31**, 120–133 (2013)
11. Szymański, J.: Mining relations between wikipedia categories. In: Networked Digital Technologies, 248—255. Springer (2010)
12. Chernov, S., Iofciu, T., Nejdl, W., Zhou, X.: Extracting semantic relationships between wikipedia categories. In: Proceedings of Workshop on Semantic Wikis (SemWiki 2006), Citeseer (2006)
13. Holloway, T., Bozicevic, M., Börner, K.: Analyzing and visualizing the semantic coverage of wikipedia and its authors. Complexity **12**, 30–40 (2007)
14. Dhillon, I.S., Modha, D.S.: Concept decompositions for large sparse text data using clustering. Mach. Learn. **42**, 143–175 (2001)
15. Manning, C.D., Raghavan, P., Schütze, H.: Introduction to information retrieval, vol. 1, Cambridge University Press, Cambridge (2008)
16. Day, W.H., Edelsbrunner, H.: Efficient algorithms for agglomerative hierarchical clustering methods. J. Classif. **1**, 7–24 (1984)

17. Yang, Y.: A study of thresholding strategies for text categorization. In: Proceedings of the 24th annual international ACM SIGIR conference on Research and development in information retrieval, ACM. pp. 137–145 (2001)
18. Ioannou, M., Sakkas, G., Tsoumakas, G., Vlahavas, L.: Obtaining bipartitions from score vectors for multi-label classification. In: Tools with Artificial Intelligence (ICTAI), 2010 22nd IEEE International Conference on, vol. 1, 409–416 (2010)
19. Draszawka, K., Szymański, J.: Thresholding strategies for large scale multi-label text classifier. In: Proceedings of the 6th International Conference on Human System Interaction, IEEE. pp. 347–352 (2013)
20. Draszawka, K., Szymanski, J.: External validation measures for nested clustering of text documents. In: Ryzko D., Rybinski H., Gawrysiak P., Kryszkiewicz M. (eds.) ISMIS Industrial Session. Volume 369 of Studies in Computational Intelligence, Springer. 207–225 (2011)

Weight Learning in TRSM-based Information Retrieval

Wojciech Świeboda, Michał Meina and Hung Son Nguyen

Abstract This chapter presents a novel approach to keyword search in Information Retrieval based on Tolerance Rough Set Model (TRSM). Bag-of-word representation of each document is extended by additional words that are enclosed into inverted index along with appropriate weights. Those extension words are derived from different techniques (e.g. semantic information, word distribution, etc.) that are encapsulated in the model by a tolerance relation. Weight for structural extension are then assigned by unsupervised algorithm. This method, called TRSM-WL, allow us to improve retrieval effectiveness by returning documents that not necessarily include words from the query. We compare performance of these two algorithms in the keyword search problem over a benchmark data set.

Keywords Information retrieval · Rough sets · Document expansion

---

The authors are supported by grant 2012/05/B/ST6/03215 from the Polish National Science Centre (NCN), and the grant SP/I/1/77065/10 in frame of the strategic scientific research and experimental development program: "Interdisciplinary System for Interactive Scientific and Scientific-Technical Information" founded by the Polish National Centre for Research and Development (NCBiR).

W. Świeboda (✉) · H. S. Nguyen (✉)
Institute of Mathematics, The University of Warsaw, Banacha 2,
02-097 Warsaw, Poland
e-mail: wswieb@mimuw.edu.pl

H. S. Nguyen
e-mail: son@mimuw.edu.pl

M. Meina (✉)
Faculty of Mathematics and Computer Science, Nicolaus
Copernicus University, Toruń, Poland
e-mail: mich@mat.umk.pl

R. Bembenik et al. (eds.), *Intelligent Tools for Building a Scientific Information Platform: From Research to Implementation*, Studies in Computational Intelligence 541,
DOI: 10.1007/978-3-319-04714-0_5,

1 Introduction

Rough set theory has been introduced by Pawlak [1] as a tool for concept approximation under uncertainty. The idea is to approximate the concept by two descriptive sets called *lower and upper approximations*. The fundamental philosophy of rough set approach to concept approximation problem is to minimize the difference between upper and lower approximations (the *boundary region*). This simple but brilliant idea leads to many efficient applications of rough sets in machine learning, data mining and also in granular computing. In this chapter we investigate the Tolerance Rough Set Model (TRSM), which has been proposed in [2, 3] as a basis to model documents and terms in Information Retrieval.

Current Information Retrieval (IR) systems share a standard interaction scenario: a user formulates a query and then the system provides results based on query-to-document relevance. Retrieval efficiency is dependent upon the number of terms that overlap between the query and a document. The main issue in IR is known as *vocabulary problem*, which is a common mismatch in choosing the same terms by a user and by an indexer. One of the first evaluations of an IR system [4] concludes that formulating a proper keyword for search query demands from the user predicting existence of possible words in a document. Such prediction can be challenging and eventually can lead to scalability issues for IR systems. In addition, users may not agree with the choice of keywords while searching for the same objects. Furnas et al. [5] reports that in spontaneous word choice for objects in five domains, two people favored same term with less than 20 % frequency. Therefore, two main techniques were developed to overcome this issue: (1) Query Expansion and (2) Document Expansion.

Query Expansion (QE) is on-line process of adding additional words to the query that best describe the user's intent in searching. Document Expansion (DE), on the other hand, is a process of document vocabulary augmentation at indexing time, predicting possible queries for each document. QE is a long studied problem in Information Retrieval (see exhaustive survey [6]), but the approach of extending documents at indexing time do not get such attention. Lack of transaction constraints (as in QE) should be considered as a promise in more effective model design. On the other hand DE involves enlarging the size of the index which can lead to problems with maintaining it. In this chapter we will also investigate trade-off between retrieval effectiveness and index size problems.

Many approaches have been studied for solving the *vocabulary problem* both in QE and DE. Straightforward methods are based examination of word co-occurrence statistics in the corpus [7] outputting possible extension words. Then those words are enclosed into index or used in on-line query processing. Cluster-based Information Retrieval assumes that belonging of two documents to the same cluster carry some information about their correspondence therefore this information can be used in search process. Variety of cluster-based search methods and clustering algorithms was introduced and tested [8, 9]. Different category of techniques exploits semantic information from external sources. Expanding term by theirs synonyms after word sense disambiguation using WordNet Ontology (along with other relation) [10] reports good performance boost. Although we see major differences between all of the techniques the common feature of term correspondence is exploited in the final extension. In this chapter we present a document extension method which may encapsulate different tolerance relations between terms. The method is a variant of a Tolerance Rough Set Model for Documents [2, 3] (TRSM). We supplement TRSM by a weight learning method in an unsupervised setting and apply the model to the problem of extending search results. This chapter is an extension of [11].

The outline is as follows: We begin this chapter by reviewing the Vector Space Model (a basic model widely used in Information Retrieval), and fundamentals of Rough Set Theory based both on indiscernibility relation and based on tolerance relation. Afterwards, we review TRSM model and introduce a weight learning scheme which we validate in the context of document retrieval.

2 Basic Notions

Rough Set Theory, developed by Pawlak [1] is a model of approximation of sets. An Information System I is a pair $\mathrm{I} = (U, A)$. U is called the *Universe* of objects and is the domain whose subsets we wish to represent or approximate using attributes, i.e. elements of A. Each attribute $a_i \in A$ is a function $a: U \to V_a$, where V_a is called the *value set* of attribute a.

For a subset of attributes $B \subseteq A$ we define a *B-indiscernibility* relation $IND(B) \subseteq U \times U$ as follows:

$$(x, y) \in IND(B) \Leftrightarrow \forall_{a \in A} a(x) = a(y) \tag{1}$$

$IND(B)$ is an equivalence relation and defines a partitioning of U into equivalence classes which we denote by $[x]_B$ $(x \in U)$. B-Lower and B-upper approximations of a concept $X \subseteq U$ are defined as follows:

$$\mathcal{L}(X) = \{x \in U : [x]_B \subseteq X\} \tag{2}$$

$$\mathcal{U}(X) = \{x \in U : [x]_B \cap X \neq \emptyset\} \tag{3}$$

2.1 Tolerance Approximation Spaces

Indiscernibility relation in standard Rough Set Model is an equivalence relation. This requirement is known to be too strict in various applications. Skowron et al. [12] introduced Tolerance Approximation Spaces (and Generalized Approximation Spaces), relaxing conditions on the underlying relation. In this framework, indiscernibility of objects is defined by a tolerance relation.

An *Approximation Space* is defined as a tuple $\mathcal{R} = (U, I, \nu, P)$, where:

- U is a non-empty universe of objects,
- An *uncertainty function* $I{:}U \rightarrow \mathcal{P}(U)$ is any function such that the following conditions are satisfied:
  - $x \in I(x)$ for $x \in U$,
  - $y \in I(x)$ iff $x \in I(y)$
- A *vague inclusion function* $\nu{:}\mathcal{P}(U) \times \mathcal{P}(U) \rightarrow [0,1]$, such that $\nu(X, \cdot)$ is a monotone set function for each $X \in P(U)$.
- A *structurality function* $P{:}I(U) \rightarrow \{0,1\}$, where $I(U) = \{I(x){:}x \in U\}$.

A *vague membership function* $\mu(I,\nu){:}U \times \mathcal{P}(U)$ is defined as $\mu(I,\nu)(x,X) = \nu(I(x), X)$ and lower and upper approximations of $X \subseteq U$ are defined as:

$$L_A(X) = \{x \in U{:}P(I(x)) = 1 \wedge \mu(I,\nu)(x,X) = 1\} \tag{4}$$

$$U_A(X) = \{x \in U{:}P(I(x)) = 1 \wedge \mu(I,\nu)(x,X) > 1\} \tag{5}$$

We will further refer to I as to the *tolerance relation* and to $I(u)$ as to the *tolerance class*.

2.2 Standard TRSM

Let $D = \{d_1, \ldots, d_N\}$ be a corpus of documents Assume that after the initial processing documents, there have been identified N unique terms (e.g. words, stems, N-grams) $T = \{t_1, \ldots, t_M\}$.

Tolerance Rough Set Model for Documents [2, 3], or briefly TRSM, is an approximation space $\mathcal{R} = (T, I_\theta, \nu, P)$ determined over the set of terms T where:

- The parameterized **uncertainty function** $I_\theta{:}T \rightarrow \mathcal{P}(T)$ is defined by

  $$I_\theta(t_i) = \{t_j \mid f_D(t_i, t_j) \geq \theta\} \cup \{t_i\}$$

  where $f_D(t_i, t_j)$ denotes the number of documents in D that contain both terms t_i and t_j and θ is a parameter set by an expert. The set $I_\theta(t_i)$ is called the *tolerance class* of term t_i.
- **Vague inclusion function** $\nu(X, Y)$ measures the degree of inclusion of one set in another. The vague inclusion function is defined as $\nu(X, Y) = \frac{|X \cap Y|}{|X|}$. It is clear that this function is monotone with respect to the second argument.
- **Structural function**: All tolerance classes of terms are considered as structural subsets: $P(I_\theta(t_i)) = 1$ for all $t_i \in T$.

In TRSM model $\mathcal{R} = (T, I, \nu, P)$, the membership function μ is defined by

$$\mu(t_i, X) = \nu(I_\theta(t_i), X) = \frac{|I_\theta(t_i) \cap X|}{|I_\theta(t_i)|}$$

where $t_i \in T$ and $X \subseteq T$. The lower and upper approximations of any subset $X \subseteq T$ can be determined by the same maneuver as in approximation space [12]:

$$L_{\mathcal{R}}(X) = \{t_i \in T \mid \nu(I_\theta(t_i), X) = 1\}$$
$$U_{\mathcal{R}}(X) = \{t_i \in T \mid \nu(I_\theta(t_i), X) > 0\}$$

2.3 Application of TRSM in Information Retrieval

Information Retrieval systems in the vector space model, often represent texts as feature vectors, that is, tuples of values $d = (w_1, w_2, \ldots, w_m)$, where w_j is the numeric value that term j takes on for this document, and m is the number of terms that occur in the document collection.

Thus any text d_i in the corpus D can be represented by a vector $[w_{i1}, \ldots, w_{iM}]$, where each coordinate $w_{i,j}$ expresses the significance of jth term in this document. The most common measure, called *tf-idf* index (term frequency-inverse document frequency) [13], is defined by:

$$w_{i,j} = tf_{i,j} \cdot idf_j = \frac{n_{i,j}}{\sum_{k=1}^{M} n_{i,k}} \cdot \log\left(\frac{N}{|\{i : n_{i,j} \neq 0\}|}\right) \tag{6}$$

where $n_{i,j}$ is the number of occurrences of the term t_j in the document d_i.

Both standard TRSM and extended TRSM are the conceptual models for the Information Retrieval. Depending on the current application, different extended weighting schema can be proposed to achieve as highest performance as possible. Let us recall some existing weighting schemes for TRSM:

1. The extended weighting scheme is inherited from the standard TF-IDF by:

$$w^*_{ij} = \begin{cases} \left(1 + \log f_{d_i}(t_j)\right) \log \frac{N}{f_D(t_j)} & \text{if } t_j \in d_i \\ 0 & \text{if } t_j \notin U_{\mathcal{R}}(d_i) \\ \min_{t_k \in d_i} w_{ik} \frac{\log \frac{N}{f_D(t_j)}}{1 + \log \frac{N}{f_D(t_j)}} & \text{otherwise} \end{cases}$$

This extension ensures that each term occurring in the upper approximation of d_i but not in d_i itself has a weight smaller than the weight of any term which appears in d_i. Normalization by vector's length is then applied to all document vectors:

$$w^{new}_{ij} = w^*_{ij} / \sqrt{\sum_{t_k \in d_i} \left(w^*_{ij}\right)^2}$$

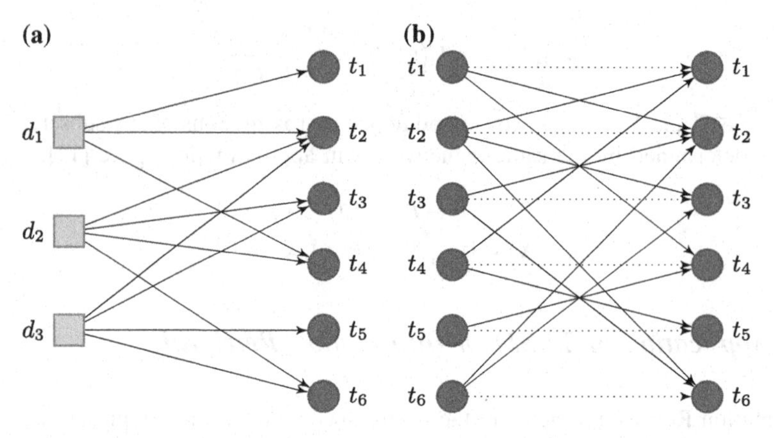

Fig. 1 Bag-of-words document representation **a** Bag-of-words document representation of a simple corpus with three documents and six terms, **b** Structural part of TRSM model for parameter $\theta = 2$. *Arrows* describe tolerance classes of terms

(see [2, 3]). Figures 1 and 2 illustrate the model on a document corpus from [14]. Weights in this example and in experiments further discussed in this chapter differ slightly from original TRSM weights: we use term frequency $f_{d_i}(t_j)$ instead of $log\left(1 + f_{d_i}(t_j)\right)$ and do not apply normalization.

2. Explicit Semantic Analysis (ESA) proposed in [15] is a method for automatic tagging of textual data with predefined concepts. It utilizes natural language definitions of concepts from an external knowledge base, such as an encyclopedia or an ontology, which are matched against documents to find the best associations. Such definitions are regarded as a regular collection of texts, with each description treated as a separate document. The original purpose of ESA was to provide means for computing semantic relatedness between texts. However, an intermediate result—weighted assignments of concepts to documents (induced by the term-concept weight matrix) may be interpreted as a weighting scheme of the concepts that are assigned to documents in the extended TRSM. This method has been proposed in [16] as follows.

   Let $W_i = \left[w_{i,j}\right]_{j=1}^{N}$ be a bag-of-words representation of an input text d_i, where $w_{i,j}$ is a numerical weight of term t_j expressing its association to the text d_i. Let $s_{j,k}$ be the strength of association of the term t_j with a knowledge base concept c_k, $k \in \{1, \ldots, K\}$ an inverted index entry for t_j. The new vector representation, called a *bag-of-concepts* representation of d_i, is denoted by $\left[u_{i,1}, \ldots u_{i,K}\right]$, where:

$$u_{i,k} = \sum_{j=1}^{N} w_{i,j} s_{j,k}.$$

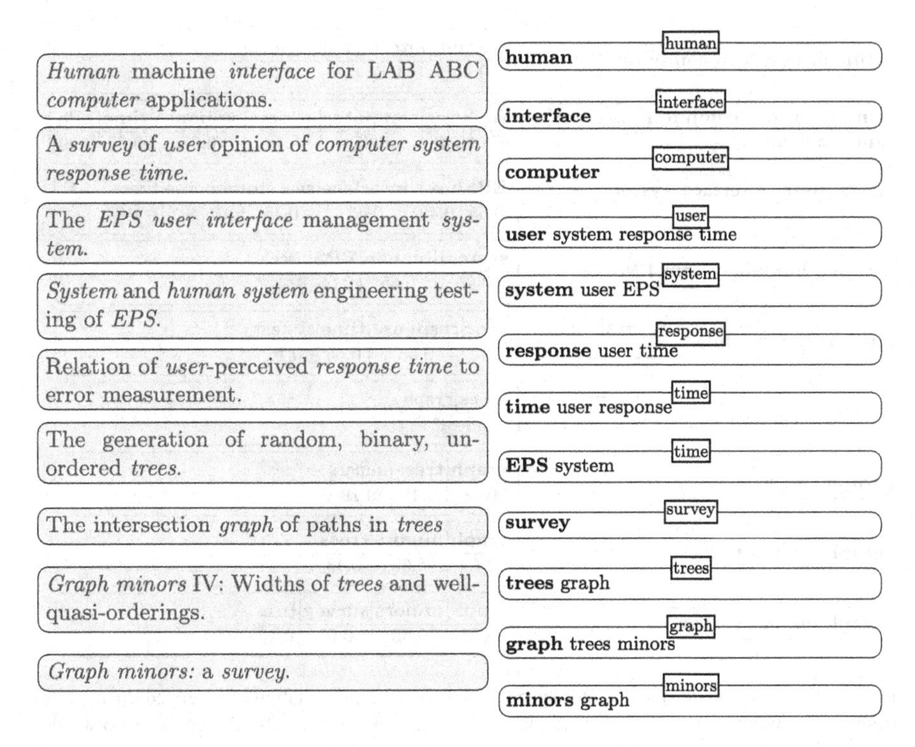

Fig. 2 A document corpus (*left*) and the corresponding tolerance classes (*right*) for $\theta = 2$. The document corpus was introduced by Deerwester et al. in [14]

For practical reasons it is better to represent documents by the most relevant concepts only. In such a case, the association weights can be used to create a ranking of concept relatedness. With this ranking it is possible to select only top concepts from the list or to apply some more sophisticated methods that involve utilization of internal relations in the knowledge base.

3 TRSM with Weight Learning

The purpose of our research is to explore an alternative framework for weight assignment in TRSM model. We define the underlying term-document structure in the (extended) Vector Space Model using TRSM—in other words, we assume that $w^*_{ij} = 0$ for $t_j \notin U_{\mathcal{R}}(d_i)$. We will speak of the model structure, the set of permitted (i.e., nonzero) weights and the set of tolerance classes interchangeably. We further propose an alternative method of determining w_{ij} for $t_j \in U_{\mathcal{R}}(d_i)$. An example of such weightening on Deerwester corpus is illustrated on Fig. 3.

Human interface computer

Human	interface	computer
0.65	0.65	0.65

survey user computer system response time EPS

survey	user	computer	system	response	time	EPS
0.65	0.48	0.65	0.48	0.65	0.65	0.19

EPS user interface system response time

EPS	user	interface	system	response	time
0.65	0.48	0.65	0.48	0.19	0.19

system human system EPS user

system	human	EPS	user
0.95	0.65	0.65	0.21

user response time system

user	response	time	system
0.48	0.65	0.65	0.16

trees graph

trees	graph
0.48	0.15

graph trees minors

graph	trees	minors
0.48	0.48	0.19

graph minors trees

graph	minors	trees
0.48	0.65	0.48

graph minors survey trees

graph	minors	survey	trees
0.48	0.65	0.65	0.16

Fig. 3 Bag of words (*left*) and vector space model (*right*) representation of extended documents from Fig. 1. We only show words that appear in at least two documents and remove stop words "a", "in", "for", …

The model that we propose aims to approximate original TF-IDF weights by a conical combination of TF-IDF weights of related terms. The set of terms related to t_i is the tolerance class of t_i in TRSM model (excluding t_i itself). In other words:

$$w^*_{ij} = \sum_{k=1}^{N} \delta(i,k,j)\beta_{kj}w_{ik} \tag{7}$$

where

$$\delta(i,k,j) = \begin{cases} 1 & \text{for } t_k \in d_i \wedge t_j \in I_\theta(t_k) \\ 0 & \text{otherwise} \end{cases} \tag{8}$$

We will further demand that $\beta_{kj} > 0$ to stress the fact that tolerance classes in TRSM aim to capture similarities rather than dissimilarities between terms. We can thus rewrite Eq. 7 as follows:

$$w^*_{ij} = \sum_{k=1}^{N} \delta(i,k,j)e^{\alpha_{kj}}w_{ik} \tag{9}$$

In what follows, we propose a framework for determining weights α_{kj}. The underlying idea is to train a set of linear neurons (with trivial transfer functions) whose inputs are determined by TRSM. One can think of the problem as of "document recovery" and wish to approximate hidden w_{ij} by w_{ij}^*, i.e. try to assign weights w_{ij}^* so as to minimize error

$$E = \sum_{i=1}^{N} \sum_{j=1}^{M} L\left(w_{ij}, w_{ij}^*\right)$$

for a convenient loss function L. For simplicity, we pick the square loss function. Since the choice of a weights α_{kj} has no bearing on $L(w_{ij}, w_{ij}^*)$ for $t_j \notin U_{\mathcal{R}}(d_i)$, we can further restrict the summation to $i = 1, \ldots, N$ and j such that $t_j \in U_{\mathcal{R}}(d_i)$.

A natural additive update (proportional to negative gradient) is:

$$\Delta\alpha_{kj} \propto \sum_{i=1}^{N} \left(w_{ij} - w_{ij}^*\right) \delta(i,k,j) e^{\alpha_{kj}} w_{ik} \tag{10}$$

A commonly used mode of training is on-line learning, where the algorithm iterates the corpus document-by-document. In this approach, the corresponding updates during processing of document i are: $\Delta\alpha_{kj} \propto \left(w_{ij} - w_{ij}^*\right) \delta(i,k,j) e^{\alpha_{kj}} w_{ik}$.

Please note that as soon as the model structure is determined, perceptron weights are updated independently of each other. However, typically the model itself (whose size is determined by the set of nonzero weights) can be stored in computer memory, whereas the document corpus needs to be accessed from a hard drive. Processing the corpus sequentially document-by-document and updating all relevant weights is beneficial for technical purposes due to a smaller (and sequential) number of reads from a hard disk. Thus, training all perceptrons simultaneously is beneficial (strictly) for technical reasons.

Let us further call this model TRSM-WL (TRSM with weight learning). In principle it is an unsupervised learning method. The learned model can be applied to new documents and thus is inferential in nature.

Algorithm 1 shows pseudo-code for the weight update procedure. For simplicity, we assume that we iterate only once over each document (document d_i is processed in step i). In practice (and in experiments that follow) we iterated over the document corpus several times. The damping factor $\eta(i)$ used in experiments was picked inversely proportional i.

Algorithm 1: Weight update procedure in TRSM-WL.

```
Input: W = (w_ik)_{i,k} (the document-term matrix), Tol : T → T (tolerance class
       mapped to each term).
Output: α = (α_kj)_{k,j}.
1 for k, j such that t_j ∈ I_θ(t_k) do
      /* The initializing distribution of α_kj (implicitly - of
         β_kj = e^{α_kj}) is a modeling choice.                           */
2     α_{k,j} = RandNorm(μ = 0, σ^2 = 1)
3 for i in 1, ..., |D| do
      /* d̃: the set of terms in the extension of document d.              */
4     d̃ = d ∪ ⋃_{t_k ∈ d_i} I_θ(t_k);
      /* Determine weights w*_ij for t_j ∈ d̃_i.                            */
5     for j in 1, ..., |d̃_i| do
6         w*_ij = Σ_{k: t_k ∈ d_i} e^{α_kj} w_ik
7     for j in 1, ..., |d̃_i| do
8         for k in 1, ..., |d_i| do
              /* Apply updates to weights α. η(i) is a damping factor
                 which determines the proportionality ratio of
                 consequent updates.                                        */
9             α_kj = α_kj + η(i)(w_ij − w*_ij) e^{α_kj} w_ik
```

4 Experimental Results

We conducted experiments on the ApteMod version of Reuters-21578 corpus. This corpus consists of 10,788 documents from Reuters financial service. Each document is annotated by one or more categories. The distribution of categories is skewed with 36.7 % of the documents in the most common category and only 0.0185 % (2 documents) in each of the five least common categories.

We applied stemming and stop word removal in order to prepare a bag-of word representation of documents. Test queries were chosen among all words in corpus using Mutual Information (MI) coefficient:

$$I(C,T) = \sum_{c \in \{0,1\}} \sum_{t \in \{0,1\}} p(C=c,\, T=t) \log_2 \left(\frac{p(C=c, T=t)}{p(C=c)p(T=t)} \right), \tag{11}$$

where $p(c=0)$ represents the probability that randomly selected document is a member of particular category and $p(c=1)$ represents probability that it isn't. Similarly, $p(t=1)$ represents the probability that a randomly selected document contains a given term, and p(T $=$ 0) represents the probability that it doesn't. Next for each category five terms with highest MI was taken and used as query.

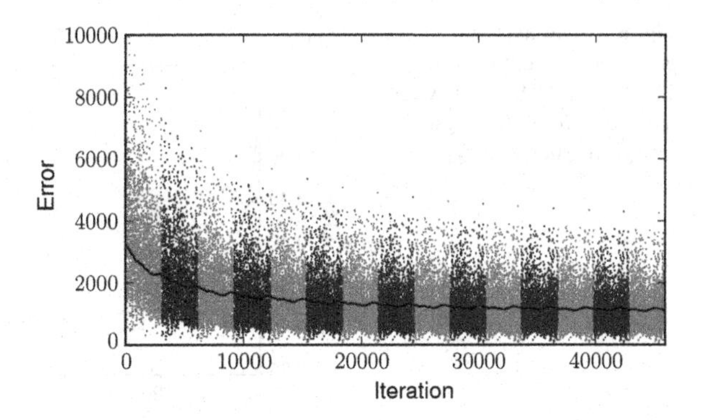

Fig. 4 Model convergence

Fig. 5 Recall of document search based on TF-IDF and TRSM extended document representations

Query-Set	TF-IDF	TRSM / TRSM-WL
All	0.457 ± 0.317	0.764 ± 0.335
Q_1	0.329 ± 0.252	0.733 ± 0.364
Q_2	0.325 ± 0.200	0.993 ± 0.006

4.1 Model Convergence

Figure 4 shows the squared loss for each document in the training set when the algorithm is executed. The y axis on the plot shows the contribution of each document i to error E, i.e.

$$E_i = \sum_{j=1}^{M} \left(w_{ij} - w_{ij}^*\right)^2.$$

Each colored block corresponds to a single iteration over the entire document repository, and thus the error corresponding to each document appears several times, once in each alternating block. The graph shows that the first two iterations provide the biggest contribution to overall model fit.

4.2 Searching Using Extended Document Representation

In order to measure the effectiveness of search we prepared Information Retrieval systems based on TF-IDF, TRSM and our TRSM-WL weighting schemes (the methodology is described in Sect. 2). We considered two test cases by dividing the query set into two subsets Q_1 and Q_2 taking into account sizes of expected result lists, e.g. number of documents in category. First one describes small categories ($n < 100$) and second large ones ($n \geq 100$).

As we see on Fig. 5 TRSM and TRSM-WL resulted in significantly better recall compared to standard TF-IDF. Both models with extended representation outputs

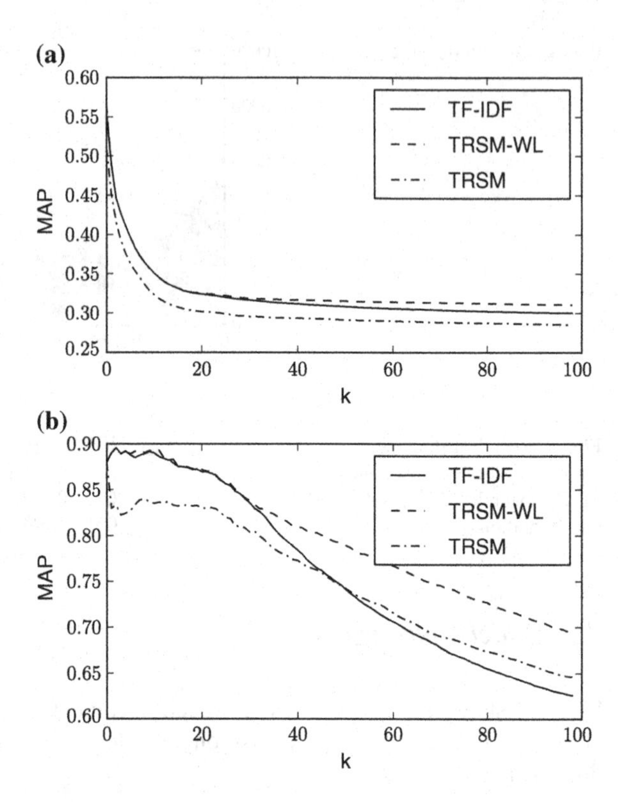

Fig. 6 Mean average precision at k cut-off rank. TRSM and TRSM-WL with parameter $\theta = 10$. **a** Query set Q_1. **b** Query set Q_2

the same recall score since TRSM-WL reassigns weights for the same co-located terms. It's worth stressing that extended representation for low θ parameters tends to be very large in most cases. Moreover the size of the extended representation correlates positively with corpora size with makes the model unusable for larger documents collection. Therefore our weighting scheme for TRSM discards some terms (by assigning low weights) that are insignificant for retrieval performance.

In order to examine top-k queries we present Mean Average Precision on Fig. 6. TF-IDF is treated as a baseline document model in search task. It is noticeable that in two test cases classic TRSM is worse that baseline method. Our weighting schemata for TRSM makes our model better or on the same level that the baseline. In second test run (Q_2) after significant drop after $k = 30$ represents the "vocabulary problem". Simply there are no terms in documents to match with the query and our model tries to match additional ones on the same level of precision baseline method.

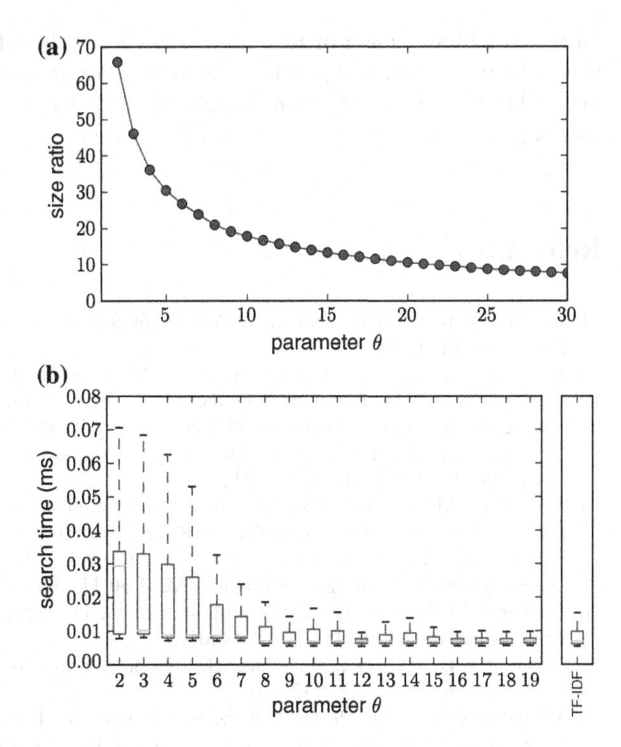

Fig. 7 Inverted index size and search time trade-off on TRSM-based information retrieval, **a** ratio of inverted index size between TRSM and TF-IDF bow model versus θ parameter, **b** inverted index scan time

4.3 Performance Analysis

Our document representation model introduces slight performance degradation in Information Retrieval system. Since the extended representation is larger (depending on θ parameter of TRSM model) than original TF-IDF representation inverted index takes up more disk space and it's scan operation is slower.

Figure 7a shows the inverted index size on disk by ratio of TRSM-based IR system and TF-IDF index (baseline). For given $\theta = 2$ index size is almost 70 times larger than the baseline. Figure 7b shows the index scan time built for both models. Index scan operation for TF-IDF model took 0.01 ms on average and it is again treated as a baseline. Nevertheless experiments presented in previous section was performed with $\theta = 10$ and shows good retrieval performance. For this model parameter index scan size is almost 20 times larger, but scan time degradation it's not noticable.

5 Conclusions

We presented an application of TRSM (tolerance rough set model) in Information Retrieval. We proposed a methods for weight learning to enhance the search effectiveness of the Information Retrieval system based on TRSM. The experiments on the ApteMod version of Reuters-21578 corpus show that TRSM enhanced by weight learning algorithm performs better than the standard TF-IDF with respect to the MAP (Mean Average Precision) measure when over 40 documents are requested.

References

1. Pawlak, Z.: Rough Sets: Theoretical Aspects of Reasoning about Data. Kluwer Academic Publishers, Dordrecht (1991)
2. Kawasaki, S., Nguyen, N.B., Ho, T.B.: Hierarchical document clustering based on tolerance rough set model. In: Proceedings of the 4th European Conference on Principles of Data Mining and Knowledge Discovery. PKDD '00, pp. 458–463. Springer, London, UK (2000)
3. Ho, T.B., Nguyen, N.B.: Nonhierarchical document clustering based on a tolerance rough set model. Int. J. Intell. Syst. **17**, 199–212 (2002)
4. Blair, D.C., Maron, M.E.: An evaluation of retrieval effectiveness for a full-text document-retrieval system. Commun. ACM **28**(3), 289–299 (1985)
5. Furnas, G.W., Landauer, T.K., Gomez, L.M., Dumais, S.T.: The vocabulary problem in human-system communication. Commun. ACM **30**(11), 964–971 (1987)
6. Carpineto, C., Romano, G.: A survey of automatic query expansion in information retrieval. ACM Comput. Surv. **44**(1), 1:1–1:50 (2012)
7. Manning, C.D., Raghavan, P., Schtze, H.: Introduction to Information Retrieval. Cambridge University Press, New York (2008)
8. Voorhees, E.M.: The cluster hypothesis revisited. In: Proceedings of the 8th Annual International SIGIR Conference on Research and Development in Information Retrieval, SIGIR '85, pp. 188–196. ACM, New York, NY, USA (1985)
9. Leuski, A.: Evaluating document clustering for interactive information retrieval. In: Proceedings of the Tenth International Conference on Information and Knowledge Management, CIKM '01, pp. 33–40. ACM, New York, NY, USA (2001)
10. Agirre, E., Arregi, X., Otegi, A.: Document expansion based on wordnet for robust IR. In: Proceedings of the 23rd International Conference on Computational Linguistics: Posters. COLING '10, Association for Computational Linguistics, pp. 9–17. Stroudsburg, PA, USA (2010)
11. Świeboda, W., Meina, M., Nguyen, H.S.: Weight learning for document tolerance rough set model. In: Lingras, P., Wolski, M., Cornelis, C., Mitra, S., Wasilewski, P. (eds.) Eight International Conference on RSKT. Lecture Notes in Computer Science, vol. 8171, pp. 385–396. Springer, Berlin (2013)
12. Skowron, A., Stepaniuk, J.: Tolerance approximation spaces. Fundam. Inform. **27**(2/3), 245–253 (1996)
13. Feldman, R., Sanger, J.: Text Mining Handbook: Advanced Approaches in Analyzing Unstructured Data. Cambridge University Press, New York (2006)
14. Deerwester, S., Dumais, S.T., Furnas, G.W., Landauer, T.K., Harshman, R.: Indexing by latent semantic analysis. J. Am. Soc. Inform. Sci. **41**(6), 391–407 (1990)
15. Gabrilovich, E., Markovitch, S.: Computing semantic relatedness using wikipedia based explicit semantic analysis. In: Proceedings of the 20th International Joint Conference on Artificial Intelligence, pp. 1606–1611, (2007)
16. Janusz, A., Slezak, D., Nguyen, H.S.: Unsupervised similarity learning from textual data. Fundam. Inform. **119**(3–4), 319–336 (2012)

An Analysis of Contextual Aspects of Conceptualization: A Case Study and Prospects

Krzysztof Goczyła, Aleksander Waloszek and Wojciech Waloszek

Abstract In this chapter we present a new approach to development of modularized knowledge bases. We argue that modularization should start from the very beginning of modeling, i.e. from the conceptualization stage. To make this feasible, we propose to exploit a context-oriented, semantic approach to modularization. This approach is based on the Structural Interpretation Model (SIM) presented earlier elsewhere. In the first part of thischapter we present a contextualized version of the SYNAT ontology developed using the SIM methodology. For the approach to be useful in practice, a set of tools is needed to enable a knowledge engineer to create, edit, store and perform reasoning over contextualized ontologies in a flexible and natural way. During our work on the SYNAT project, we developed such a set of tools that are based on a mathematical ground of tarset algebra (also introduced elsewhere). In the second part of this chapter we give a deeper insight into some aspects of using these tools, as well as into ideas underlying their construction. The work on contextualization of knowledge bases led us to further theoretical investigation of hierarchical structure of a knowledge base systems. Indeed, in a system of heterogeneous knowledge sources, each source (a knowledge base) can be seen in its own separate context, as being a part of a higher level contextual structure (a metastructure) with its own set of context parameters. Higher levels of the knowledge hierarchy do not substantially differ

This work was partially supported by the Polish National Centre for Research and Development (NCBiR) under Grant No. SP/I/1/77065/10 within the strategic scientific research and experimental development program: "SYNAT-Interdisciplinary System for Interactive Scientific and Scientific-Technical Information".

K. Goczyła (✉) · A. Waloszek (✉) · W. Waloszek (✉)
Gdańsk University of Technology, Gdańsk, Poland
e-mail: kris@eti.pg.gda.pl

A. Waloszek
e-mail: alwal@eti.pg.gda.pl

W. Waloszek
e-mail: wowal@eti.pg.gda.pl

R. Bembenik et al. (eds.), *Intelligent Tools for Building a Scientific Information Platform: From Research to Implementation*, Studies in Computational Intelligence 541,
DOI: 10.1007/978-3-319-04714-0_6,

from basic SIM structure, so that all context-aware services, including reasoning, can be performed. The theoretical background of this conception is presented in the third part of this chapter.

Keywords Knowledge base · Ontology · Modularization · Tarset · Contextualization

1 Introduction

1.1 Foreword

In this chapter we described the first systematic practical case study for deployment of our original methods of knowledge base modularization (the methods are described in details in [1]; they are tarset approach [2] and SIM [3]—Structured Interpretation Model—knowledge base organization; a short introduction to these methods is presented in Sects. 1.2 and 2.4). This case study allowed us to illustrate these methods and resulted in abundance of conclusions and recommendations for further development.

In the first part of this chapter (first part of Sect. 2) we report our work on contextualizing a fragment of SYNAT ontology, taking into account all aspects differentiating the process of conceptualization characteristic for contextual knowledge bases from the process for non-contextual ones. The main aspect we put stress on was the manner of implementing social roles and activities. To analyze the issue systematically, we distinguish the role- and concept-centric approaches, the former more often used during non-contextual, and the latter during contextual conceptualization. We illustrate the role-centric approach showing knowledge bases founded on the c.DnS ontology.

In the main part of Sect. 2 we describe our work on contextualizing a fragment of SYNAT ontology. This case study should be perceived as an experiment, whose goal was to gather as much experience as possible in order both to verify and in future develop modularization methods proposed by our group. Its results show that the proposed methods allow a knowledge engineer to build structures that significantly ease formulating queries and inserting data. These results are also very beneficial for our further research, because they let us to draw some important conclusions. Some of them we already managed to bring into life, e.g. optimization and improvement of the algorithm of reasoning. The others let us to formulate new ideas, for example, a proposal of controlled incomplete reasoning (see Sect. 3).

In our opinion, the most valuable result is the idea of a truly universal framework for all contextual approaches, where contexts would be substantial elements of designed models from the very commencement of the conceptualization stage. This framework would enable one to design knowledge bases in many levels with use of a recurrent mechanism embracing every layer of knowledge representation structure. The last part of this chapter (Sect. 4) introduces the idea and provides evidence that it is feasible due to properties of our previous proposition.

1.2 Preliminaries

In our works in SYNAT project we concentrated on the *tarset theory of knowledge bases* and the *SIM* model. Both of them were applied in the practical tool called *CongloS* (see http://conglos.org) In this section we shortly describe these results. For more details we refer the Reader to our previous publications [1–3].

Tarset knowledge bases consist of modules called *tarsets* (originally named *s-modules*, see [2]). Every tarset T is a pair $(\mathbf{S}, \mathbf{W})$ where $\mathbf{S}$ is a *signature* (or a *vocabulary*, i.e. a set of names), and $\mathbf{W}$ is a set of Tarski style interpretations of the vocabulary $\mathbf{S}$.

To formally define a tarset knowledge base we introduce a *tarset variable* or *a module* (the notion similar to a *relation variable* in relational databases). Tarset variables are basically labels. We assign each tarset variable a value, i.e. a pair $(\mathbf{S}, \mathbf{W})$, this value may be perceived as a single point from the space of all possible tarsets $\mathbf{T}$.

In the space of tarset we define a set of operations, which together constitute tarset algebra (see [1, 2] for more details). Tarset algebra is a variant of a cylindric algebra, one of the simplest operations is intersection, defined as $(\mathbf{S}, \mathbf{W}_1) \cap (\mathbf{S}, \mathbf{W}_2) = (\mathbf{S}, \mathbf{W}_1 \cap \mathbf{W}_2)$; since this operation reduces the set of possible models, effectively all the conclusions from both the tarsets being intersected can be drawn from the intersection.

We can constrain values of tarset variables using a *coupler*. A coupler syntactically is an *inequality* of two tarset algebra expressions. The *arity* of a coupler is the number of tarset variables contained in both expressions of the inequality. For example, a coupler containing three tarset variables T_1, T_2 and T_3 (a *ternary coupler*) is the inequality: $T_1 \leqslant T_2 \cap T_3$ (this coupler means that all conclusions inferred from the tarsets T_2 and T_3 hold in the tarset T_1). We define an *instance of a tarset knowledge base* as a set of tarset variables (together with their assignments) and a set of couplers containing these variables.

Tarset variables and couplers are sufficient to express semantic properties of modular bases created according to existing modularization techniques, as for example DDL [4], but they still lack very vital information concerning admissible changes of the structure of a knowledge base. This gap is filled by the notion of a *schema of a tarset knowledge base*. A schema consists of definitions of *tarset types* and *coupler types*. To satisfy a schema, an instance of a knowledge base has to contain tarset variables and couplers such that every tarset variable has to be assigned a tarset type defined in the schema, and every coupler has to be assigned a coupler type defined in the schema.

Due to the notion of schema, tarset model gains flexibility which allows for describing within it also other models of modular knowledge bases. In our work we performed this task for SIM contextual knowledge bases [1], expressing its types of modules (context types and context instances) as tarset types and its three types of relationships between modules (instantiation, aggregation, and inheritance) as coupler types. (More information on SIM model is presented in Sect. 2.).

The CongloS system is a tool allowing a user to create tarset knowledge bases accordingly to the SIM method. Technically it is a set of plug-ins for Protégé—the most popular editor of ontologies in the OWL standard (http://protege.stanford.edu/). These plug-ins enrich the interface of the editor by adding elements necessary from the point of view of the SIM method and provide the system with mechanism for management of the structure of the knowledge base being edited. One of the most important plug-ins is an inference engine capable of reasoning from contextual knowledge bases.

While focused on Protégé, ConglоS is a flexible tool that may be used in different configurations and consists of subsystems suitable also for handling general tarset knowledge bases (and for reasoning from them). ConglоS can be adjusted to different target configurations by replacing specially distinguished adaptation layer.

2 Contextualizing SYNAT Ontology

2.1 SYNAT Ontology as a Case Study for Contextualization

Originally, our choice of SYNAT ontology (see [5]) as an object of the case study was simply justified by our participation in this project. However, SYNAT ontology turned out to have some features that make it a remarkable experimental material for such research. Firstly, it is an ontology with a very broad domain of interest which embraces different but strictly related problem sub-domains. Examples of the sub-domains are various aspects of organization of institutions, their structure, activities, effects of activities, events, work of people, their career paths, geographical locations, etc. Diversity of the domains alone suggests the possibility of distinguishing some subsystems, which, while being autonomous, are connected with others by various relationships.

Secondly, SYNAT ontology bears many characteristic features of constructive stance (discussed in Sect. 2.2, introduced in [6]), that is to say some specific set of solutions directed towards multi-aspectual description of diverse phenomena within a non-contextual ontology. Such a non-contextual ontology must simultaneously display attributes of high-level ontology and of operational knowledge base, so it must be able to be expanded by facts in order to make reasoning over them possible (according to Gruber [7], the former should be characterized by the weakest possible ontology commitment, while the latter should strengthen the commitment, making it possible to particularize meaning of terms to one selected context). As a result predicates of high arity are used there to precisely capture all the contextual information needed.

The main purpose of the SYNAT ontology is to gather information about objects and events concerning academic and scientific society. It contains 472 classes and 296 properties.

The ontology is internally divided into five parts determined by five generic concepts (see Fig. 1):

1. *Agent*—the concept being an ancestor of human agents, organizations and groups.
2. *InformationResource*—the extension of this concept contains different carriers of information, mainly physical like documents, or electronic like audio and video resources.
3. *Event*—the concept represents all kinds of events associated with scientific and academic activity.
4. *Project*—concerns scientific projects.
5. *Characteristic*—this concept describes all kinds of abstract entities reifying properties of instances of all other concepts.

These five concepts are roots of separated trees of taxonomies. First four concepts embrace entities existing in the reality. The last one is the essence of the ontology, typical for the constructive approach. It is the root for the richest tree of classes defining reified properties of different kinds and arities for all the other classes of individuals. The best example is characteristic of information resources gathered under the class *InformationResourceCharacteristic*. While the class *Information-Resource* describes only physical carriers, its characteristic contains information about the content. First, there is a group of concepts describing content types, e.g. Catalogue, Guide, Encyclopedia, Norm, Patent, News, Report, etc. Then there are concepts for periodical resources, for describing structure of documents, for access conditions, or a very rich group of notions concerning web resources.

Another example is characteristic of persons. The main concept for this branch of conceptualization is *PersonCharacteristic*. It contains basic personal data, like names and addresses, as well as all data connected with the career in science and education. Of course, the organization of the data relies on reification of relations (see Fig. 2).

The ontology is divided into four modules, however this division is not realized as modularization in its common sense, i.e. as a way allowing to reason from smaller parts of knowledge. The purpose of the division is to separate information taken from two external ontologies, GIO (Geographic Information Objects, see [8]) and SCPL (Science Classification elaborated by the Polish government), and to adapt them to the needs of designed structure of concepts. This design decision allows a user to import the original ontologies without harming existing classification.

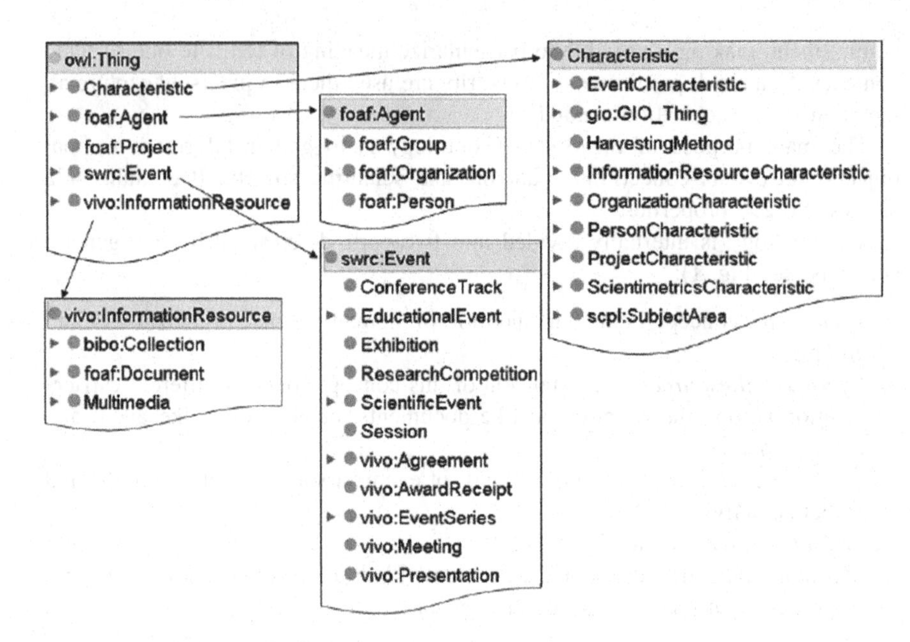

Fig. 1 The main branches of taxonomy in the SYNAT ontology (from [5])

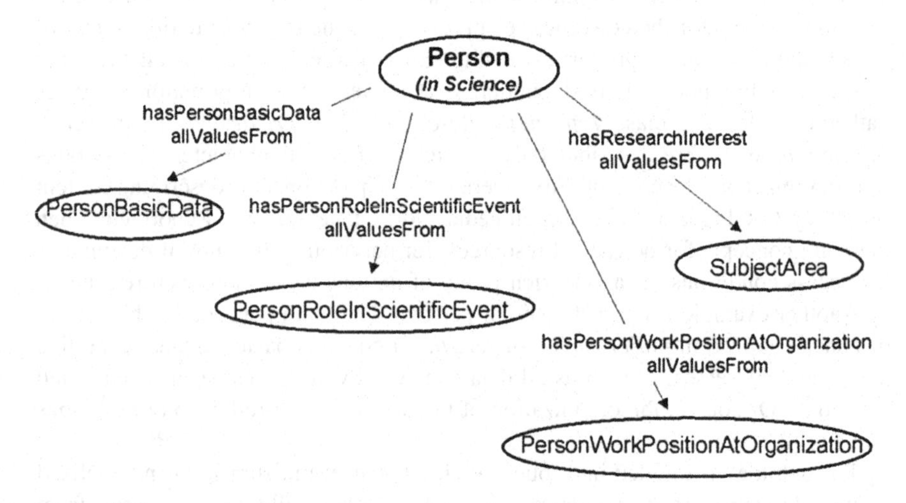

Fig. 2 Reified *n*-ary properties describing persons (from [5])

The SYNAT ontology is also ready to reuse some other globally known ontologies (available under "free" licenses, like CC and BSD), like for example FOAF [9], VIVO [10], SWRC [11] or BIBO [12]. This is realized by relating local terms to the terms defined by those ontologies with axioms (the ontologies are imported by the SYNAT ontology).

2.2 c.DnS as a Classical Example of Constructive Approach

Inclusion of contextual information within description of facts (added to an ontology in order to enable factual reasoning) is connected with constructive approach and is mainly due to the requirement of monotonicity of an ontology. This requirement prevents introduction to an ontology sentences that are in contradiction with its current contents. For example it is infeasible to state that "John Doe is a rector" and then "John Doe is not a rector" to model a situation when John Doe ends his term of office and is not elected a rector for the next period. This is the reason why an operational knowledge base cannot constrain the meaning of the concept *Rector* to "current rector".

Non-contextual knowledge bases must therefore conform to the requirement of monotonicity, but without constraining the meaning of the used terms. Consequently, descriptions of various phenomena within such bases have to contain large amounts of additional contextual information. For instance, the fact of employment of a person in some institution has to be expressed as a predicate of very high arity containing information about the time of the employment, work post, position etc.

c.DnS ontology is an extreme example of constructive approach. It was described in [6]. The abbreviation means Constructive Descriptions and Situations and it reflects the intension of its authors to describe situations in a constructive way. According to a philosophical stance called constructivism, reality is described as a mental structure depending on a context and for a specific reason with use of properly chosen system of notions. The authors, recalling this stance, propose a system containing very few classes describing the most general objects of a domain of interest (e.g. *Entity*, *Social Agent*, *Situation*, *Collection*, *Information Object*, *Time*), and also some meta-classes (e.g. *Description*, *Concept*) for objects that represent elements of a description of a domain, not a domain itself.

The meaning of the main classes is as follows:

1. *Description*—individuals of this class represent a kind of conceptualization, e.g. laws (government regulation between others), plans, projects, etc.; they are mainly related to individuals representing the classes *Concept* and *Entity*.
2. *Situation*—individuals represent things which have to be described, e.g. facts, legal cases, actions (realizing plans or projects), etc.; the relations connect this individuals with instances of *Description*, other situations (as sub- or super-situations) and entities.
3. *Concept*—individuals represent concepts defined by a description; concepts are related to collections and entities.
4. *Entity*—the extension represents anything what exists. This concept is a superclass of all others. The individuals are divided into two groups: schematic and non-schematic. They are related to instances of *Collection*, *Information Object* and other entities.

5. *Social Agent*—the only role played by an individual being an instance of this class is to describe situations or share it with other agents.
6. *Collection*—is the super-class for groups, teams, collectives, associations, collections, etc. A special case is a collective of agents sharing one description of a given situation.
7. *Information Object*—an information object may express a description and be about an entity.
8. *Time*—individuals represent time intervals and are properties of instances of the *Description* and *Situation* classes.

Every knowledge base is understood as a *c.DnS* relation, i.e. a set of tuples, each of which contains eight elements:

$$c.DnS(d, s, c^*, e^*, a^*, k^*, i^*, t^*) \rightarrow D(d) \wedge S(s) \wedge C(c^*) \wedge E(e^*) \wedge A(a^*) \wedge K(k^*) \\ \wedge I(i^*) \wedge T(t^*)$$

The asterisk means that a given variable is an ordered list of values of this same type.

The meaning of a tuple is as follow: a social agent a (A = *Social Agent*) as a member of knowledge communities k^* (K = *Collection*) perceives a situation s (S = *Situation*) and describes it using a description d (D = *Description*) and assigning entities e^* (E = *Entity*) with concepts c^* (C = *Concept*). Information objects i^* (I = *Information Object*) are supposed to express the description d. Time intervals t^* (T = *Time*) play two roles: first, they are temporal attributes for s, informing when the situation occurred; secondly they should describe a time interval when the description was carried out.

An example of a c.DnS tuple is:

$$\begin{aligned} c.DnS(&KnowledgeOfPreviousCases\#1, KillingSituation\#1, \\ &\{Precedent, Killer, Tool, HypotheticalIntention\}, \\ &\{Event\#1, PhysicalAgent\#1, Tool\#1, Plan\#1\}, \\ &Detective\#1, InvestigationTeam\#1, CriminalCode\#1, \\ &\{TimeOfEvent\#1, TimeOfInterpretation\#1\}) \end{aligned}$$

The tuples may be *projected* on relations of lower arity, the main projections are depicted in Fig. 3. The bold gray arrows in the figure represent ternary relations that include time as an additional parameter. The dashed arrows represent derived relations (e.g. a description unifies a collection iff the description defines a concept which covers the collection).

The real power of cDnS ontology is *redescription*. Redescription allows us to describe this same situation from another points of view. For example, while the above tuple describes a situation of intentional murder, during an investigation the interpretation of the same event may be changed to self-defense:

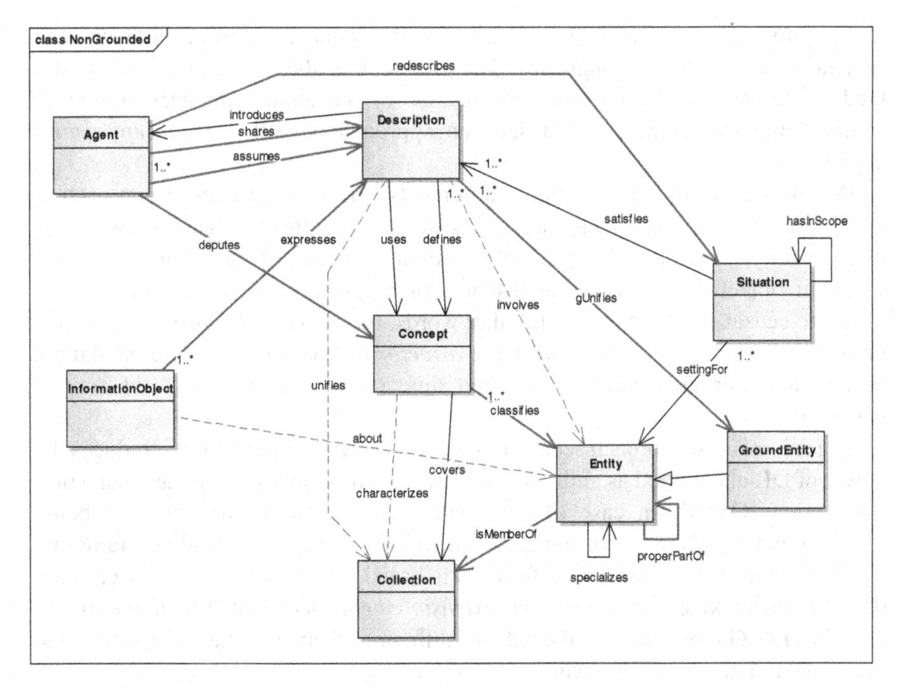

Fig. 3 The basic concepts and roles from c.DnS ontology in the form of UML diagram

$$
\begin{aligned}
&c.DnS(KnowledgeOfPreviousCases\#1,\ KillingSituation\#1,\\
&\quad \{Precedent,\ Killer,\ Tool,\ HypotheticalIntention\},\\
&\quad \{Event\#1,\ PhysicalAgent\#2,\ Tool\#1,\ SelfDefence\#1\},\\
&\quad Detective\#2,\ InvestigationTeam\#2,\ CriminalCode\#1,\\
&\quad \{TimeOfEvent\#1,\ TimeOfInterpretation\#2\})
\end{aligned}
$$

2.3 Role-Centric Versus Concept-Centric Approach

The presented extreme case of the constructive approach enables us to characterize it more precisely. For purpose of our argumentation we call it a *role-centric approach*. It is justified by the fact that the main property of this approach is cessation of designing hierarchic taxonomies of classes towards graph solutions. For example, if we want to model the property of *being a rector*, we introduce (in the simplest case) a binary predicate relating a given person to an abstract object called for example *rector's post* instead of declaring this person as an instance of a concept (e.g. *Rector*).

Assigning an individual to an extension of a concept is in accordance with specific character of DL languages. The grammar of these languages bases on a kind of descriptions, what therefore enables to formulate sentences similar to natural language utterances. Let such an approach be called a *concept-centric approach*.

Despite of the mentioned above arguments, the concept-centric approach is rarely used by knowledge engineers. The reason is, already recalled above, the monotonicity principle. This principle eliminates ability of modeling a situation when an object ceases to be an instance of a given class, for example, when someone ceases to act as rector. In other words, the concept *Rector* has too strong ontological commitment, because *being a rector* means someone who is taking a post of a rector in a certain place and time (i.e. for given values of context parameters).

The role-centric approach seems to have big expressive power but it reduces the power of DL understood as ability of terminological reasoning with the open world assumption (OWA). In case of the role-centric approach this power is being replaced by the power of reasoning similar to the one used in logical programming based on Horn rules—reasoning from facts based on the close world assumption (CWA). Such a kind of reasoning has *extensional character* rather than *intensional* one. In [13] Guarino has explained the difference between the extensional and intensional character of reasoning.

The extensional reasoning has its origin in the definition of conceptualization given in [14]. By Genesereth and Nilsson conceptualization is defined as a structure (Δ, R) where Δ is a domain and R is a set of relations on Δ. Guarino criticized this definition arguing that it is extensional while the process of conceptualization has intensional character. Calling this definition extensional Guarino meant the mathematical understanding of the set of relation R. Every element of this set is an n-ary relation on Δ, i.e. a subset of Δ^n. During conceptualization a human mind takes into account not only currently existing relationships, but also all possible relationships in all possible worlds.

Such kind of relations Guarino defines as *conceptual relations*. To consider this fact we should first define a *domain space* as a pair (Δ, W) where W is a set of all possible worlds. Formally, a conceptual relation ρ of arity n is a function $W \to \mathcal{P}(\Delta^n)$, from all possible worlds to the powerset of Δ^n. The image $\rho(W)$ is, by Guarino, the set of all *admittable extensions* of ρ, i.e. all possible (mathematical) relations of one kind in all possible worlds.

Finally, conceptualization is a triple $\mathbf{C} = (\Delta, W, \mathcal{R})$ where $\mathcal{R}$ is a set of conceptual relations defined in a domain space (Δ, W). A conceptualization is, according to this definition, much richer phenomenon because every ρ from the set $\mathcal{R}$ assigns to it a lot of its admittable extensions (mathematically defined relations).

The concept-centric approach gives us much better ability to implement intensional reasoning in the sense given by Guarino. Describing someone's profession by a concept, i.e. unary predicate, allows us to utilize all advantages of DL languages in describing relations between extensions of given concepts. For example, it is easy to say that someone who is a *Rector* has always to be a *Professor* using a construct stating subsumption between them. This statement restricts the set of possible interpretations to only those corresponding to the set of possible worlds. Formulating statements uttering similar meanings in role-centric approach is much more difficult.

Entering data is also much more complicated. As an example let us say, according to the SYNAT ontology, that an individual *johnDoe* is a Rector. To do it we have to state the following statements:

Person(johnDoe)

PersonWorkPositionAtOrganization(personPositionAtOrganization_JD)

Rector(rector_JD)

hasPersonPositionAtOrganization(johnDoe, personPositionAtOrganization_JD)

hasRoleAtOrganization(personPositionAtOrganization_JD, rector_JD)

All these sentences replace one simple statement *Rector(johnDoe)* used in the concept-centric approach. In consequence, the role-centric approach produces a lot of additional individuals, that have no equivalent in the reality, and a lot of sentences relating these individuals to each other.

Sometimes ontology designers try to reduce the aforesaid faults. The good example is the ontology of fraudulent disbursement described in [15, 16]. Its architecture is depicted in Fig. 4. This is a modular ontology, and the only mapping used to connect modules is the OWL import statement. In the modular structure the cDnS ontology is used as a top level ontology.

The core layer of the structure is formed by four modules: $\text{c.DnS}_{\text{Crime}}$ contains description of crime situations, $\text{c.DnS}_{\text{Inquiry}}$ contains information about inquiries managed by public prosecutors, $\text{c.DnS}_{\text{Investigation}}$ describes investigations conducted by the police. The last module, $\text{c.DnS}_{\text{Workflow}}$, contains description of situations that cannot be classified to any of the first three modules. The lowest layer contains modules specializing $\text{c.DnS}_{\text{Crime}}$.

Of course, every information inserted into the ontology makes a proper c.DnS tuple. However, the authors created a parallel taxonomic structure of domain concepts inheriting from the c.DnS concept *Entity*. A set of specifically defined axioms make a reasoner to infer extensions of these concepts basing of binary relations they are involved in. As a result we can easier find a given individual and its properties. This parallel structure mitigates some of the drawbacks of the role-centric approach, however at the cost of making the ontology somewhat more complex. In addition, modularization of the ontology divides information into smaller portions. This division helps to keep better control over quickly growing amount of data and to avoid reasoning from the entire ontology.

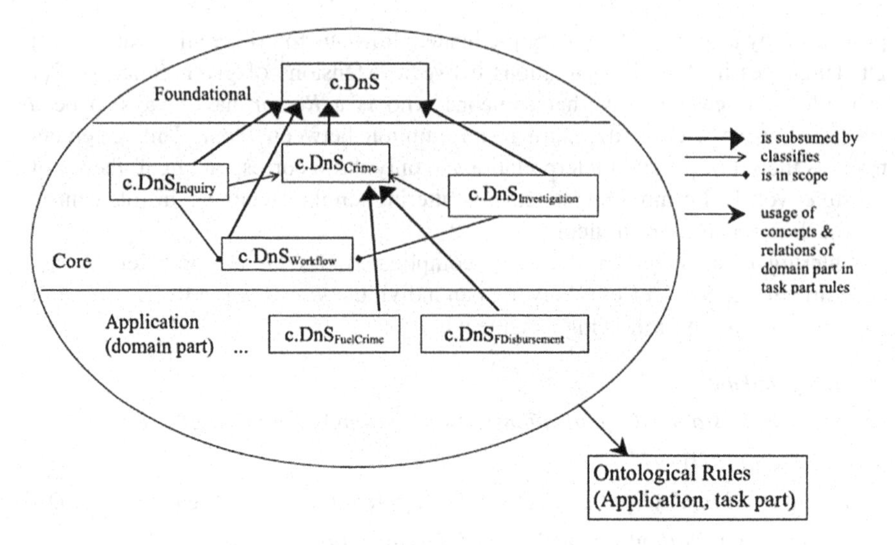

Fig. 4 The architecture of the fraudulent disbursement ontology (from [16])

2.4 SIM Model as a Contextual Approach

SIM model gives hope for more extensive use of concept-centric approach without losing the operability of a knowledge base. It is possible due to the fact that a SIM base is divided into modules-contexts, and every context is characterized by its own set of contextual parameters (not necessarily made explicit). As a result, the meaning of each concept can be particularized without restriction simply by using such a term in a proper place within knowledge base structure.

Since SIM model has been thoroughly described in other publications (e.g. [1, 3]), here we only very briefly review the basics of the method.

According to the SIM method TBox and ABox of an ontology constitute two structures organized in different manners. Division of TBox is based on a relation very similar to OWL *import*. Contextualized TBox $T = T(\{T_i\}_{i \in I}, \trianglelefteq)$ is a set of *context types* $\{T_i\}$ (partial TBoxes) connected with *inheritance* relation $\trianglelefteq$ established on a set of indexes I. $\trianglelefteq$ is a poset with a minimal element m, T_m being a *top context type*.

A contextualized TBox cannot have a model by itself. What determines the structure of the interpretation is a division of ABox. Contextualized ABox is defined as $A = (\{A_j\}_{j \in J}, inst, \ll)$ and consists of *context instances* $\{A_j\}$ (partial ABoxes), function $inst$: $J \rightarrow I$ (called *instantiation function*) assigning each partial ABox to its partial TBox and the *aggregation relation* $\ll$ established on a set of indexes J; $\ll$ is a poset with a minimal element n, A_n being a *top context instance*, $inst(n) = m$. Each partial ABox has its local interpretation defined as $\mathcal{I}_j = (\Delta^{\mathcal{I}j}, \bullet^{\mathcal{I}j})$, $\bullet^{\mathcal{I}j}$ assigns elements of the domain to concepts and roles defined in $T_{i:\ inst(j)\ =\ i}$ and its ancestors. Contextualized interpretation $\mathcal{I} = (\{\mathcal{I}_j\}_{j \in J}, \ll)$ is a set of local interpretations connected with the relation $\ll$ (the same as in the case of ABoxes). The flow of conclusions is assured by aggregation conformance rules obligatory for all models:

(1) $\bigcup_{k \in \{k:j \ll k\}} \Delta^{\mathcal{I}k} \subseteq \Delta^{\mathcal{I}j}$,
(2) $\bigcup_{k \in \{k:j \ll k\}} C^{\mathcal{I}k} \subseteq C^{\mathcal{I}k}$,
(3) $\bigcup_{k \in \{k:j \ll k\}} R^{\mathcal{I}k} \subseteq R^{\mathcal{I}k}$,
(4) $a^{\mathcal{I}k} = a^{\mathcal{I}j}$ for $j \ll k$.

Generally, the nodes of a SIM knowledge base structure may be perceived as tarsets, and the relationships are couplers. The transformation is straightforward. There are only two tarset types predefined: the first one for context types and the second one for context instances. And, correspondingly, there are three coupler types: the first one for expressing inheritance, the second one for instantiation, and the third one for aggregation relationships.

2.5 Contextualization Process

During our work we conducted a case study: contextualization, in the form of SIM knowledge base of a fragment of the domain of interest covered by SYNAT ontology. Within the case study we set ourselves a goal to exploit concept-centric approach and to investigate whether its use allows for preserving operability of a knowledge base.

As the fragment of the domain we picked subjects covered by the factual part (ABox) of SYNAT ontology. These subjects embrace publishing of articles, their editors, and employment of persons in institutions.

As an indicator of operability we chose conformance to the two sets of requirements:

1. Ability to answer the set of 21 competency queries, examples of which are presented below:
   (a) Which articles were written by person *x*?
   (b) By which institutions person *x* was employed?
   (c) Which persons were co-authors of the articles written by person *x*?
2. Ability to perform the set of 18 update operations to the base, examples of which are presented below:
   (a) addition of a new person,
   (b) addition of a new working post for a person,
   (c) addition of a new publisher.

As mentioned above, we took as a starting point the exemplary ABox provided with SYNAT ontology. A major part of the ABox is presented in Fig. 5. For the

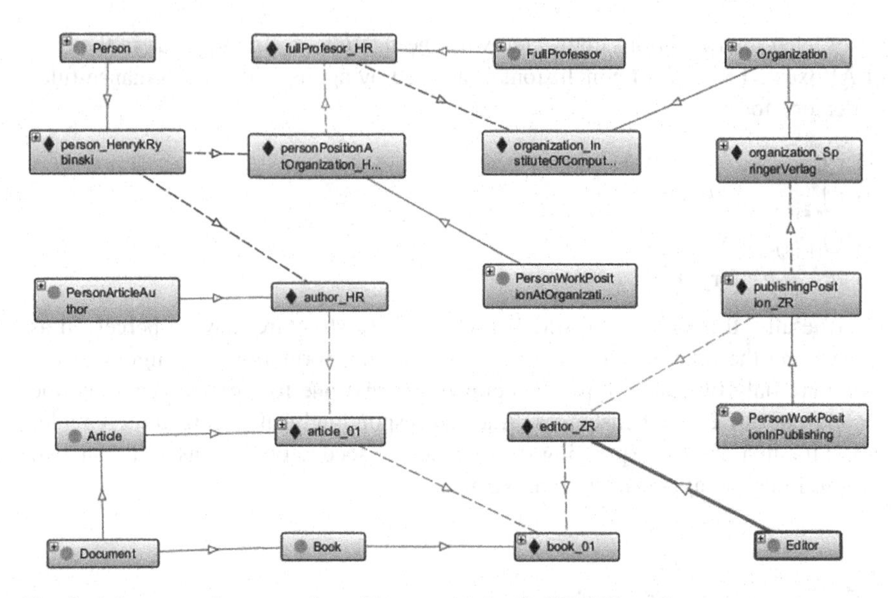

Fig. 5 A fragment of an exemplary ABox provided with SYNAT ontology

sake of this presentation we chose a fragment which allows for an answer to a query about which employees of Institute of Computer Science published articles in Springer-Verlag (in the further part of this chapter this query is denoted as Q). The picture was generated from *OntoGraf* plug-into Protégé editor.

The graph structure presented in Fig. 5 is typical for role-centric approach. It gives an ontology user considerable flexibility of specifying complex contextual information, however at the cost of complication of the structures that need to be inserted to the base. Ground objects (Agents and Information Resources) in the picture are *person_henrykRybinski*, *organization_InstituteOfComputerScience*, *organization_SpringerVerlag*, article_*01*, *book_01*. The remaining individuals are reifications of characteristics of the objects and specify relationships between them. For instance individual *author_HR* describes the authorship of the *article_01* by *person_henrykRybinski* (and can be associated with attributes one would like to relate to this authorship—e.g. its date). One of the cost of this flexibility is complexity of queries: the query Q is in fact query about instances of the following concept:

$$\exists hasPersonWorkPositionAtOrganization.\exists hasRoleAtOrganization.\exists holdsPersonRole AtOrganization.\{organization\_InstituteOfComputerScience\} \sqcap \exists hasPersonAuthorship. \exists isAuthorOfArticle.\exists isIncludedIn.\exists isEditorOf^{-}\exists hasPersonRoleInPublishing^{-}. \exists hasWorkPositionInPublishingOrganization.\{organization\_SpringerVerlag\}$$

As a first step to contextualization of the base we divided the domain of interest of the ontology into two fragments: objects we wanted to model as individuals (like in the original ontology) and objects we wanted to reflect as elements of

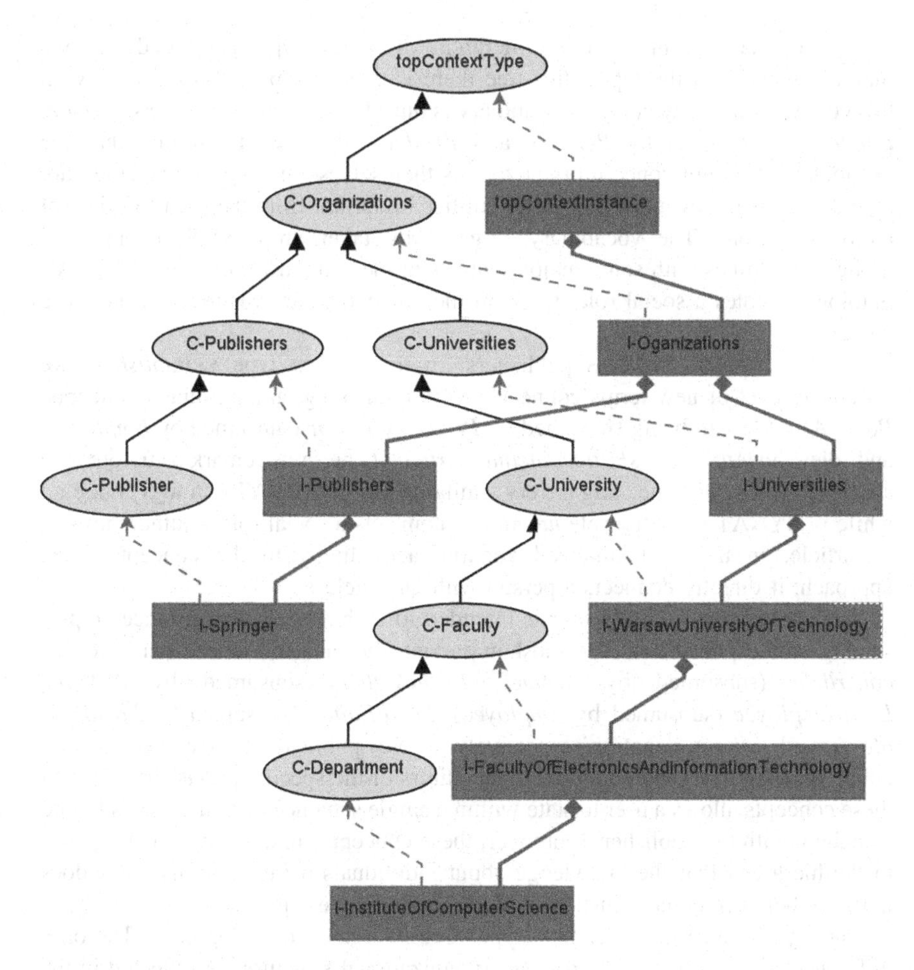

Fig. 6 Structure of contextualized SYNAT ontology

contextual structure. This decision has been made on the basis of analysis of competency queries and updates. As a result we decided to reflect in the contextual structure institutions like publishers, faculties, and institutes.

In the effect we obtained the modular structure of the knowledge base depicted in Fig. 6. In the figure context types are denoted with light ovals, and context instances with darker rectangles. The names of the context types are prefixed with *C-* and should be interpreted as referring to the contents of their context instances, e.g. the name *C-Universities* means that each instance of this context type represents a group of universities, similarly, the name *C-University* indicates that each context instance of this type represents a single university. The names of context instances are prefixed with *I-* and may carry additional information, e.g. about which university is represented by this instance.

As it can be seen in Fig. 6, the hierarchy of context types is divided into two major branches. At the top of this tree there is context type *C-Organizations*. In this contexts only very general vocabulary is introduced, namely concepts: *Person*, *Employee* (subsumed by *Person*), and *Product*. (It is worth noting, that the vocabulary does not concern organizations themselves—they are represented not as individuals but as modules—other entities important from the point of view of an organization.) The vocabulary is generally taken from SYNAT ontology, though sometimes with some major changes in meaning, as *Employee* in SYNAT ontology denotes a social role, while in the contextualized version it relates to a person.

The left branch concerns publishers: in the context type *C-Publishers* we introduce a set of new terms connected with publishing, among others concepts: *Book*, *Article* (both being subsumed by *Product*), *Editor* (subsumed by *Employee*) and roles: *hasAuthor*, *hasEditor*, *includesArticle*. (The same remark as before also applies to roles: their meaning is very similar to that in SYNAT ontology, but e.g.: while in SYNAT ontology role *hasAuthor* connects a social role—authorship—to an article, in the contextualized version, accordingly to the concept-centric approach, it directly connects a person with an article.)

The context type *C-Publisher* is intended to embrace context instances representing single publishers. The most important concepts introduced here are *LocalArticle* (subsumed by *Article*), *LocalAuthor* (subsumed by *Author*), *LocalEmployee* (subsumed by *Employee*), *LocalEditor* (subsumed by *LocalEmployee* and *Editor*), denoting respectively articles published by the specific publisher, their authors, and employees and editors of the specific publisher. Defining these concepts allows a user to state within a single sentence that an entity is bound somehow with the publisher. Moreover, these concepts are defined in such a place in the hierarchy that the knowledge about individuals being their instances does not flow between context instances representing single publishers.

The right branch of the context types tree is organized analogously. The only difference is that also some aspects of organizational structure are reflected in the tree: as a consequence, here a university is understood as both a single university and a group of faculties it includes (analogous relation takes place between a faculty and its institutes).

At the top level of the branch, in the context type *C-Universities*, several concepts are introduced, however, for our discussion the important one is *Professor*, subsumed by *Employee*. At the level of *C-University* new concepts are introduced: *UniversityEmployee* (subsumed by *Employee*) and *UniversityProfessor* (subsumed by *UniversityEmployee* and *Professor*). The concepts are analogous to concepts *LocalEmployee* and *LocalEditor* defined for publishers, and have their counterparts in the lower levels of the hierarchy: *C-Faculty* introduces *FacultyEmployee* and *FacultyProfessor*, and *C-Institute* introduces *InstituteEmployee* and *Institute Professor*.

In order to reflect the ABox fragment from Fig. 5 in this contextual structure, following assertions should be formulated in the context instance *I-Springer*:

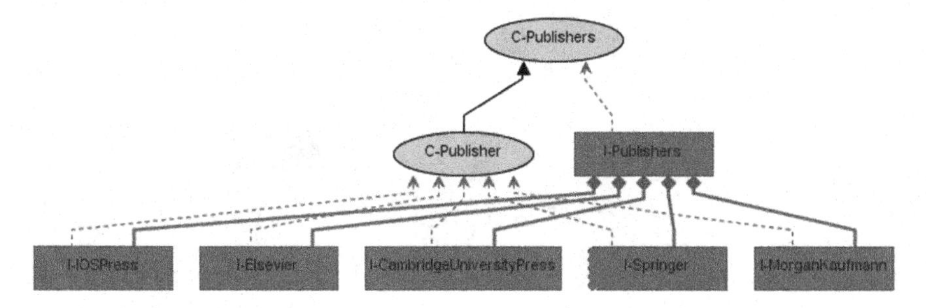

Fig. 7 A fragment of contextualized SYNAT ontology after addition of new publishers

$$LocalAuthor(person\_henrykRybinski)$$
$$LocalArticle(article\_01)$$
$$hasAuthor(article\_01,\ person\_henrykRybinski)$$
$$LocalBook(book\_01)$$
$$includesArticle(book\_01,\ article\_01)$$
$$LocalEditor(person\_ZbigniewRas)$$
$$hasEditor(person\_ZbigniewRas)$$

Additionally, in the context instance *I-InstituteOfComputerScience*, the single assertion has to be formulated: *InstituteProfessor(person_henrykRybinski)*.

This information suffices e.g. to answer query *Q*, which in the contextualized knowledge base has to be issued in two steps: first, one has to ask tarset-algebraic query for the module *I-InstituteOfComputerScience* ∩ *I-Springer*, and then ask the resulting module about instances of concept *InstituteEmployee* ⊓ *LocalAuthor*.

The resulting contextual knowledge base has been verified against all the competency queries and updates. All of the queries were possible to be issued to the base, although some of them have to be formulated as queries about a set of context instances. An example of such a query is "Which publishers published articles written by person *x*?", which has be expressed as a query about a list of context instances in which it is true that *LocalAuthor*(*x*) (such a form of query is covered by KQL language we use in CongloS).

Similarly, all the updates were possible to be performed, and some of them needed to be done by altering the contextual structure of the knowledge base. For instance, adding new publishers requires to insert new context instances: Fig. 7 presents an appropriate fragment of the knowledge base structure after such an update.

To sum up, the case study showed that it is possible to create a contextualized knowledge base in SIM model with the assumption of concept-centric approach which conforms to the specified requirements. This approach may lead to substantial simplification of inserting new knowledge and querying a base.

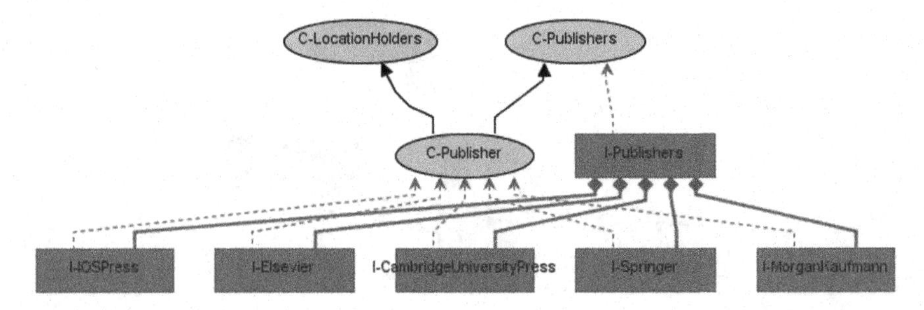

Fig. 8 A fragment of contextualized SYNAT ontology after addition of a new context type

The study was also an opportunity for observing a worrisome effect that may constitute an obstacle with use of the methods in more complex engineering solutions, namely the necessity of redefining *Professor* at each level of context hierarchy, which we plan to alleviate by introducing a new kind of coupler.

2.6 Conclusions

Analysis of the results obtained during the work allows us to draw some conclusions. The introduced structure is strictly related to the specified list of requirements. Extension of the list may lead to rendering the structure infeasible. In such a situation, a user of the knowledge base would find himself lacking of some important information, including perhaps implicit contextual parameters intended by a designer of the base.

Naturally, lack of this knowledge is simply a consequence of an incompatibility between specified requirements and user expectations, and as such might be neglected. Nonetheless, the ability to adjust the base to changing user needs is an important feature influencing the operability of the base. During the course of contextualization we identified the kind of changes in requirements that cause difficulties with the adjustment. Specifically, these are the changes that force to assign some attributes to the fragments of the original domain that have been modeled as context instances (or more generally, elements of the structure of the base) rather than individual objects.

An example of such a situation may be an addition of a new competency question about articles published by publishers from Berlin. It requires us to extend stored information by geographical locations of publishers, however there seems to be no natural way of doing this.

One of the possibilities here is to introduce a new context type *C-LocationHolders* containing a concept like *GeographicalLocation* and to inherit *C-Publisher* from this context type (see Fig. 8). A user may then assign each publisher a location by specifying an assertion like *GeographicalLocation*(*Berlin*) in *I-Springer*. However effective, this solution may easily lead to perplexity, as it mixes different levels of descriptions.

Analysis of this example leads to the conclusion that the most natural course of action is to assign attributes directly to publishers, which means directly to context instances. It in turn shows the direction of refining the description of contextual structure towards a manner similar to describing individual objects of the basic domain of interest. In the case of SIM method such a refinement might consist in treating context types in an analogous way to concepts and context instances to individuals. As a result we could gain a possibility of flexible adjustments of the knowledge base structure to changing user needs, without necessity of utilizing unnatural constructs.

Moreover, introduction of this analogy would open new possibilities for reasoning. The description of knowledge base structure could be then treated as conforming to OWA (Open World Assumption) and the relationships between modules could be inferred from some characteristic of the structure. Further elaboration of this idea led us to formulation of postulates for a new theory of information contexts described in Sect. 4.

Another issue identified during the experiment was the problem of complexity of reasoning in the relation to both the size of the structure and the contents of a knowledge base. The conducted case study embraces only a small fragment of real-world domain of interest. It is not easy to notice that the size of the base, assuming it indeed covers the majority of universities and their employees in Poland, would grow enormously.

It should be stressed that this problem also exists for non-contextual ontologies. Reasoning with OWA is very expensive and reasoning engines conforming to the assumption have difficulties with efficient handling of relatively small numbers of individuals (some attempts to overcome these limitations have been undertaken in [17], however, current trend involves rather sacrificing OWA, like e.g. in [18]).

Deployment of contextual approach should improve the effectiveness of reasoning: after all, contextualization means constraining only to relevant information, and as such should result in faster inferences. Unfortunately, with currently exploited algorithms and methods of knowledge representation, we do not observe the improvement of reasoning speed. This problem is discussed in much more details in Sect. 3, however we may briefly recall here the main reason of its occurrence: the manner of representing knowledge with the requirement of maintaining completeness of reasoning require to gather all of the sentences from the whole knowledge base in order to perform inferences from any point of view.

There are two possible ways of solving this problem. The first one consists in optimizing algorithms and improving the methods of representing contextual knowledge. Though we recorded some successes along this path, we find the second way much more promising. This way consists in supplying the inference engine with knowledge about elements of the structure of a base, embracing contextual parameters important for the process of reasoning. Such knowledge would allow the engine to neglect some of the portions of the base in some kinds of inferences (we call this *controlled incompleteness of reasoning*). This idea harmonizes very well with the aforementioned theory of information contexts and is elaborated upon in Sects. 3 and 4.

3 Reasoning Over Contextualized KB

Commencement of practical work over contextualizing SYNAT ontology was an opportunity to review and assess the principles of work of internal mechanisms which comprise CongloS system. One of the most important mechanisms to assess was the contextual inference engine.

In general, contextual approach was assumed to reduce the complexity of reasoning. Contexts are created to ease the process of inference by constraining the domain and the level of details of its description only to interesting aspects. However, sound and complete reasoning in full accordance with SIM definitions requires that the information from all the contexts in the base should be used.

The currently utilized methods employ description logics sentences (axioms and assertions) to represent contents of contexts. With this assumption, the only way to consider in the target context (the one representing the point of view from which we reason) conclusions from other contexts is by transferring sentences from one representation to another. Such a transfer is conducted basically by executing operations of tarset algebra described by couplers defined between modules of a knowledge base.

During our work over contextualizing SYNAT ontology we identified two fundamental problems related to this approach. The first problem consists in the fact that the transfer of sentences has to be all-embracing, i.e. has to embrace all the sentences from all the contexts, which is the consequence of the inference engine lacking information about relationships between knowledge in different contexts. The second problem stems from complicated structure of connections between modules (e.g. aggregation relation is represented by two cyclic couplers, i.e. the presence of aggregation between two modules M_1 and M_2 requires transfer of the knowledge—in the form of sentences—both: from M_1 to M_2 and from M_1 to M_2). Consequently, very often redundant knowledge was being transferred to the target module representation, resulting in vast growth of the number of sentences that had to be considered during reasoning.

The second problem is mainly of technical nature, and has already been mostly alleviated by introducing an optimization to the process of reasoning. The optimization consists in identification of redundant subsets of sentences. This task is carried out by performing graph simulations [19]. This solution is described in more details in Sect. 3.1.

The first problem, however, is much more significant. It can be partially solved by abandoning the sentential representation in favour of some other approach. Other kind of representation may be better suited for expressing knowledge at different levels of detail and its use may result in reducing the amount of information being transferred. This idea is illustrated by use of *cartographic representation* in Sect. 3.2.

Nonetheless, it seems that much more natural way to solve the first problem is to provide a user with a means of describing contents of specific contexts in intensional manner, thus specifying their influence on inference process. This information may be provided in various forms: by simple heuristics or by more systematic ontological description of the knowledge base structure. This family of solutions is discussed in Sect. 3.3.

3.1 Optimizations to Sentential Reasoning

In this subsection we describe the optimization to reasoning over tarsets represented in sentential manner. This optimization is crucial for knowledge bases with complicated structure of couplers between modules (tarset variables).

First, we have to briefly recall the principles of sentential representation of tarset contents, included in the theory of $\mathcal{L}$-(2)-**D** representability. Given the set of *dummy* names **D**, a tarset T is $\mathcal{L}$-(2)-**D**-*representable* iff there exists a (finite) set of sets of sentences $\mathcal{S} = \{\mathcal{O}_i\}_{i\in[1..k]}, k \in N$, being its $\mathcal{L}$-(2)-**D** *representation* (denoted $T \sim \mathcal{S}$), which means that all the conclusions that can be drawn from T can also be drawn from $\mathcal{S}$ and, conversely, all the conclusions that can be drawn from $\mathcal{S}$ which do not concern terms from **D**, can also be drawn from T.

The contents of modules of SIM knowledge base can be easily shown to be $\mathcal{L}$-(2)-**D** representable, since (1) the user initially fills the modules with sentences, and (2) the SIM couplers consist only of three tarset algebra operations: intersection, rename and projection, and all of them preserve $\mathcal{L}$-(2)-**D** representability. A constructive algorithm for calculation of $\mathcal{L}$-(2)-**D** representation (Algorithm A_1) for the results of these operations is sketched below:

1. If $T_1 \sim \mathcal{S}_1, \mathcal{S}_1 = \{\mathcal{O}_{1:i}\}_{i\in[1..k]}$ and $T_2 \sim \mathcal{S}_2, \mathcal{S}_2 = \{\mathcal{O}_{2:j}\}_{j\in[1..m]}$, then $T_1 \cap T_2 \sim \{\mathcal{O}_{1:i} \cup \mathcal{O}_{2:i} : i \in [1..k], j \in [1..m]\}$.
2. If $T \sim \mathcal{S}, \mathcal{S} = \{\mathcal{O}_i\}_{i\in[1..k]}$, then $\rho_\gamma(M) \sim \{\gamma\mathcal{O}_i : i \in [1..k]\}$.
3. If $T \sim \mathcal{S}, \mathcal{S} = \{\mathcal{O}_i\}_{i\in[1..k]}$, then $\pi_{\mathbf{S}}(T) \sim \{\gamma_{T,\mathbf{S},\mathbf{D}}(\mathcal{O}_i) : i \in [1..k]\}$; where by $\gamma_{T,\mathbf{S},\mathbf{D}}$ we understand a function renaming terms from T not included in **S** to terms from **D**.

In the case when no non-standard couplers with union have been defined by the user, the above formulas can be greatly simplified, by the observation that all the representations may contain only a single set of sentences:

1. If $T_1 \sim \{\mathcal{O}_1\}$ and $T_2 \sim \{\mathcal{O}_2\}$, then $T_1 \cap T_2 \sim \{\mathcal{O}_1 \cup \mathcal{O}_2\}$.
2. If $T \sim \{\mathcal{O}\}$, then $\rho_\gamma(M) \sim \{\gamma(\mathcal{O})\}$.
3. If $T \sim \{\mathcal{O}\}$, then $\pi_{\mathbf{S}}(T) \sim \{\gamma_{T,\mathbf{S},\mathbf{D}}(\mathcal{O})\}$; $\gamma_{T,\mathbf{S},\mathbf{D}}$ is defined as above.

In the case of a standard SIM knowledge base, the source of complexity of reasoning over sentential representation is twofold: it is a consequence of cyclic couplers and of projection operation (formula 3 in the above list) which produces new sentences.

Let us illustrate this effect with a simple example. We consider a knowledge base K with four modules (see Fig. 9): context types M_1 and M_2, and context

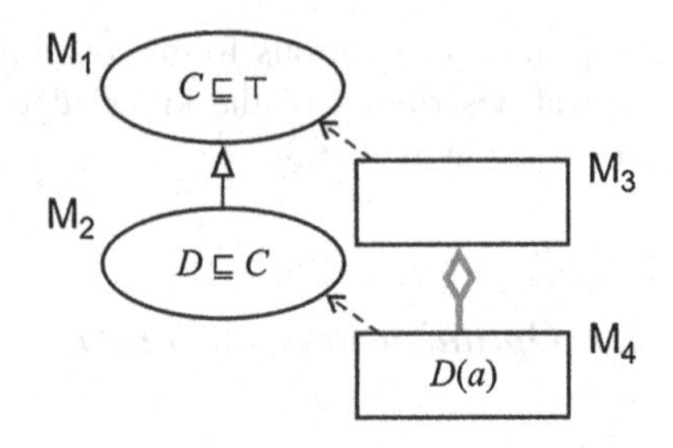

Fig. 9 An exemplary knowledge base K

instances M_3 (the only instance of M_1) and M_4 (the only instance of M_2). M_2 inherits from M_1, M_3 aggregates M_4, and initial contents of modules is: $M_1 := \{C \sqsubseteq \top\}$, $M_2 := \{D \sqsubseteq C\}$, $M_3 := \{\}$, $M_4 := \{D(a)\}$. Let us moreover assume that the set of dum*X*my names **D** consists of infinite (but countable) number of names of the form $_i$ (for concepts) and x_i (for individuals), $i \in N$.

Execution of a naïve implementation of algorithm A_1 for K would result in infinite growth of representations of modules M_3 and M_4. Below we present the results of some first iterations of the algorithm for these modules:

1. $M_4' \sim \{\{D(a), D \sqsubseteq C\}\}$
2. $M_3' \sim \{\{X_1(a), X_1 \sqsubseteq C\}\}$
3. $M_4'' \sim \{\{D(a), D \sqsubseteq C, X_1(a), X_1 \sqsubseteq C\}\}$
4. $M_3'' \sim \{\{X_1(a), X_1 \sqsubseteq C, X_2(a), X_2 \sqsubseteq C\}\}$

Numerous strategies exist that can be used to eliminate the above effect. One of the simplest ideas involves remembering which sentences, after projection, have already been transferred to other modules. This projection prevention strategy can be included into a category of pre-execution strategies as it prevents some of the superfluous transfers of sentences. However, the strategy requires a very careful control policy, as every projection has to be applied to a whole set of sentences (like in the above example, transfer of the conclusion that $C(a)$ to M_3 requires us to translate and transfer the whole set of sentences $\{X_1(a), X_1 \sqsubseteq C\}$).

After several experiments we found that the most effective solution is deployment of a simple post-execution strategy (hence removal strategy), which consists in removal of redundant subsets of sentences containing dummy terms. Post-execution strategies generally have an advantage over pre-execution ones, as they are not dependent on the set of tarset algebra operations.

The removal strategy consists of two major stages. In the first stage potentially redundant subsets are identified. To identify the subsets we used a procedure for calculating graph simulation [19], after changing the sentences in module representation to a graph form.

A simulation relation is defined for a labeled directed graph G understood as a quadruple (V, E, L, l) where V is a set of vertices, E set of edges, L set of labels and l is a function which assigns each vertex from V to a label from L. A relation $\leqslant \subseteq V \times V$ is a simulation iff for every $u \leqslant v$ it follows that (1) $l(u) = l(v)$ and (2)

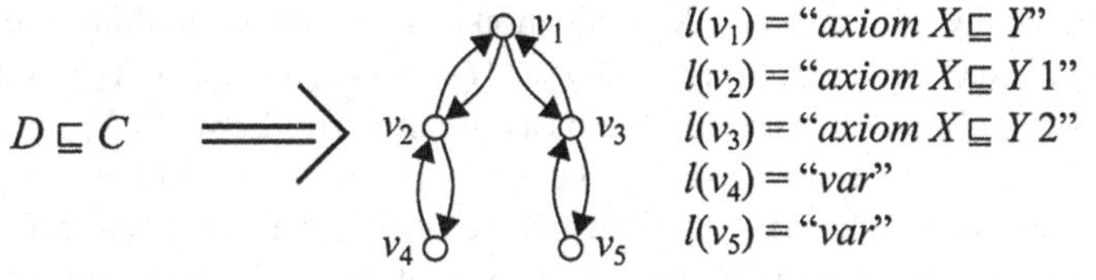

$D \sqsubseteq C \Longrightarrow$ v_1, v_2, v_3, v_4, v_5

$l(v_1) = \text{“}axiom\ X \sqsubseteq Y\text{”}$
$l(v_2) = \text{“}axiom\ X \sqsubseteq Y\,1\text{”}$
$l(v_3) = \text{“}axiom\ X \sqsubseteq Y\,2\text{”}$
$l(v_4) = \text{“}var\text{”}$
$l(v_5) = \text{“}var\text{”}$

Fig. 10 An illustration of converting a sentence to a graph (removal strategy)

for all vertices u' such that $(u, u') \in E$, there exists a vertex v' such that $(v, v') \in E$ and $u' \leqslant v'$.

A simulation calculation algorithm utilized for the first stage of removal strategy is an elimination algorithm, i.e. it calculates the next approximation $\leqslant^{i+1}$ of the $\leqslant$ on the basis of the current approximation $\leqslant^i$ by eliminating from $\leqslant^i$ pairs which do not conform to the conditions (1) and (2) from the previous paragraph, and stops after reaching a fixpoint. In the original algorithm the first approximation $\leqslant^1$ was calculated as $\{(u, v) \colon l(u) = l(v)\}$.

In order to use the simulation calculation algorithm for identifying redundant subsets of sentences, the $\mathcal{L}$-(2)-**D** representation is transformed into a tri-partite graph, where labels are strings of characters. The first partition consists of vertices representing axioms. Only axioms of exactly the same grammatical structure (e.g. $C \sqcap D \sqsubseteq E$ and $E \sqcap F \sqsubseteq D$) are assigned the same labels. Vertices in the second partition represent possible reassignments, i.e. i-th position in the grammar structure of each axiom. Their label is the concatenation of the label of the axiom and the number of the position. The third partition embraces vertices representing actual terms used in grammar structure, and are all assigned the same label “*var*”. Edges in the graph are set as follows: we connect bi-directly a vertex for every axiom with vertices for each of its assignments and, also bi-directly a vertex for every assignment (see Fig. 10 for the illustration of the process).

Such manner of constructing the graph allows us to almost directly use the simulation calculation algorithm to identify similar sets of axioms and terms, i.e. axioms of the same grammatical structure which hold similar terms in the same positions, and terms which occur in the same positions in similar axioms of the same grammatical structure. The only (small) modification to the algorithm is a slight change of the first approximation $\leqslant^1$, from which there are removed edges directed from vertices representing non-dummy names to vertices representing dummy names, to reflect the assumption that we cannot replace a non-dummy name with dummy one.

In the second stage of the algorithm, the resulting simulation $\leqslant$ is being analyzed, and the subsets of axioms that can be reduced to other subsets by replacing assignments to those of similar terms are calculated. A simple heuristic is used here, according to which more frequently used terms are replaced with similar terms in the first place (in a greedy fashion). After these replacements the set of axioms is reduced.

In the example with the base K application of the strategy modifies the 4th step of algorithm execution. The representation $\{\{D(a), D \sqsubseteq C, X_1(a), X_1 \sqsubseteq C\}\}$ of M''_4 is reduced to $\{\{D(a), D \sqsubseteq C\}\}$ on the basis of similarity between X_1 and D. X_1 is identified as a term used in exactly the same positions of similar sentences as D and is therefore replaced with D, which results in the reduction. Further consequence of this reduction is reaching the fixpoint in the 4th step and termination of the algorithm.

Ideologically, the removal strategy can be easily explained by the observation that dummy terms can be interpreted as "some examples" of concepts, individuals etc. So, for instance, the representation $\{\{X_1(a), X_1 \sqsubseteq C\}\}$ of M'_3 can be read as "there exists some concept being subsumed by C, and a is a member of the concept." Such interpretation allows us to draw a (correct) conclusion that a is also a member of C. Thus sentences $X_1(a)$ and $X_1 \sqsubseteq C$ are clearly superfluous in the representation $\{\{D(a), D \sqsubseteq C, X_1(a), X_1 \sqsubseteq C\}\}$ of M''_4 as there already exists such a concept, and it is D.

The deployment of the removal strategy allowed for faster identification of fixpoints and for vast decrease of the time of reasoning in comparison with projection prevention strategy. (Preliminary tests indicate that the decrease is the higher the more complicated is the structure of the knowledge base—reaching two orders of magnitude in the case of knowledge bases comprising of more than 20 modules; more systematic experiments are planned to be executed in the nearest future.)

3.2 Non-sentential Representation of Tarsets

Another approach that may be assumed in order to avoid problems with transferring vast numbers of sentences is a change of the method of representing tarsets. New representation should be more deeply rooted in semantics and allow for easier change of the level of details of expressing knowledge (to facilitate movement of knowledge between different levels of SIM structure).

A good candidate for the new representation is knowledge cartography. Knowledge cartography (hence KC) is a technique developed by authors in their earlier works. It focuses on simplified description of the space of ontology models, yet aiming at capturing the largest number of possibly useful conclusions.

The key notion in KC is a *map of concepts*. A map of concepts m for ontology $\mathcal{O}$ is a function which assigns concepts from a chosen set $\mathbf{C}$ a subset of *set of regions* $\mathbf{R} = \{r_1, r_2, \ldots r_n\}$ ($n \in N$ is a *size of the map*) in such a way that $\forall C, D \in \mathbf{C} : m(C) \subseteq m(D) \Leftrightarrow \mathcal{O} \models C \sqsubseteq D$. Putting it in other words, relations between sets of regions assigned to concepts reflect interrelationships (like inclusion or disjointness) between interpretation of such concepts in ontology models.

An extension of the notion of map of concepts is an *individual placement*, a function r which assigns each individual from an ontology (let us denote their set $\mathbf{I}$) a subset of $\mathbf{R}$ in such a way that $\forall C \in \mathbf{C}, a \in \mathbf{I} : r(a) \subseteq m(C) \Leftrightarrow \mathcal{O} \models C(a)$.

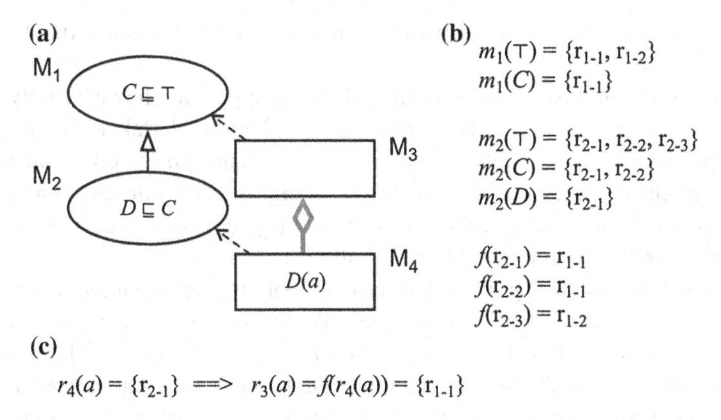

Fig. 11 Use of KC for inferences: **a** shows the KB structure, **b** maps m_1 and m_2 for M_1 and M_2 resp. and mapping f between them, **c** placements r_4 and r_3 for M_3 and M_4 and transfer of knowledge from M_4 and M_3 with use of the function f

The main strength of KC is its ability to unify different syntactic forms of the same concepts, e.g. concepts like $C \sqcap D$, $C \sqcap D \sqcap D$ and $\neg(\neg C \sqcup \neg D)$ have to be assigned the same set regions (by definition of the map). Moreover, this unification goes further to considering relations between concepts introduced by the ontology itself: e.g. if $C \sqsubseteq D$ is an axiom in $\mathcal{O}$ it has its consequences for instance in $m(C \sqcap D) = m(C)$.

Due to this feature KC can be used to transfer conclusion in the unified form. The idea here is to create a map of concepts for each context type and to establish a mapping between regions of different maps. The mapping function f may be then used to recalculate placement of individuals in different context instances. The idea of such a mapping and KC-based transfer of conclusions is illustrated in Fig. 11 (it is based on the example of the knowledge base K from Fig. 9).

The idea of using KC seems very appealing, and has vast potential to reduce the amount of information necessary to be transferred between contexts. However, KC also has drawbacks, one of the main being lack of full OWL 2 expressiveness. Moreover, use of KC does not eliminate the necessity of examination of contents of all the context from a KB in order to answer a query issued toward any target context.

3.3 Controlled Incomplete Reasoning Strategies

While the technique proposed in Sect. 3.2 has its potential in reducing necessary transfers of knowledge between modules of a SIM knowledge base, it still requires to carefully check all the contexts in search of all the relevant information. Though such kind of reasoning has its uses, it seems more natural for contextual approach to be able to abstract from information stored in some of the contexts at least in some of inference tasks.

This direction of development seems to be very appealing, however only a very introductory conceptual research in this area has been undertaken by our group. The simplest strategy that can be exploited here is *user-guided* query answering. With use of this strategy during solving reasoning problem inference engine asks the user to explicitly decide upon the set of context considered during reasoning (knowledge from other contexts is not analysed).

The drawback of user-guided approach is that its effectiveness relies on the familiarity of the user with advanced issues, like the structure of a knowledge base. A viable alternative may be to use a *heuristic-guided* strategy. This strategy assumes that a set of heuristics is employed to determine the set of relevant contexts. Heuristics may vary in complication: the simplest ones may put a threshold on the distance between modules (understood as a path distance in a graph of knowledge base structure), the more sophisticated may count the numbers of common terms within two modules.

Both of the strategies introduced above bring in the serious danger of omitting important knowledge in the process of reasoning, for critical queries. To eliminate this danger it seems necessary to introduce a strategy which involves explicit specification of the knowledge about meaning of specific modules (contexts and context instances). Such idea of specifying "semantics of structure" harmonizes very well with the theory of information contexts described in the next section. This semantics may be used to perform a preliminary inferences whose result will determine the set of contexts to be used in answering a query. However, this would require major development of both knowledge base structure and contextual theory, according to the assumptions outlined in Sect. 4.

4 Hierarchical Structure of Knowledge Bases

The conducted experiments give us valuable hints and argument for the support of our main goal, which is to develop a novel, universal and holistic theory of knowledge representation. This theory of information contexts (TIC) is focused on contexts perceived as one of the main kinds of elements forming the structure of knowledge, and is not constrained in its possible range of application to knowledge bases.

What we perceive an important argument in favor of undertaking this task is that knowledge management techniques developed so far are becoming more and more inadequate for handling enormously growing heterogenic repositories available on the Internet. In the situation when the size of the repositories reaches petabytes and above, retaining ACID properties becomes impossible and relational databases gradually lose their status of a primary information sources to emerging NoSQL databases. Research on new methods and tools for managing large information sets (*big data*) is becoming necessity.

Our concept lies in engaging contexts as a model for substantial elements from the beginning of the conceptualization stage of a designing process. Conceptualization and contextualization should be realized together and affect each other. Both levels of abstraction contain mutually related fragments of a single model. The contextual level delivers elements allowing to create types and instances of contexts. Similarly the conceptual level delivers elements allowing to create classes of objects and their instances.

This similarity of the elements allows us to believe that there exists a way of unification of rules that order their creation and usage. The unification gives a basis for elaboration of a recurrent mechanism embracing every layer of knowledge representation structure. We propose to distinguish the following four layers in the structure.

Domain layer embraces ontological description of the universe of discourse. Assuming Description Logics as the basis, we can notice the well-known TBox/ABox distinction: terminological (general) knowledge is expressed with *concepts* and *roles*, while factual knowledge assigns individual objects to concepts, and pairs of objects to roles. Contextual aspects (assumed point of views) are most often neglected in this layer and need some external description (like OMV metadata [20]).

Module arrangement layer concerns modular knowledge bases, though we may also treat standard import OWL relation as inter-modular. The essential function of the layer is to determine a role of each module in the structure and its contribution to the process of inference. One of the major problems within this layer is version management: possible influence of updating one module on other modules.

System management layer describes relationships between whole knowledge bases. Due to this specificity, the main problem within the layer (apart from inherited version management) is knowledge integration.

Paradigm layer concerns different methods of encoding knowledge and relationships between modules. It embraces problems of transformation between diverse methods of modularization of knowledge bases (like e.g. DDL [4], $\mathcal{E}$-connections [21]).

One may notice the internal similarity of different layers: in each of the layers we may see a dichotomy very similar to well-known TBox/ABox relation. This observation, supported by the encouraging results of the case study, led us to formulation of a hypothesis that the use of the hierarchical structure of knowledge may be beneficial at every layer. Moreover, the possibility of treating various elements of such structures alternatively and simultaneously as the elements of conceptualization (objects and classes) or contextualization (context instances and context types) would open significant possibilities of improvement of managing knowledge.

As a consequence we see substantial reason in creation of hierarchical and recurrent model of knowledge organization. If constructed properly, such a model can offer unified view on many important problems of knowledge engineering, thus broadening the range of use of many existing techniques.

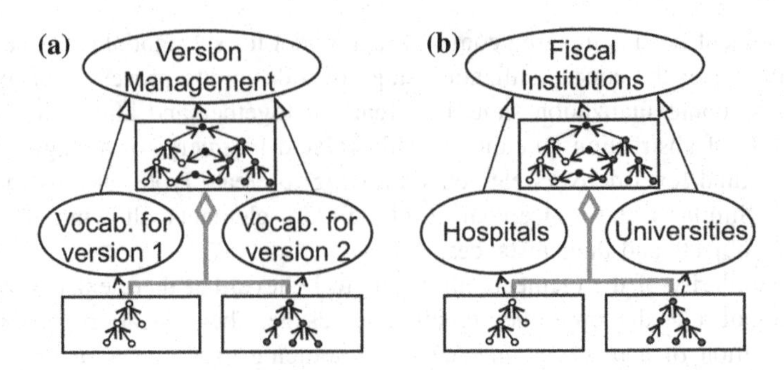

Fig. 12 Version management (**a**) and knowledge integration (**b**) in recursive knowledge model

An example of use of such a model covers a very important problem of *version management*. The issue here is the possibility that ontologies that import the one being updated may easily become unsatisfiable or incoherent, which is the consequence of the fact that the author of the original ontology being imported cannot easily foresee all of its uses.

In the recursive knowledge model this problem has an interesting solution. For two versions of a knowledge base we can create two descriptions in the form of two context instances in the system management layer. These two instances can be then aggregated in an instance which gathers knowledge about both versions. By placing appropriate couplers in the aggregating instance, we can establish a flow of facts and conclusions between the two versions. Consequently, it is possible to maintain only the newer version and to assure that only "safe" conclusions are transferred to the older version, which prevents the importing knowledge bases from becoming unsatisfiable.

The illustration of such a situation is presented in Fig. 12a. The structure of the older version is described with use of individuals depicted with white center, and the newer version with gray center. The aggregating context instance builds a new structure by adding a new top context (black individual at the top) and new couplers between chosen elements of both structures (black individuals below).

The couplers are basically algebraic operation on tarsets. Within version management projection and selection seem particularly well fit for the task. The former allows to ignore new terms and thus reduce the risk of name conflicts, while the latter allows to focus only on a subset of facts (e.g. from a specified date range). The end users, though, do not have to worry about mathematical formalities, as instead of algebraic terms, they use vocabulary defined in Version Management context type.

A remarkable feature of the model is the fact that exactly the same course of action can be undertaken to integrate knowledge from two knowledge bases. In Fig. 12b we can see exploitation of the same pattern for connecting two institutions. Lower instances represent individual knowledge bases of a hospital and a university, for the sake of discussion we may assume Medical University Hospital in Gdańsk, and Gdańsk University of Technology. Since both institutions are fiscal, we may aggregate their bases at the higher level. In such an approach the fact that *Fiscal institutions* context type provides vocabulary plays two roles: facilitates building more specific contexts (like *Universities*), and allows for integration of knowledge bases in the whole sub-tree with use of common terms.

Lifting the same technique further allows for describing the way of *migration* from one representation model (like DDL) to another (like SIM). The aggregating context instance has to be placed on the highest level of system management layer and specify the connections between contents of appropriate module in the basic language of couplers and tarsets.

5 Summary

In this chapter we described our experiences and conclusions from performing contextualization of SYNAT ontology. This contextualization is a first systematic practical case study for deployment of our original methods of knowledge base modularization.

Contemporary knowledge bases grow larger and larger. The vast amount of information that has to be stored and managed in an intelligent way motivated intensive research in the field of management of large knowledge bases, especially their modularization.

In our research we consequently assume the stance that contextualization (decomposition of knowledge in accordance with various points of view) should be considered an integral part of conceptualization (the mental process which leads to creation of a description of some domain of interest). Participation in SYNAT project gave us the opportunity to practically verify this thesis.

The course of the experiments confirmed the usefulness and generality of the proposed approaches and allowed us to make many valuable observations. The main conclusion is that by further development of the contextual methods, it is possible to create a truly universal framework for capturing and effective management of knowledge contained on the Internet, which would cover aspects of Big Data management and Context-Oriented computing, consequently allowing for intelligent handling of vast amount of information and data currently produced on everyday basis.

Acknowledgments We want to thank Professor Wierzbicki for his comment to our presentation during the last SYNAT Workshop, which became a valuable inspiration for our further work.

References

1. Goczyla, K., Waloszek, A., Waloszek, W., Zawadzka, T.: OWL API-based architectural framework for contextual knowledge bases. In: Bembenik, R., Skonieczny, Ł., Rybiński, H., Niezgódka, M. (eds.) Intelligent Tools for Building a Scientific Information Platform: Advanced Architectures and Solutions. Springer, Berlin (2013)

2. Goczyła, K., Waloszek, A., Waloszek, W.: A semantic algebra for modularized description logics knowledge bases. In: Proceedings of DL 2009, Oxford, United Kingdom (2009)
3. Goczyła, K., Waloszek, A., Waloszek, W.: Contextualization of a DL knowledge base. In: Proceedings of DL Workshop DL'2007, pp. 291–299 (2007)
4. Borgida, A., Serafini, L.: Distributed description logics: assimilating information from peer sources. J. Data Semantics **1**, 153–184 (2003)
5. Wróblewska, A., Podsiadły-Marczykowska, T., Bembenik, R., Rybiński, H., Protaziuk, G.: SYNAT system ontology: design patterns applied to modeling of scientific community, preliminary model evaluation. In: Bembenik, R., Skonieczny, Ł., Rybiński, H., Niezgódka, M. (eds.) Intelligent Tools for Building a Scientific Information Platform: Advanced Architectures and Solutions. Springer, Berlin (2013)
6. Gangemi, A., Lehmann, J., Catenacci, C.: Norms and plans as unification criteria for social collectives. In: Proceedings of Dagstuhl Sem. 07122, vol. II, pp. 48–87 (2007)
7. Gruber, T.R.: Towards principles for the design of ontologies used for knowledge sharing. Int. J. Hum. Comput. Stud. **43**, 907–928 (1995)
8. Paliouras, G., Spyropoulos, C. D., Tsatsaronis, G. (eds.): Knowledge-driven multimedia information extraction and ontology evolution: bridging the semantic gap. Series: LNCS, vol. 6050 (2011)
9. FOAF Vocabulary Specification 0.98. Namespace Document 9 August 2010 - Marco Polo Edition. Available at: http://xmlns.com/foaf/spec/
10. Krafft D.B., Cappadona N.A., Caruso B., Corson-Rikert J., Devare M., Lowe B.J., VIVO Collaboration: VIVO: Enabling National Networking of Scientists, Web-Sci10: Extending the Frontiers of Society On-Line. (Raleigh, NC, April 26-27, 2010). VIVO model available at: http://vivoweb.org/download
11. Sure, Y., Bloehdorn, S., Haase, P., Jens Hartmann, J., Oberle, D.: The SWRC ontology—semantic web for research communities. In: 12th Portuguese Conference on Artificial Intelligence-Progress in Artificial Intelligence EPIA 2005. LNCS, vol. 3803, pp. 218 – 231. Springer, Heidelberg (2005). SWRC model available at: http://ontoware.org/swrc/
12. BIBO Ontology, http://bibotools.googlecode.com/svn/bibo-ontology/trunk/doc/index.html
13. Guarino, N.: Formal ontology and information systems. In: Proceedings of the 1st International Conference on Formal Ontologies in Information Systems FOIS'98. Trento, Italy, IOS Press, pp. 3–15 (1998)
14. Genesereth, M.R., Nilsson, N.J.: Logical Foundation of Artificial Intelligence. Morgan Kaufmann, Los Altos (1987)
15. Cybulka, J.: Applying the c.DnS design pattern to obtain an ontology for investigation management system. In: Computational Collective Intelligence. Semantic Web, Social Networks and Multiagent Systems. LNCS, vol. 5796, pp. 516–527 (2009)
16. Jędrzejek, C., Cybulka, J., Bąk, J.: Towards ontology of fraudulent disbursement. In: Agent and Multi-Agent Systems: Technologies and Applications. LNCS, vol. 6682, pp. 301–310 (2011)
17. Fokoue, A., Kershenbaum, A., Li Ma, Schonberg, E., Srinivas, K., Williams, R.: SHIN ABox reduction. In: Proceedings of DL Workshop DL'2006, pp. 135–142 (2006)
18. Horst, H.J.: Completeness, decidability and complexity of entailment for RDF Schema and a semantic extension involving the OWL vocabulary. J. Web Semantics **3**(2–3), 79–115 (2005)
19. Henzinger, M.R., Henzinger, T.A., Kopke, P.W.: Computing simulations on finite and infinite graphs. In: Proceedings of the 36th Annual Symposium on Foundations of Computer Science, pp. 453–462 (1995)
20. Hartmann J., Palma R., Sure Y.: OMV—Ontology Metadata Vocabulary, ISWC 2005
21. Kutz, O., Lutz, C., Wolter, F., Zakharyaschev, M.: $\mathcal{E}$-connections of abstract description systems. Artif. Intell. **156**(1), 1–73 (2004)

Part III
Research Environments

Creating a Reliable Music Discovery and Recommendation System

Bożena Kostek, Piotr Hoffmann, Andrzej Kaczmarek and Paweł Spaleniak

Abstract The aim of this chapter is to show problems related to creating a reliable music discovery system. The SYNAT database that contains audio files is used for the purpose of experiments. The files are divided into 22 classes corresponding to music genres with different cardinality. Of utmost importance for a reliable music recommendation system are the assignment of audio files to their appropriate genres and optimum parameterization for music-genre recognition. Hence, the starting point is audio file filtering, which can only be done automatically, but to a limited extent, when based on low-level signal processing features. Therefore, a variety of parameterization techniques are shortly reviewed in the context of their suitability to music retrieval from a large music database. In addition, some significant problems related to choosing an excerpt of audio file for an acoustic analysis and parameterization are pointed out. Then, experiments showing results of searching for songs that bear the greatest resemblance to the song in a given query are presented. In this way music recommendation system may be created that enables to retrieve songs that are similar to each other in terms of their low-level feature description and genre inclusion. The experiments performed also provide basis for more general observations and conclusions.

Keywords Music information retrieval · Music databases · Music parameterization · Feature vectors · Principal component analysis · Music classification

B. Kostek (✉)
Audio Acoustics Laboratory, Gdańsk University of Technology,
80-233 Gdańsk, Poland
e-mail: bokostek@audioacoustics.org; bozenka@sound.eti.pg.gda.pl

P. Hoffmann · A. Kaczmarek · P. Spaleniak (✉)
Multimedia Systems Department, Gdańsk University of Technology,
80-233 Gdańsk, Poland
e-mail: papol@sound.eti.pg.gda.pl

R. Bembenik et al. (eds.), *Intelligent Tools for Building a Scientific Information Platform: From Research to Implementation*, Studies in Computational Intelligence 541,
DOI: 10.1007/978-3-319-04714-0_7,

1 Introduction

In order to enable the music service user to browse song databases and allow creation of effective music recommendation systems, music files are usually parameterized with low-level descriptors, which are usually based on the MPEG 7 standard, Mel-frequency Cepstral Coefficients (MFCC's) or, finally, parameters suggested by researchers [1, 3– 5, 7, 10, 12– 16, 18, 20, 28–30, 31]. Although it has been several years since the MPEG 7 standard, dedicated primarily to audio–video signal parameterization, was created, in practice, each music service uses their own solution for music parameterization [17, 25]. On the other hand, some on-line services (i.e. social music networking systems) as well as record labels and record companies utilize metadata such as for example: music genres, album name, date of the album release, names of artists, length of a particular song, lyrics, etc., to make searching for artists and categorizing music resources easier. Finally, the most important research in the Music Information Retrieval (MIR) domain is related to the content-based retrieval. In particular, *Query-by-Category* utilizes musical style, genre, mood/emotion of a musical piece in music retrieval [27].

With the expansion of the social networking systems represented by music services such as iTunes [23], Amazon [22], Lastfm [17], Pandora [25] and others, automatic precise classification of music genres is becoming more and more imperative. The problem with comparing various algorithms and the effectiveness of low-level descriptors in this field is that authors/services use different music databases, different taxonomy and music excerpts of different lengths (a song may be classified to a given genre for 30 % of its duration, while for another song, its entire playback time will be used).

Under the Synat project carried out in the Multimedia Systems Department, Faculty of Electronics, Telecommunications and Informatics, Gdańsk University of Technology, a music service was created with a database of more than 50,000 songs divided into 22 genres. Music genre classification is based on low-level feature vectors and, on decision algorithms.

This chapter first discusses the music database which was made for the Synat music service. As part of the experiments undertaken, an analysis of the effectiveness of music genre classification was made, taken into account the redundancy check employing t-Student's test and the Principal Component Analysis. The tests were performed with an external application that uses the algorithms implemented in the Synat service.

2 Synat Music Service

2.1 Description of the System

The Synat music service encompasses approximately 52,000 30-s song excerpts allocated to 22 music genres: Alternative Rock, Blues, Broadway & Vocalists,

Children's Music, Christian & Gospel, Classic Rock, Classical, Country, Dance & DJ, Folk, Hard Rock & Metal, International, Jazz, Latin Music, Miscellaneous, New Age, Opera & Vocal, Pop, Rap & Hip-Hop, Rock, R & B, and Soundtracks. The database contains additional metadata, such as: artist name, album name, genre and song title. In addition to the items listed in the database, songs include also track number, year of recording and other descriptors typically used for annotation of recordings.

The music service utilizes a number of technologies, among which are: Debian 6.0 Squeeze AMD64—for server software and basic programming libraries; Ip-tables Guarddog—protects the system against unauthorized access from the Internet; Apache httpd FastCGI mod_fcgid—the service is implemented as a fcgid application running in the popular Apache http server environment; Firebird 2.5/ 3.0—database server stores information on song metadata, as well as their features and metrics; FFmpeg—converts audio files to wav format. This is required for the libsndfile library to read samples for subsequent song parameterization, etc.

The architecture of the Synat service (Fig. 1) consists of three layers: the user interface, web server with implemented algorithms and a database containing a list of users and audio tracks. The interface layer includes a module to handle the service via a web browser and applications intended for use on mobile platforms. The internet service employs Web Toolkit. A detailed description of functionalities provided by the user interface is contained in Sect. 2.2. Applications for mobile devices that are planned in near future will enable users to send music file and get answers about the song genre. The user sends the track via a TCP/IP protocol to the application layer where a query is processed. The main task of the application layer is to manage all functionalities of the service. The service enables the user to browse the whole database, search for songs and genres assigned to audio files. This part is linked to the database which contain information about the songs (ID3 tags and descriptors), songs themselves, and also users' accounts.

2.2 User Interfaces

This Section presents graphical user interfaces of the music recognition system with the associated music genres. Home page (Fig. 2), available to all users visiting the website, allows for:

- getting acquainted with the idea of the project/service,
- signing in for users with an active account,
- signing up for new users,
- going to the SYNAT project website.

The signed/logged-in user will be redirected to the subpage where they can select an audio file for music classification analysis (Fig. 3). One can choose a file either from their own computers or from the Synat music database. After loading,

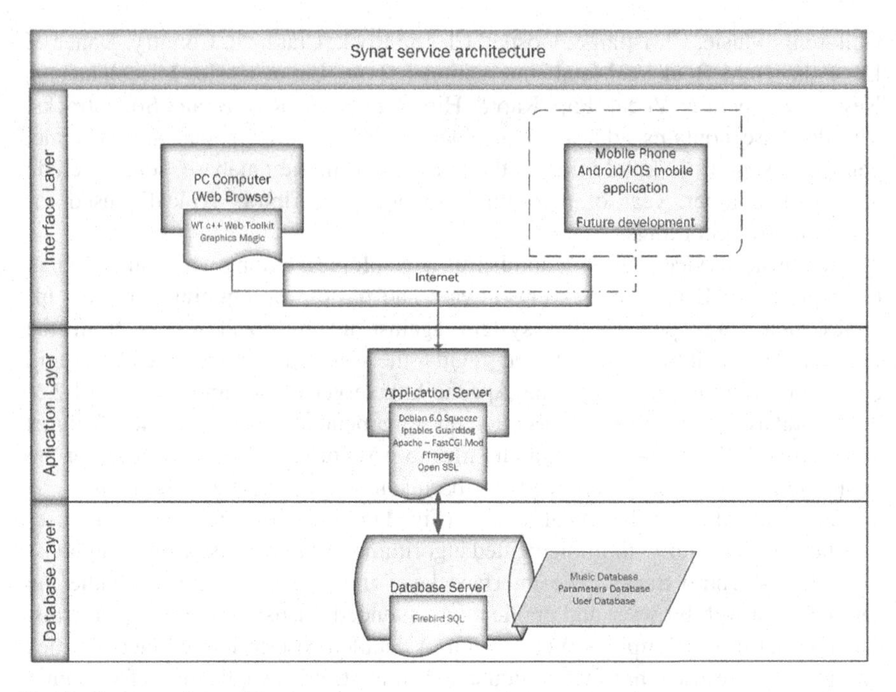

Fig. 1 Synat service architecture

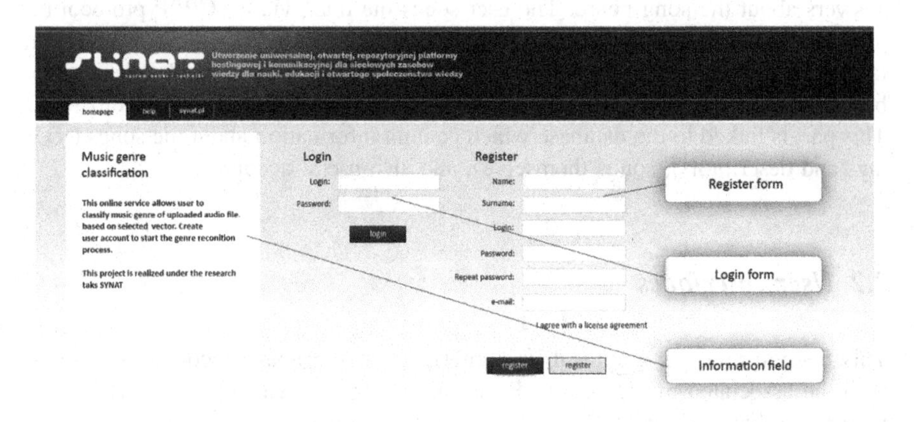

Fig. 2 Home page of the website

the file can be played with a simple player supporting playback. Before the search analysis result is displayed, one should specify the desired feature vector. On the right side of the interface, there is information about the recommended audio file format to be uploaded.

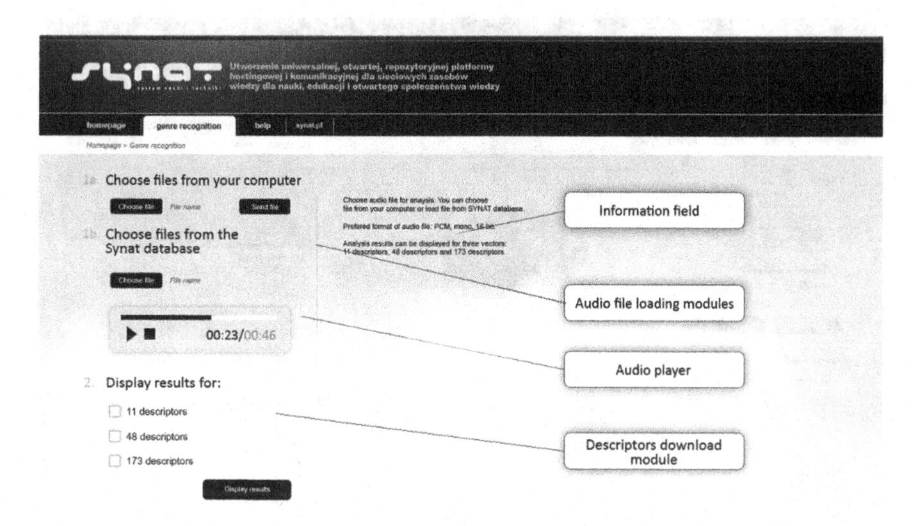

Fig. 3 Song uploading/adding subpage

On the result page, in addition to playing the song, one can read the pie chart showing the degree to which the song belongs to the given genre, given as a percentage (Fig. 4). The page displays also a list containing the recommended songs. This list is built on the previously selected size of the feature vector. The table at bottom of Fig. 4 contains a list of songs arranged according to their distance (based on feature vector similarity) from the song which is analyzed. Similarity is determined by means of the *k*NN (*k*-Nearest Neighbor) algorithm on the basis of feature vectors containing 11 or 48 parameters. The list is composed of 10–40 tracks according to the value of *k* in the *k*NN algorithm. The interface shown in Fig. 4, in addition to showing the artist's name, song title and album and the parametric distance from the track searched, has the ability to download 30-s excerpts. Also, advanced options are available (Fig. 5) which adhere to the subjective impression of the user attributed to the listened music expressed in terms of distance.

On the advanced options subpage (Fig. 5) one can view and edit the file data. One can also download the available feature vectors. At the bottom of the page, there is an audio player.

2.3 Parameterization

The original version of the system includes the following parameters [9, 11, 16]: 127 descriptors of the MPEG-7 standard: Audio Spectrum Centroid (ASC), Audio Spectrum Spread (ASS), Spectral Flatness Measure (SFM), Audio Spectrum Envelope (ASE), Mel-Frequency Cepstral Coefficients (MFCC) (40 descriptors: mean values and variances calculated from 20 MFCC values) and 24 dedicated

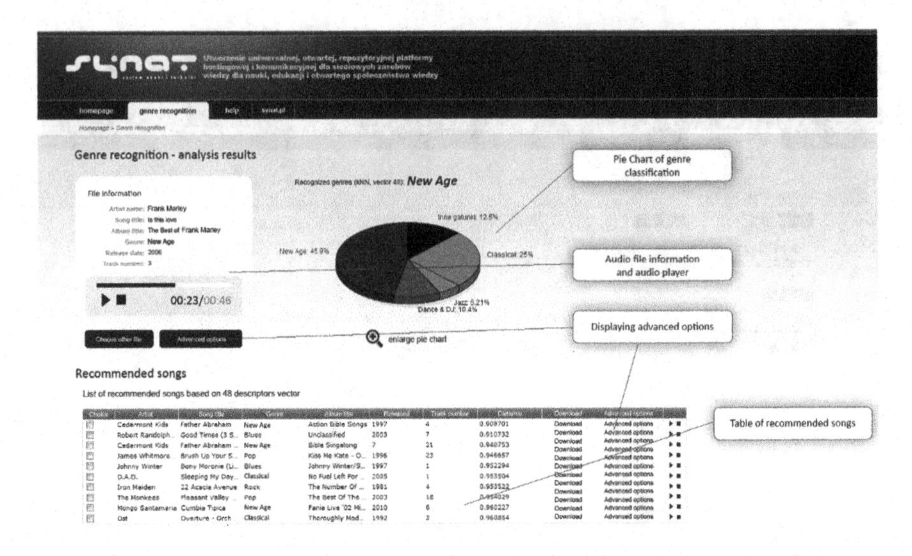

Fig. 4 Subpage with the result of music genre recognition and analysis

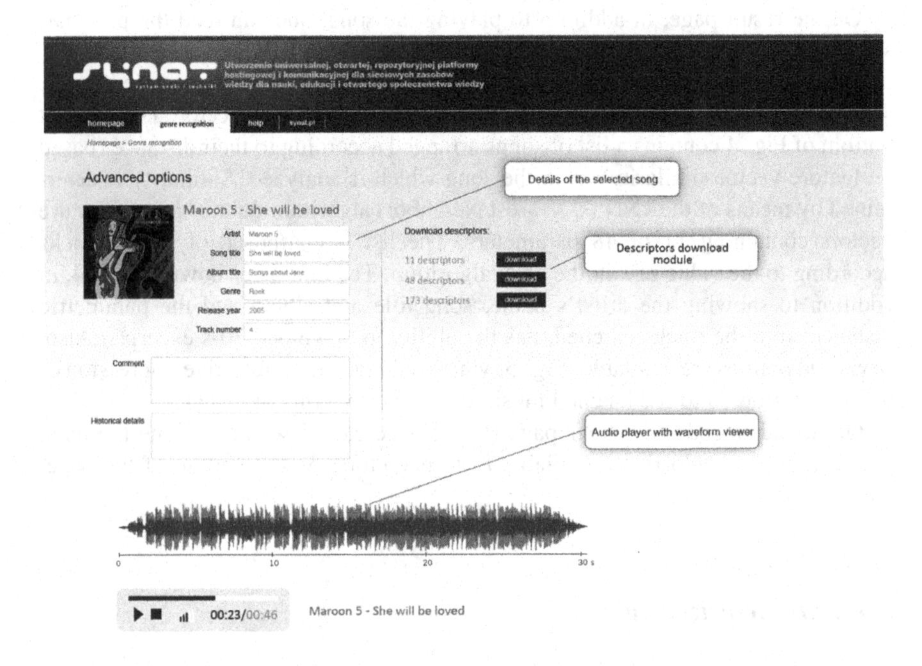

Fig. 5 “Advanced options” subpage

parameters: zero-/threshold-crossing rates, i.e., beat histogram. Parameter vector includes 191 descriptors. It should also be noted that the parameters are normalized for range (−1, +1) [9, 11, 16].

MFCC parameters utilize linear and logarithmic frequency scales that perceptually approximates frequency scales. It is based on the division of the range from 0 to approx. 6,700 Hz into 40 sub-bands with the first 13 being linear sub-bands with an equal bandwidth of 66.6667 Hz, and the remaining 27 being the logarithmic part of the scale with bandwidth ascending by 1.0711703 and increasing centre frequencies (typically approx. linear $f < 1$ kHz; logarithmic above that). The number of MFCC coefficients in this system is 40: 20 PAR_MFCC parameters are calculated as arithmetic means, while 20 PAR_MFCCV parameters are calculated as variances of these coefficients calculated in all the segments.

The vector having 191 features was thoroughly tested on a smaller, carefully selected database—ISMIS [16]. The ISMIS music database contains songs by 60 composers/performers. For each musician/composer, 15–20 songs from several different albums were selected, but all having the same style. All of them were checked by listening to them to make sure each song correspond to the genre to which it had been assigned. Music genres in this database were narrowed down to basic styles such as: classical music, jazz, blues, rock, heavy metal and pop.

An important aspect of the analysis of the effectiveness of parameterization is reducing the feature vector redundancy, therefore, the resulting parameter vector was examined for separability. Feature vector redundancy was checked based on two approaches. In the preliminary phase the correlation analysis based on t-Student's test was utilized. In the second phase of experiments the Principal Component Analysis (PCA) method was employed to reduce the dimensionality of a data set.

The basic correlation analysis involves calculating the covariance matrix, then the correlation matrix, and interpreting the individual coefficients based on t-Student's statistics. With this one can determine which descriptors can be considered redundant. By denoting the correlation coefficient calculated for x_1, $x_2, \ldots x_n$ of parameter x and for y_1, $y_2, \ldots y_n$ of parameter y with R_{xy}, it is possible calculate the statistics having the decomposition of t-Student's test with $n-2$ degree of freedom, using the following equation:

$$t = \frac{R_{xy}}{\sqrt{1 - R_{xy}^2}} \cdot \sqrt{n - 2} \tag{1}$$

where:

n the number of parameter vectors.

With proper selection of the parameters in the feature vector, as well as checking them for parameter separability, we were able to enter the ISMIS database for the ISMIS 2010 conference contest. The contest results for classification of music genres were encouraging, with effectiveness over 80–90 % [16].

The next step was to test the effectiveness of this vector (containing a number of features reduced to 173 because of the lower sampling frequency of 22,050 Hz for the Synat database. In the first tests, the effectiveness was low. However, when we listened to random song excerpts in the Synat database, it turned out that the collection of songs collected by the music robot is not uniform, which means that the songs are often not properly assigned to the genre.

It was therefore decided to optimize database parameterization. To this end, we prepared a database of 11 of the most distinctive/characteristic music genres (Blues, Classical, Country, Dance & DJ, Jazz, Hard Rock & Metal, NewAge, Pop, Rap & Hip-Hop, R & B, and Rock) with three new parameter vectors proposed for them. Altogether we used 32,000 audio files.

Parameterization included the following steps:

- downloading an excerpt of a recording with a duration of approx. 26 s,
- conversion to a 22,050 Hz monophonic signal (right, left channel summing),
- segmentation into 8,192 samples, i.e., 2 to the power of 13, due to the need for the FFT algorithm for spectral analysis,
- calculating Fourier spectra with a sample rate of 4,410 samples (time equal to 0.2 s), and using a Blackman window, hence the overlap size equal to 3,782 samples,
- calculating the power spectrum using a logarithmic scale for each segment.

Due to the sampling frequency of 22.050 Hz, the frequency resolution for the Fourier analysis is 2.692 Hz and the analysis band reaches 9986.53 Hz. The entire available frequency band is divided into sub-bands of increasing width directly proportional to the centre frequencies. The first sub-band has a centre frequency of 100 Hz. We used the nonlinearity coefficient = 1.194 borrowed from the König scale that defines the ratio of the width of the subsequent sub-bands in the entire analyzed frequency range.

- calculating the spectrograms in the above-described scale followed by cepstrograms by calculating cepstral coefficients C_i according to the discrete cosine transform [6]:

$$C_i = \sum_{j=1}^{N} logE_j \cos\left(\frac{\pi i}{N}(j - 0,5)\right) \tag{2}$$

where:

i number of a cepstral coefficient,
E_j energy of the jth sub-band of the filter bank,
N number of (filter) channels in the filter bank.

- calculating statistical moments to 3, inclusive of individual cepstral coefficients representing sub-vectors being parts of the full 2048-long parameter vector.

Mean value m_i, i.e. the moment of the first order for the ith cepstral coefficient is calculated with the following equation [2]:

$$m_i = \frac{1}{K}\sum_{k=0}^{K} C_i(k) \tag{3}$$

where:
K number of segments,
$C_i(k)$ value of the ith cepstral coefficient for segment no. k.

Variance and skewness, i.e. moments of the second and third order (variance and skewness) for the ith cepstral coefficient are calculated with the following equation [2]:

$$M_i(n) = \frac{1}{K}\sum_{k=0}^{K} [C_i(k) - m_i]^n \tag{4}$$

where:
i number of a cepstral coefficient,
n order of the moment (2nd or 3rd).

For cepstral parameters, each sub-vector has a length equal to the number of segments. However, their number is equal to the number of cepstral coefficients (cepstrum order) = 16. The vector length is 2,048 which consists of 16 sub-vectors with 128 values each.

The resulting cepstrogram can be converted into a shorter vector in several ways, such as by delta, trajectory or statistical transformation. The pilot studies showed that the transformation based on the statistical analysis proved to be the most effective. It should be noted that the full vector with a length of 2,048 was also tested, but proved to be ineffective in the classification process.

Therefore, based on the cepstrograms, we determined the statistical parameters of individual cepstral coefficients: statistical moments. We used three statistical parameters: average value (arithmetic mean), variance and skewness. This number is the result of optimization. In this way, each sub-vector is described by 3 numbers, resulting in a major reduction of the data, that is, a shortening of parameter vectors. Finally, parameter vectors have a length of $48 = 3 \times 16$.

We have also developed a parameterization vector based on fuzzy logic for assessing the degree to which a particular song belongs to particular genres. This system is designed for searching for songs that bear the greatest resemblance to the song in question, throughout the entire SYNAT database. Each database object is described by x numbers, which are the distances from the centroids of all x classes. This is further data reduction, which allows songs to be described in terms of the genres they belong to. Approximation of membership functions is based on the histograms of the distances of objects from the intrinsic class centroid, according to the following equation:

$$\begin{aligned} Fp(x) &= 1 dla\, x < g \\ Fp(x) &= \exp[-d(x-g)] dla\, x > g \end{aligned} \tag{5}$$

where:

g limit value that can be interpreted as the value of distance of overall (almost 100 %) objects in a particular class from the centroid of this class,

d coefficient dependent (inversely proportional) on standard deviation of these distances.

In this way, one can obtain the intrinsic membership function for each class. The value of the membership function can be interpreted as the quotient of the number of objects more distant from a given argument by the number of all objects of a given class.

Multidimensional data are generally not evenly scattered along all axes of the coordinate system, but are concentrated in certain subspaces of the original space. The purpose of Principal Components Analysis (PCA) is to find the subspace in the form of main components, i.e., the method produces linear combinations of the original variables to generate the axes. The principal components are the eigenvectors of the covariance matrix of the data set. The eigenvectors are related to the corresponding eigenvalues that are positive real numbers. By choosing vectors corresponding to several largest eigenvalues, we get the wanted collection of new directions of the coordinate system. These directions (which is the essence of the method) are directions which maximize variability in the data in terms of variance. These directions are uncorrelated with each other. All of the attributes describing the initial data set are 100 % of the variation of these data [26].

The PCA is a well known method for reducing the dimensionality of data. It was decided in experiments to use this method and reduce the number of parameters used in the classification process after obtaining low effectiveness based on t-Student's test. The results of this experiment are described in Sect. 3.

3 Experiments

The aim of the experiments was to examine the effectiveness of classification and to come up with conclusions and solutions that would support automatic recognition of music genres. The tests were performed with an external application that uses decision algorithms implemented in the Synat service:

- Fuzzy Logic,
- *k* of the nearest neighbors using an 11-element vector parameter (kNN11) and a 48-element vector parameter (kNN48).

Table 1 Number of songs in the databases: "1,100" and Synat

Genre	Size of song databases	
	1100	Synat
P–Pop	100	5,976
Ro–Rock	100	4,957
Co–Country	100	3,007
RB–R & B	100	2,907
Ra–Rap & Hip-Hop	100	2,810
Cl–Classical	100	2,638
J–Jazz	100	2,543
D–Dance & Dj	100	2,289
N–NewAge	100	2,122
B–Blues	100	1,686
H–Hard Rock & Metal	100	1,175
$\sum$	1,100	3,2110

The *k*NN algorithm is the well accepted classification method due to its ease and practical efficiency. It finds a group of *k* objects in the training set that are nearest to the test object, and bases the assignment of a label on the predominance of a particular class in this neighborhood. This property is used to build the lists of recommendation. Thus, the main reason of using minimum-distance classifier, i.e. *k*NN algorithm in the service is to build the recommendation list based on inter object distance in feature space. As a result of *k*NN classification the distances from database objects are calculated and after sorting is performed, the recommendation list can be obtained. The other advantages of *k*NN classifier are: in general high classification quality and the ease of implementation and application of modifications, which in the music service may be an important issue. Commonly known shortcomings, such as long time computing and large memory requirements, i.e. the need to store the whole reference set in memory are not too troublesome in the context of technological progress. Furthermore, as checked, for the 11 classes database containing 32110 objects containing vectors of the length of 173, creating the recommendation list takes (on SYNAT computer) around 10 ms.

The recognition effectiveness was tested on a group of the eleven most uniform music genres. For the experiments, we used a 1,100-song database (composed of songs that are not present in the Synat database) containing 100 songs from each of the 11 selected genres + the Synat database. Table 1 shows the number of songs in the 1,100 and the Synat databases for the analyzed genres. In experiments we tested the effectiveness of music genre recognition based on two approaches to feature redundancy checking. As mentioned before, to reduce redundancy of feature vectors the t-Student's test and PCA method were used. The main reason of using t-Student's method was to check to what extend the classification effectiveness depends on music excerpt, as well as on *k*—values in *k*NN classifier and whether it is profitable to use weighted feature vectors in the classification process.

These results are treated as the preliminary ones. The main results of the effectiveness of music genre recognition were obtained with the PCA method described at end of this Section.

Pilot tests were carried out using the "*leave-one-out*" procedure and the NN classifier, following prior normalization of parameters <−1, +1>. The tests focused on the selected parameters from vector 173. We tested selected parameter groups (ASE, SFM, MFCC, dedicated parameters: beat histogram), as well as other individual parameters (such as centroids and spectrum centroid variances).

The results of pilot tests showed that the best classification performance is displayed by the group of MFCC parameters. These were the mean values and variances of 20 Mel-Frequency Cepstral Coefficients calculated from 40 power spectrum sub-bands distributed over the frequency scale in a combined linear and logarithmic manner. The linear part related to the lower part of the band, while the logarithmic part concerned the upper part of the frequency band.

3.1 Preliminary Results

This led us to checking how a division of the given band influences the classification performance of mel-frequency cepstral parameters. Furthermore, it should be verified whether the parameters of higher statistical moments such as skewness and kurtosis can be included in the resulting vector. Pilot tests confirmed that the best division of the frequency band is a fully logarithmic one, and the maximum order of statistical moment should be 3 (skewness). Therefore, for further research we use the 48-element vector containing 16 frequency sub-bands.

Further experiments were designed to identify which of 30-s fragments of a song provides the most characteristic information for the entire song for effective genre recognition. We analyzed four fragments of songs: intro, the middle of the first half (middle1), the middle of the second half (middle2) and the middle of the song (middle). Doing parameterization for the final section of the song was deemed irrelevant since songs are faded out coming to the end. It should be noted that this is a very important part of the research due to the fact that the music databases store 30-s fragments of songs (the copyright aspect). For this part of the experiments, we used a database with 1,100 songs.

Figure 6 shows preliminary results of the effectiveness of music genre recognition within the collection of 1,100 songs for the analyzed song fragment. For comparison of results, the *k*NN algorithm was tested using two parameter vectors: 11 and 48 reduced with the t-Student's test.

The best results were achieved for the middle of the first half of songs. The maximum increase in the classification efficiency was 56 % compared with the song intro. All classifiers achieve higher scores when testing further song fragments in relation to the initial section. Using the shorter parameter vector for the *k*NN algorithm reduces the classification efficiency by 7 % points, on average. In the best case, the song classification efficiency was 45 %, so in comparison with

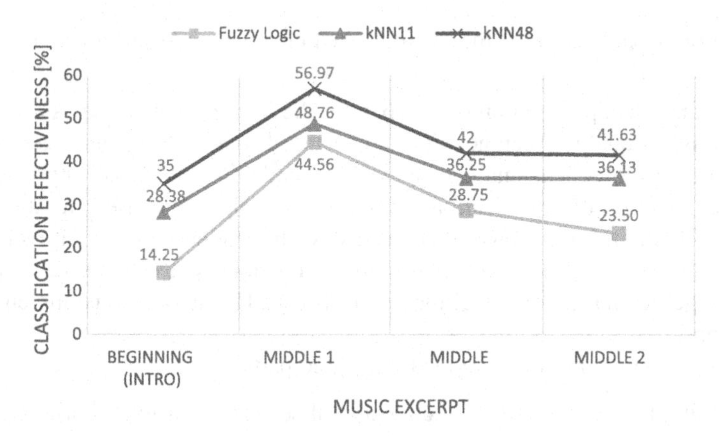

Fig. 6 Genre classification results for fragments of the song

common test sets the score should be considered low [21, 24]. For further experiments, we selected a 30-s fragment of the middle of the first part of the song.

Another aspect leading to improving the classification effectiveness is the optimization of the resulting vector due to the minimum-distance classifiers used. As mentioned earlier, in the first run, we normalized mini-max parameters for range <−1, +1>. This means that the weights of parameters are aligned, but this assumption should be regarded only as a starting point for optimization activities.

Proper selection of parameter weights is key to maximizing the effectiveness of *k*NN classifier. Therefore, we developed an optimization procedure for weights of the parameter vector. It involves multiple use of the *k*NN classifier on the basis of the parameters with simultaneous systematic changes of weight values.

Weight vector at the starting point of this procedure is aligned, and all weights are equal to 1. The next optimization steps involve changing the values of the individual weights by *in plus* and *in minus* increments assumed *apriori*.

As a result, *k*NN classifier forms a confusion matrix, and in each optimization step it is necessary to assess whether there was an *in plus* or an *in minus* change. In the described optimization system, these criteria were applied to the value of the *kappa* coefficient calculated with the following equation [8]:

$$\kappa = \frac{N \cdot \sum_{i=1}^{r} x_{ii} - \sum_{i=1}^{r} (x_{i+} \cdot x_{+i})}{N^2 - \sum_{i=1}^{r} (x_{i+} \cdot x_{+i})} \tag{6}$$

where:

N the number of all the objects in the database,
r number of classes,
x_{ii} values on the main diagonal,
x_{i+} sum in rows (TP + FN),
x_{+i} sum in columns (TP + FP),
TP correct indication of the distinguished class (TP—*true positive*),

FP incorrect indication of the distinguished class (FN—*false positive*).

The aim of the optimization is to maximize the *kappa* value, which is a measure of the classifier quality. Weight values are changed so as to bring about a situation in which each subsequent small change of the weight does not cause improvement. Therefore, a series of optimization is used where weight increases and decreases become smaller in the consecutive steps. Optimization begins with changes +100 % and −50 %, and successively after reduction by a factor of 2/3 in each step, one get to change values of the order of 3 %. The algorithm is implemented in 10 steps.

Further experiments were divided into three parts:

- examining the classification effectiveness of selected decision algorithms using the Synat database,
- examining the effects of *k* parameter changes in the *k*NN algorithm for the 1,100-song database,
- examining the classification effectiveness for 6 music genres for: Synat and 1,100-song databases.

All tests were carried out taking into account the optimization of parameter vectors. For the classifiers training phase, 70 % of the collections in Table 1 was used, while for the testing phase 30 %. Song fragments used in the study were taken from the middle of the first part of the song.

Below are the test results for the following classifiers: *Fuzzy Logic*, *k*NN11, *k*NN48. The results are provided in the following tables containing classification matrices. For each stage of the study, we prepared a summary table with the defined genre classification efficiency in the analyzed aspect.

The first one of the analyzed algorithms, Fuzzy Logic, draws on the membership function. During the tests, each of the 11 classes (music genres) received its membership function by which the elements of test sets are assigned. Classification matrices (before and after optimization) for the *Fuzzy Logic* classifier are presented in Table 2. Average gain resulting from the optimization is 2.5 %. For most genres, the average gain ranges between 2 and 3 %. The best recognition performance is obtained for Classical (64.44 %), and the worst for Blues (16.61 %). The biggest gain after optimization was recorded for Hard Rock & Metal, and is 5.5 %.

Another algorithm, *k*NN, is a minimum-distance classifier, where *k* value is the number of items included in the decision-making process. For the tests, we used $k = 15$. The algorithm was tested in two variants of the parameter vector: 11 and 48. Table 3 shows test results for vector 11, while Table 4 shows test results for vector 48.

The total classification efficiency of *k*NN11 algorithm is 38 %, which makes it comparable with the *Fuzzy Logic* algorithm. Gain after optimization is very low, at a negligible level of 0.5 %. For the short parameter vector, optimization gives less

Table 2 Confusion matrix for Fuzzy Logic classifier before (top) and after optimization (below) for the Synat database

	P	Ro	Co	RB	Ra	Cl	J	D	N	B	H
P	*555*	314	274	173	103	57	118	58	28	68	45
Ro	294	*525*	227	85	64	32	68	43	24	51	75
Co	152	98	*472*	59	20	11	32	8	9	39	3
RB	177	84	85	*320*	81	11	49	28	9	27	2
Ra	90	48	58	75	*501*	0	11	34	0	21	5
Cl	84	27	24	17	3	*485*	56	9	76	11	0
J	131	90	49	75	21	40	*258*	23	39	34	3
D	110	84	43	81	76	20	40	*154*	28	22	29
N	59	46	22	21	5	117	59	15	*286*	7	0
B	89	79	85	61	20	14	52	8	12	*82*	4
H	46	95	23	13	13	6	3	12	2	10	*130*

	P	Ro	Co	RB	Ra	Cl	J	D	N	B	H
P	*602*	334	237	173	97	52	109	50	41	56	42
Ro	292	*575*	197	83	44	39	49	43	29	68	69
Co	147	96	*498*	48	18	9	33	9	10	33	2
RB	173	88	74	*340*	73	10	42	28	4	35	6
Ra	97	44	46	77	*521*	1	10	30	0	16	1
Cl	82	28	22	10	6	*510*	43	6	71	11	3
J	123	79	52	85	18	47	*269*	21	33	33	3
D	109	90	43	57	79	18	46	*164*	34	26	21
N	71	33	26	18	4	109	53	17	*295*	10	1
B	88	75	80	60	29	21	44	7	10	*84*	8
H	40	96	20	8	14	3	3	5	4	8	*152*

gain. The best recognition performance is obtained for Classical (63.68 %), and the worst for Pop (14.89 %) and Blues (15.03 %).

The classifier variant with the longer parameter vector is the most effective decision algorithm among the ones tested. The total classification efficiency is 45 %. The optimizations give percentage gain in genre recognition performance. As for *k*NN11 and Fuzzy Logic, the best results are obtained for Classical (71.9 %), and the worst for Pop (19.02 %) and Blues (19.57 %).

The first stage of tests on the optimization of parameter vectors confirmed that changing the weights of parameter vectors was appropriate. The gain in relation to the main results is most visible for the best classification results. Summary results showing the classification effectiveness of the tested algorithms before and after optimization are shown in Table 5.

The best of the tested algorithms is *k*NN48. Its characteristic feature is a large spread of values. There are genres that are recognized with the efficiency of 20 and 70 % in a single test set. The genre that is recognized best of all in the Synat database is Classical. For further work on the optimization of parameter vectors, *k*NN48 algorithm will be used.

Table 3 Confusion matrix for *k*NN11 classifier $k = 15$ before (top) and after optimization (below) for the Synat database

	P	Ro	Co	RB	Ra	Cl	J	D	N	B	H
P	*280*	287	202	187	145	88	198	98	67	136	105
Ro	183	*371*	162	97	46	46	112	83	48	142	198
Co	102	118	*337*	71	31	18	65	15	16	104	26
RB	99	61	79	*251*	129	9	90	58	12	74	11
Ra	26	21	38	99	*518*	1	18	74	3	30	15
Cl	26	13	6	6	2	*491*	57	16	149	15	11
J	95	51	38	78	10	62	*238*	48	70	64	9
D	50	46	10	56	90	19	49	*215*	56	33	63
N	20	16	12	8	2	173	50	17	*322*	14	3
B	36	80	61	69	25	24	66	15	39	*80*	11
H	18	52	6	4	10	6	5	29	9	8	*206*
	P	**Ro**	**Co**	**RB**	**Ra**	**Cl**	**J**	**D**	**N**	**B**	**H**
P	*267*	265	243	176	155	103	190	114	74	126	80
Ro	171	*343*	177	86	50	53	116	101	50	136	205
Co	106	101	*370*	68	36	22	62	15	13	97	13
RB	82	61	88	*252*	151	16	99	47	16	46	15
Ra	39	18	34	90	*508*	1	20	91	4	26	12
Cl	14	23	9	8	2	*504*	48	14	153	8	9
J	69	59	39	89	19	75	*226*	40	77	63	7
D	58	39	13	50	97	28	39	*217*	66	27	53
N	19	8	9	2	2	143	62	25	*356*	8	3
B	54	71	69	63	18	30	66	13	30	*76*	16
H	18	60	7	1	9	5	4	20	8	4	*217*

Subsequent tests were designed to determine the optimal value for k parameter in *k*NN algorithm with the 48-element parameter vector. The tests were carried out for the 1,100-song database for three k values: 4, 11, 15. Using the reduced song database allowed for comparing optimization performance for smaller collections of songs that we listened to. Table 6 shows classification matrices for this test.

Using the reduced training database improved the optimization performance by 52.12 %, a 7 % improvement in relation to the Synat database. In relation to the Synat database, the best recognizable music genres changed. This may indicate improper assignment of music genres in the Synat database. During the experiments, we examined the effects of the k parameter on the optimization performance result. As the value of the parameter decreases, the music genre classification performance increases after parameter vector optimization. For $k = 4$, the performance is 56.97 %.

The results show that optimization of parameter weights is correct. Genres recognized more effectively achieve better scores after results optimization. The Synat database contains over 32,000 songs, making it impossible to check the accuracy of the assignment of all of the songs to corresponding genres. This is reflected in the training process conducted incorrectly for classifiers in the decision algorithms. The large size of the set has a negative impact on the optimization of

Table 4 Confusion matrix for kNN48 classifier $k = 15$ before (top) and after optimization (below) for the Synat database

	P	Ro	Co	RB	Ra	Cl	J	D	N	B	H
P	*329*	252	448	217	152	60	126	25	30	80	74
Ro	191	*405*	365	92	65	24	65	16	29	80	156
Co	61	44	*640*	40	24	7	30	4	6	41	6
RB	98	44	121	*341*	141	6	51	17	3	46	5
Ra	35	26	68	63	*609*	1	12	16	0	7	6
Cl	26	22	32	12	1	*553*	47	7	74	13	5
J	70	47	83	96	24	35	*317*	11	32	46	2
D	86	75	49	80	122	16	38	*123*	30	20	48
N	38	21	29	18	9	126	56	7	*324*	6	3
B	46	44	126	70	28	17	67	3	10	*86*	9
H	18	54	18	7	15	6	5	4	2	7	*217*
	P	Ro	Co	RB	Ra	Cl	J	D	N	B	H
P	*341*	235	409	226	136	70	129	50	38	85	74
Ro	169	*412*	343	95	66	25	62	32	28	91	165
Co	71	47	*630*	37	26	9	23	5	8	43	4
RB	96	53	124	*331*	133	4	59	18	6	44	5
Ra	25	22	73	80	*602*	0	7	22	1	9	2
Cl	28	20	35	11	2	*569*	32	8	69	12	6
J	68	52	66	92	30	48	*316*	10	34	43	4
D	69	66	42	78	112	18	42	*163*	42	27	28
N	32	20	27	20	7	110	69	8	*334*	9	1
B	55	53	125	62	20	18	50	2	12	*99*	10
H	14	54	14	5	12	5	5	4	2	12	*226*

Table 5 Percentage recognition efficiency for music genres in the Synat database

Genre	Fuzzy logic		kNN11		kNN48	
	Before	After	Before	After	Before	After
Pop	30.96	*33.58*	15.62	*14.89*	18.35	*19.02*
Rock	35.30	*38.67*	24.95	*23.07*	27.23	*27.70*
Country	52.32	*55.20*	37.36	*41.02*	70.95	*69.84*
R & B	36.69	*38.99*	28.78	*28.90*	39.10	*37.95*
Rap_&_Hip-Hop	59.43	*61.80*	61.45	*60.26*	72.24	*71.41*
Classical	61.28	*64.44*	62.04	*63.68*	69.88	*71.90*
Jazz	33.82	*35.26*	31.20	*29.62*	41.55	*41.42*
Dance_&_DJ	22.43	*23.88*	31.31	*31.60*	17.91	*23.74*
New_Age	44.93	*46.34*	50.58	*55.92*	50.90	*52.47*
Blues	16.21	*16.61*	15.82	*15.03*	17.00	*19.57*
Hard_Rock_&_Metal	36.88	43.12	58.44	*61.56*	61.56	*64.11*
$\sum$	39.11	41.63	37.96	*38.69*	44.24	*45.38*

parameter vectors. It should be mentioned at this point that there are no reference music databases available of this size. Some conferences, such as e.g. ISMIR, MIREX provide databases but they are not that large.

Table 6 Confusion matrix for kNN48 classifier $k = 15$ before (top) and after optimization (below) for the 1,100-song database

	P	Ro	Co	RB	Ra	Cl	J	D	N	B	H
P	*7*	0	15	0	0	1	2	0	1	2	2
Ro	1	*24*	1	0	1	1	2	0	0	0	0
Co	0	0	*18*	0	0	0	0	2	1	0	9
RB	0	0	2	*6*	0	2	0	10	2	2	6
Ra	6	4	4	0	*7*	0	1	1	2	2	3
Cl	0	0	4	0	0	*7*	0	3	0	0	16
J	3	6	5	0	0	0	*9*	2	0	1	4
D	0	0	2	0	0	0	0	*17*	4	4	3
N	0	0	2	1	0	0	0	2	*21*	3	1
B	0	0	0	0	0	0	0	8	10	*11*	1
H	0	0	3	0	0	2	0	1	0	0	*24*
	P	Ro	Co	RB	Ra	Cl	J	D	N	B	H
P	*8*	1	13	0	0	1	2	0	1	2	2
Ro	0	*28*	1	0	0	0	1	0	0	0	0
Co	0	0	*20*	0	0	0	0	3	1	1	5
RB	0	0	2	7	0	2	0	11	2	2	4
Ra	6	3	4	0	*11*	0	4	0	1	1	0
Cl	0	0	4	0	0	*10*	0	2	0	1	13
J	5	6	5	0	0	0	*9*	3	0	1	1
D	0	0	2	2	0	0	0	*18*	2	4	2
N	0	0	2	1	0	0	0	1	*22*	3	1
B	0	0	0	0	0	0	0	7	7	*15*	1
H	0	0	3	0	0	3	0	0	0	0	*24*

Table 7 shows a summary of results for variable value of k parameter. The genre that is recognized best of all is Rock and Hard Rock & Metal. The performance over 90 % is comparable to the results which were achieved for common test sets [24]. This shows that there is great potential in genre classification solutions that use minimum-distance classifiers coupled with optimization of parameter weights. The optimal value of the k parameter which yields the best weight optimization results is 4.

Next tests aimed at investigating the effects of reducing the number of classes (music genres) on the classification effectiveness. The initial set of 22 genres had been reduced during the pilot tests to 11. Further tests were conducted using 6 genres. The size of sets is presented in Table 1.

For the tests, we selected the following genres: Classical, Dance & DJ, Hard Rock & Metal, Jazz, Rap & Hip-Hop, and Rock. We compared 1,100-song and Synat databases using the kNN48 classifier. The results presented as a confusion matrix for 1,100-song and Synat databases are shown in Tables 8 and 9.

Table 7 Percentage of recognition efficiency for music genres in the 1,100-song database using *k* parameter of *k*NN algorithm

Genre	$k = 15$		$k = 11$	$k = 4$
	Before	After	After	After
Pop	23.33	*26.67*	40.00	*50.00*
Rock	80.00	*93.33*	96.67	*93.33*
Country	60.00	*66.67*	70.00	*66.67*
R & B	20.00	*23.33*	36.67	*50.00*
Rap_&_Hip-Hop	23.33	*36.67*	40.00	*53.33*
Classical	23.33	*33.33*	40.00	*40.00*
Jazz	30.00	*30.00*	36.67	*36.67*
Dance_&_DJ	56.67	*60.00*	60.00	*66.67*
New_Age	70.00	*73.33*	70.00	*66.67*
Blues	36.67	*50.00*	46.67	*36.67*
Hard_Rock_&_Metal	80.00	*80.00*	73.33	*66.67*
$\sum$	45.76	*52.12*	55.45	*56.97*

Table 8 Confusion matrix for *k*NN48 classifier $k = 15$ before (top) and after optimization (below) for the Synat database for a reduced number of classes (genres)

	Ro	Ra	Cl	J	D	H
Ro	*971*	158	37	139	27	156
Ra	69	*730*	0	19	19	6
Cl	49	7	*638*	86	4	8
J	157	48	65	*472*	19	2
D	167	182	39	83	*162*	54
H	90	16	6	4	10	*227*
	Ro	Ra	Cl	J	D	H
Ro	*1014*	101	40	134	40	159
Ra	76	*712*	0	19	30	6
Cl	48	4	*657*	71	6	6
J	179	51	74	*435*	23	1
D	162	166	42	74	*193*	50
H	81	12	7	9	8	*236*

Classification effectiveness for songs in the Synat database including 6 genres is 19 % higher after optimization than for 11 genres with optimization. Rock—poor recognition performance in the Synat database (27.23 %); after optimizing the number of classes and parameter weights, its recognition performance amounts to 70 %. The genres that were effectively recognized before changing the number of classes continue to be properly classified after optimization.

Similar results as for the Synat database can be observed in tests carried out for the 1,100-song database. The average classification effectiveness improved in relation to the tests without optimization by 32 % and amounts to 77.22 %. The

Table 9 Confusion matrix for *k*NN48 classifier $k = 15$ before (top) and after optimization (below) for the 1,100-song database with 6 genres

	Ro	Ra	Cl	J	D	H
Ro	*25*	3	0	0	0	2
Ra	3	*27*	0	0	0	0
Cl	0	0	*29*	0	0	1
J	2	4	3	*19*	1	1
D	9	5	0	0	*15*	1
H	22	1	0	0	0	*7*
	Ro	Ra	Cl	J	D	H
Ro	*27*	2	0	0	1	0
Ra	2	*28*	0	0	0	0
Cl	0	0	*30*	0	0	0
J	2	2	1	*24*	0	1
D	4	4	1	0	*21*	0
H	19	1	0	0	1	*9*

Table 10 Percentage recognition efficiency for music genres in the 1,100-song and Synat databases with 6 genres

Genre	Synat		1100	
	Before	After	Before	After
Rock	65.29	*68.19*	83.33	*90.00*
Rap_&_Hip-Hop	86.60	*84.46*	90.00	*93.33*
Classical	80.62	*83.02*	96.67	*100.00*
Jazz	61.87	*57.02*	63.33	*80.00*
Dance_&_DJ	23.59	*28.11*	50.00	*70.00*
Hard_Rock_&_Metal	25.53	*66.95*	23.33	*30.00*
$\sum$	57.25	*64.62*	67.78	*77.22*

result is similar to those obtained for common test databases [24]. After optimizations, Classical genre recognition performance is 100 %. The summary comparing the resulting classification effectiveness for 6 genres is shown in Table 10.

The experiments conducted confirm the need for optimization of data to be classified. In the tests, we were optimizing weights in parameter vectors and the number of music genres (classes). Another important aspect in developing automatic music genre recognition systems is to prepare a training set accurately.

3.2 PCA-Based Analysis Results

The PCA method is recognized as an efficient tool for reducing the number of parameters in feature space. Moreover, this method is used to reduce the dimensionality of an input space without losing a significant amount of information. For

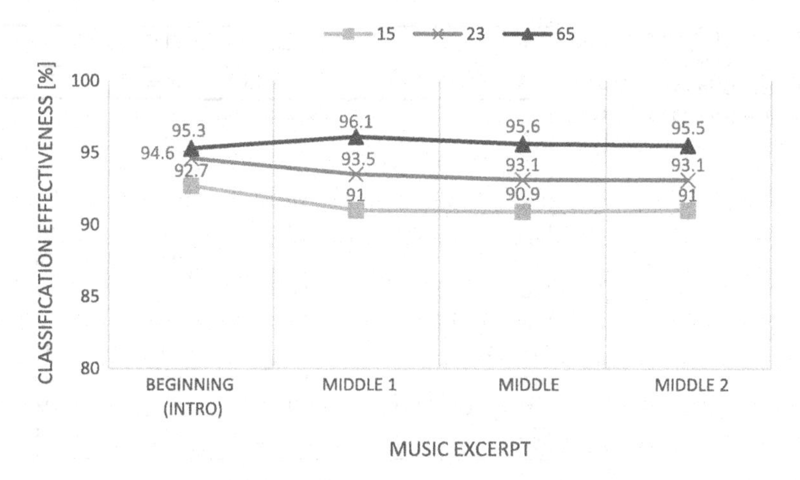

Fig. 7 Genre classification results for excerpts of the song by using the PCA method

the purpose of this phase tests based on the PCA method were performed taking into account the 173 feature vector. The feature space dimension has been reduced to 65, 23 and 15 principal components (PCs) respectively, proportional to the amount of variance (information) represented by the corresponding PCs, i.e. 95, 80 and 70 % of information. Tests were carried out on the basis of 1100 tracks using the *k*NN algorithm. The value of the parameter *k* was set to 11.

Looking at the results, it may be said that the effectiveness of the genre classification using the PCA method is very good. The effectiveness evaluated in the classification process of 11 genres raised to 90 % irrespective of the number of tested PCs in comparison to the results from the preliminary phase of experiments. In Fig. 7 the classification efficiency is shown for the four pre-selected fragments of songs. The results for all music excerpts are above 90 % which–because of the number of genres and the amount of data contained the testing set–should be considered as a very good result. Similarly to t-Student's test-based experiments the best results in genre recognition were achieved for 30 s of the middle first half of an excerpt.

All analyzed music genres were recognized with high effectiveness, i.e. above 80 %. Classical genre was recognized with the highest efficiency regardless of the number of PCs. This results is similar to the preliminary phase result in which Classical was the best recognized music genre. The worst effectiveness of classification was reported for Rap & Hip-Hop and Dance & Dj genres. For these genres it is necessary to use higher number of PCs (it was set to 65). Detailed results for the PCA –based experiments are presented in Table 11. It contains the classification results for 11 genres including three PCA factors.

Table 11 Percentage of recognition efficiency for music genres in the 1,100-song database using the PCA method with various number of factors

Genre	Number of the PCA factors		
	F = 15	F = 23	F = 65
Pop	92	93	91
Rock	96	93	94
Country	93	94	98
R & B	80	92	93
Rap_&_Hip-Hop	84	87	98
Classical	98	99	99
Jazz	90	93	97
Dance_&_DJ	81	85	95
New_Age	91	93	93
Blues	96	98	98
Hard_Rock_&_Metal	94	96	98
$\sum$	90.45	93	95.81

4 Conclusions

Music genre classification solutions cover a number of systems that when used with common test sets (GZTAN [24], ISMIR [30], MIREX [19]) achieve efficiency above 80 %; however, the size of these sets is approximately 1,000 songs. Therefore, it is difficult to compare the effectiveness of the proposed solutions with the Synat database since one needs to consider the length of an analyzed music fragment, a test set and the capacity of such a classifier to learn and improve. Moreover, in most cases, music databases contain 20- to 30-s excerpts of recordings. The studies have shown that the optimal fragment for testing the genre recognition effectiveness is the middle of the first part of a song.

The results of experiments form an important step towards a more efficient automatic music genre recognition system. Incorrect assignment of music genres in the training set leads to false readings in the test set. An overly long parameter vector leads to deterioration of the genre classification effectiveness. At the current stage of the research, the best results with the Synat database are achieved for the vector containing 48 parameters. The tests confirmed that optimizing input data is beneficial. Assigning weights to parameters and reducing the number of classes substantially increased the classification effectiveness of the *k*NN algorithm.

The most important stage in genre recognition that affect the classification effectiveness is the parameterization stage, and to be more precise—the stage of redundancy checking and reducing the dimensionality of the feature vectors. Using the PCA method allow for increasing the effectiveness of music genre classification up to 95 %.

As part of the experiment, we greatly reduced the redundant information. We selected the best fragment of a song to be analyzed, decision algorithm along with the parameter settings, parameter vector length with weights, the number of classes and the PCA method as the appropriate for the parameter redundancy checking. All these operations improved the music genre recognition effectiveness by over 100 % compared to the initial results. The current effectiveness of the automatic music genre recognition achieved in the final phase of the experiment is acceptable, and further work should aim to increase the classification effectiveness for real music databases such as for example the whole Synat database.

Acknowledgments This research was conducted and partially founded within project No. SP/I/1/77065/10, 'The creation of a universal, open, repository platform for the hosting and communication of networked knowledge resources for science, education and an open knowledge society', which is part of the Strategic Research Program, 'Interdisciplinary systems of interactive scientific and technical information' supported by the National Centre for Research and Development (NCBiR) in Poland.

The authors are very grateful to the reviewers for their comments and suggestions.

References

1. Aucouturier, J.-J., Pachet, F.: Representing musical genre: a state of art. J. New Music Res. str. **32**, 83–93 (2003)
2. Benesty, J., Mohan Sondhi, M., Huang, Y.: Springer Handbook of Speech Processing. Springer, Heidelberg (2008)
3. Bello, J.P.: Low-level features and timbre, MPATE-GE 2623 music information retrieval. New York University. http://www.nyu.edu/classes/bello/MIR_files/timbre.pdf
4. Ewert, S.: Signal processing methods for music, synchronization, audio matching, and source separation, Bonn (2012)
5. Holzapfel, A., Stylianou, Y.: Musical genre classification using nonnegative matrix factorization-based features. IEEE Trans. Audio Speech Lang. Process. **16**(2), 424–434 (2008)
6. Hyoung-Gook, K., Moreau, N., Sikora, T.: MPEG-7 Audio and Beyond: Audio Content Indexing and Retrieval. Wiley, New York (2005)
7. Jang, D., Jin, M., Yoo, C.D.: Music genre classification using novel features and a weighted voting method. In: ICME, pp. 1377–1380 (2008)
8. Kilem, G.: Inter-Rater Reliability: Dependency on Trait Prevalence and Marginal Homogeneity, Statistical Methods For Inter-Rater Reliability Assessment, No. 2, (2002)
9. Kostek, B.: Content-Based Approach to Automatic Recommendation of Music, 131 Audio Engineering Convention. New York (2011)
10. Kostek, B.: Music information retrieval in music repositories. In: Suraj, Z., Skowron, A. (eds.) Chapter in Intelligent Systems Reference Library, pp. 459–485. Springer, Berlin (2012)
11. Kostek, B.: Music information retrieval in music repositories. In: Skowron, A., Suraj, Z. (eds.) Rough Sets and Intelligent Systems, pp. 463–489. Springer, Berlin 2013
12. Kostek, B.: Perception-Based Data Processing in Acoustics, Applications to Music Information Retrieval and Psychophysiology of Hearing, Series on Cognitive Technologies. Springer, Berlin (2005)
13. Kostek B.: Soft computing in acoustics, applications of neural networks, fuzzy logic and rough sets to musical acoustics. In: Studies in Fuzziness and Soft Computing, Physica, New York (1999)
14. Kostek, B., Czyzewski, A.: Representing musical instrument sounds for their automatic classification. J. Audio Eng. Soc. **49**, 768–785 (2001)

15. Kostek, B., Kania, Ł.: Music information analysis and retrieval techniques. Arch. Acoust. Str. **33**(4), 483–496 (2008)
16. Kostek, B., Kupryjanow, A., Zwan, P., Jiang, W., Ras, Z., Wojnarski, M., Swietlicka, J.: Report of the ISMIS 2011 Contest: Music Information Retrieval, Foundations of Intelligent Systems, ISMIS 2011, pp. 715–724. Springer, Berlin (2011)
17. Lastfm: http://www.last.fm/
18. Li, T., Ogihara, M., Li, Q.: A comparative study on content-based music genre classification. In: 26th Annual International ACM SIGIR Conference on Research and Development in Information Retrieval, str. pp. 282–289. Toronto, Canada (2003)
19. Lidy, T., Rauber, A., Pertusa, A., Inesta, J.: Combining audio and symbolic descriptors for music classification from audio. In: Music Information Retrieval Information Exchange (MIREX) (2007)
20. Lindsay, A., Herre J.: MPEG-7 and MPEG-7 audio - an overview. J. Audio Eng. Soc. Str. **49**(7/8), 589–594 (2001)
21. Mandel, M., Ellis, D.: LABROSA's audio music similarity and classification submissions, Austrian Computer Society, Columbia University, LabROSA (2007)
22. Music store Amazon: http://www.amazon.com/
23. Music store Itunes: https://www.apple.com/pl/itunes/
24. Panagakis, E., Benetos, E., Kotropoulos, C.: Music genre classification: a multilinear approach. In: Proceedings of ISMIR, pp. 583–588 (2008)
25. Pandora: http://www.pandora.com
26. Shlens, J.: A tutorial on principal component analysis, Salk Insitute for Biological Studies La Jolla, New York (2005)
27. Symeonidis, P., Ruxanda, P., Nanopoulos, A., Manolopoulos, Y.: Ternary semantic analysis of social tags for personalized music recommendation. In: 9th International Conference on Music Information Retrieval str., pp. 219–224 (2008)
28. The International Society for Music Information Retrieval/Intern. Conference on Music Information Retrieval, website http://www.ismir.net/
29. Tzanetakis, G., Cook, P.: Musical genre classification of audio signal. In: IEEE Transactions on Speech and Audio Processing Str., pp. 293–302 (2002)
30. Tzanetakis, G., Essl, G., Cook, P.: Automatic musical genre classification of audio signals. In: Proceedings of International Symposium on Music Information Retrieval (ISMIR) (2001)
31. Żwan, P., Kostek, B.: System for automatic singing voice recognition. J. Audio Eng. Soc. **56**(9), 710–723 (2008)

Web Resource Acquisition System for Building Scientific Information Database

Tomasz Adamczyk and Piotr Andruszkiewicz

Abstract This chapter describes architecture and findings made as an effect of integration of complex resource acquisition system with a frontend system. Both underlying and frontend systems are mentioned briefly with reference to their root publications. The main accent has been put on data architecture, information processing, user interaction, obtained results and possible future adaptation of the system.

Keywords Information extraction · GUI · Classification

1 Web Resource Acquisition System

The number of data available through the Internet is growing rapidly among all disciplines. Among last decades, finding interesting and relevant information in reasonable time becomes easier thanks to the development of universal search systems such as Google, Bing, Yahoo etc.

Though those systems proved their usefulness, getting relevant results needs human intervention and often also query rewrite because of progressing Information Overload [1]. At the same time, the need for formation, feeding,

This work was supported by the National Centre for Research and Development (NCBiR) under Grant No. SP/I/1/77065/10 devoted to the Strategic scientific research and experimental development program: 'Interdisciplinary System for Interactive Scientific and Scientific-Technical Information'.

T. Adamczyk · P. Andruszkiewicz (✉)
Institute of Computer Science, Warsaw University of Technology, Warsaw, Poland
e-mail: P.Andruszkiewicz@ii.pw.edu.pl

T. Adamczyk
e-mail: AdamczykTM@gmail.com

R. Bembenik et al. (eds.), *Intelligent Tools for Building a Scientific Information Platform: From Research to Implementation*, Studies in Computational Intelligence 541,
DOI: 10.1007/978-3-319-04714-0_8,

maintenance and usage of comprehensive, possibly complete stores of information on a given domain of knowledge did not lost its importance. When it comes to science and scientific research, user often does not know if there is some more information available to look further for. As an example of usefulness of such attitude one may find the popularity of related search in Google Scholar and other similar solutions.

For some categories, already structured or semi-structured data stores are available through the Internet, often with already prepared interfaces for pulling data from. These are of course very valuable sources, but their scope is often limited, and we do not have any influence or information on their up-to-datedness and completeness.

Our system is aimed to fulfill the mentioned limitations in access to the most relevant and up-to-date knowledge. On one hand disadvantages of currently existing data stores will be balanced with use of multi-purpose search engines, providing up-to-date and possibly complete information. On the other hand information overload typical for the Internet and common search engines will be carried with the use of the sophisticated classification system [2].

The proposed solution is meant to be integrated with the existing University Knowledge Base [3] and focuses on acquiring possibly complete information on a given area. For example, if conference information is considered and we have information on the past conferences, it is highly desirable to automatically supplement it with other, more up-to-date details available all over the Internet. Prelegents home pages, future events, call for papers or even whole documents and presentations are only examples of what kind of details may be acquired.

The last but not least area to cover is management of the gathering process and accessibility of already gathered resources. Information that user can not access and understand gives no added value at all. If the access is limited, difficult or unintuitive the gain is very little, so the frontend of the whole system has to fulfill requirements of usefulness and ergonomics. This aspect is also covered in the presented chapter.

2 Argumentation on Choosing REPO as a Base Platform

In order to choose a base platform for this system the following requirements have been stated:

- easily maintained complex datatypes,
- web-based, eye friendly user interface,
- ease of integration with other university systems,
- reasonable and save access to stored resources.

Developed earlier as a part of SYNAT project and implemented by Warsaw University of Technology REPO system has been recognized as meeting all of the above requirements [4].

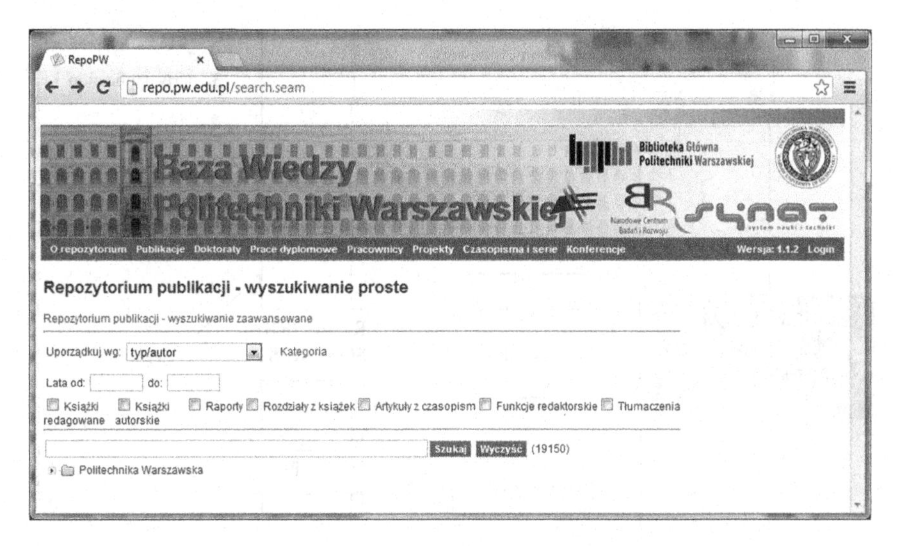

Fig. 1 Main screen of already deployed, and used REPO system

Beside the mentioned advantages REPO system has the following properties:

- flexible architecture,
- well defined and implemented access control,
- REST Web services for ease of integration with other systems,
- support for full text indexing and search.

The fact that future users of the system are already familiar with the user interface (Fig. 1) is also a nontrivial advantage of the decision made.

REPO System architecture (Fig. 2) reveals implemented MVC model. The view layer uses JSF components from the RichFaces library [4]. Its highly ergonomic user interface has been enriched with dedicated screen generator that enables automatic generation of data forms and result presentation screens for custom data types and categories. In order to provide even more flexibility, screens once generated may be customized for special visual effects or even additional functionality.

Data and the Model Layer persistence are provided with the use of JCR compliant Apache Jackrabbit. JCR standards provide, among other things, ready to use API, versioning, content observation, access control layer as well as physical data storage system independence. Lucene index aids additional efficiency in searching both in the stored objects and metadata. Controller functionality is implemented in a form of EJB 3 components, providing SOAP and REST based web service interface to stored data. More descriptive and detailed characteristic of the REPO system may be found in related publication [4].

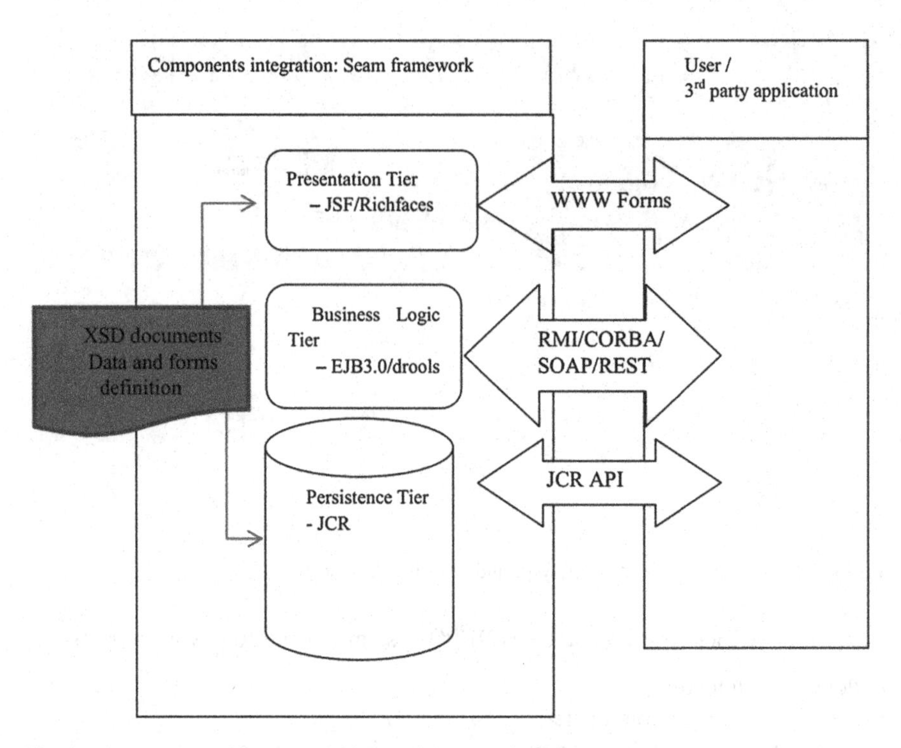

Fig. 2 The system architecture and integration scenarios [4]

3 Short Characteristic of Underlying Brokering System

REPO platform is an independent system for storage, presentation and processing of data. From the whole system architecture perspective (this will be described in the following section) it provides also an admin interface for defining new queries, controlling and monitoring the process of data acquisition. Those functionalities are quite powerful, but without external system for data facilitation it would not meet the requirements for expandable (also on-demand) knowledge base.

Underlying system developed for gaining expandability and management functions was built in SCALA programming language in the form of Multi Agent System. As the system is quite sophisticated, only a short description of the most important elements will be provided in this chapter. More detailed description may be found in references [2] and [5].

External data source connectors provide unified access for chosen data providers, such as multi-purpose web search engines, thematic repositories, science databases, etc. In order to connect a new data source, development of a dedicated connector is a sufficient action. Currently defined connectors are as follows: Google, Bing, CiteSeer. Though at the current stage of the system lifecycle free limits for using mentioned systems APIs were sufficient, the production release may need additional fee for extended access.

Classifiers module is responsible for assessment of quality of the gathered resources. It has been developed [6] in a way that enables separate development and injection of any classifier that implements the given interface. So its correctness may be improved without any intervention to other parts of the system. Also classifiers for other types of content may be developed and connected to the system in order to extend the scope of gathered data by new categories of information.

Database module provides interface for any external data storage system with adequate data structure. Currently it is configured to use REPO platform, but in general it can be any database or storage system.

Beside described functional modules the most important from the presented system perspective are strategy and task agents. The task agent is responsible for gathering resource for single REPO query. After it is called into life it uses data source connectors module in order to acquire resources and then classifiers module to assess their quality. Then the process of writing assessed resources into data storage system is called. On the other hand strategy agent is responsible for the broader perspective of gathering resources. It is able to handle multiple definitions (queries) together with the schedule of their execution. It also brings into life as many Task agents as it appears from handled definitions. In other words, it is a kind of supervisor for the Task Agents.

All modules and Agents together provide multidisciplinary, agile, intelligent tool for acquiring resources from various Internet sources, and enable REPO system to become expendable, relevant and up-to-date source of knowledge.

4 Architecture of the Process of Creating Complex Queries and their Strategies

During the process of defining the data architecture in order to obtain possibly complete, relevant and up to date resources, following requirements were stated:

- one should be able to create custom queries for a chosen category,
- once created, queries should be executed as soon as possible,
- gathered resources should be refreshed in a given time interval,
- system should support defining parameterized queries,
- resources should provide a way of further user assessment of their correctness.

Requirements stated were not trivial for modeling, as they require complex data types with the highest possible effectiveness of the user interaction, data facilitation and data transfer.

Finally the following data architecture has been developed:

Supplier stores information on external data supplier (e.g. Google).

Type describes type of the gathered resources, for which proper classifier shall be applied (e.g. PERSON, UNIVERSITY).

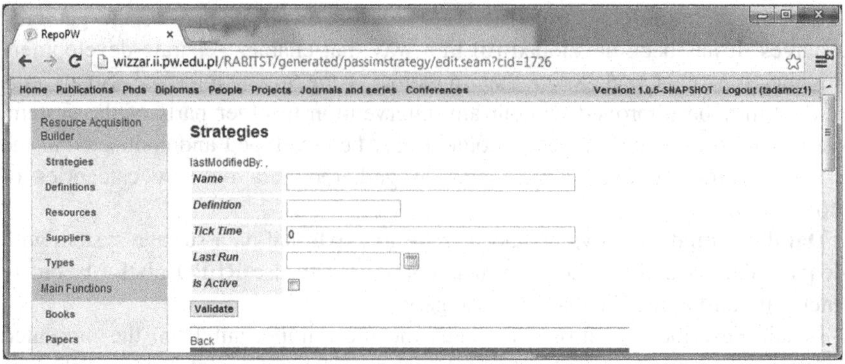

Fig. 3 New strategy object screen

Definition defines single (or generic) chunk of information to be obtained. This special type enables users also to create parameterized generic queries by using parameters file.

Strategy is a structure where the information on gathering process and refreshing gathered resources in time is defined.

Resource result of processed queries.

On the Strategy Object screen (Fig. 3) the main interface with top bar and side menu is visible. Each data structure has its own position in the menu on the left. For the backward compatibility with main REPO system other data structures have been left. Another thing visible on the screen is a form for creating new object of the Strategy. Beside such intuitive fields as Name, Last Run date and Is Active checkbox, a Definition object reference and Tick Time are visible.

Content of definition screen (Fig. 4) visualizes real-case object of Definition. Similarly to Strategy object it has a Name field which eases distinguishing Definition objects by user. Other fields are as follows: Type (category of resources to be searched), Query (search phrase description, braces defines reference to parameter names from parameter file), Supplier (defines which search engine should be used for search, also other data providers as CiteSeer may be used), Predecessor (this is the place where archive object is stored in case of a modification), File (csv format parameter file).

Resources when gathered are available through a special screen (Fig. 5) with additional controls for assessment of their relevance (user classification) and filters to narrow the scope of presented data e.g. if a user wants to see only the most dependable results or only the ones with given automatic classification result. Ovearall data structure (Fig. 6) is an effort of the whole team and has been covered more widely in related publication [2].

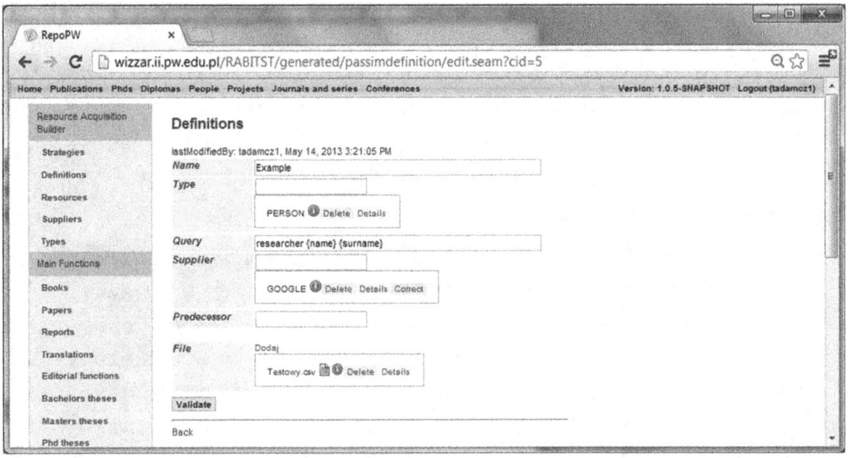

Fig. 4 Example content of a definition object

5 Use Cases and Possible Future Adaptation Possibilities

Typical use of the described system is ease of facilitation, storage and further utilization of pre-classified web resources in order to create coherent and possibly complete Knowledge Base on a given Category of Science.

To initiate this process, one should create adequate definition describing the type of requested resources as well as a query which defines more descriptively the desirable resource. One should also connect it with corresponding strategy, to define when and how often resources will be gathered and refreshed.

This is when Agents System comes into action, it queries, given by Supplier field, data source (one of preconfigured ones), fetches results from it and sends to classification module. Gathered resources, together with classification are then saved in Resource Acquisition Builder Interface (dedicated instance of REPO) system and immediately becomes visible to user with sufficient access privileges. Though automatic classification gives us quite a good idea on which resources are relevant and among other things it points out substantial part of results as rubbish, further filtration of results is needed. In the next step the user may select one or multiple resources and score them in a binary scale (True or False).

A few cycles of such feedback may be used for re-learning of classifiers or for defining probability threshold from which resources may be automatically recognized as trustworthy, relevant resources and passed for the next stage of the Knowledge Base building process. Either to provide a source for Scientific Information Database or for browsing for users with basic privileges (catalogued scientific information database).

From the description of the system provided so far, one may conclude that user intervention is necessary at least two times in the process of data facilitation. A user, in order to expand knowledge base, has to first define the Definition and Strategy of its

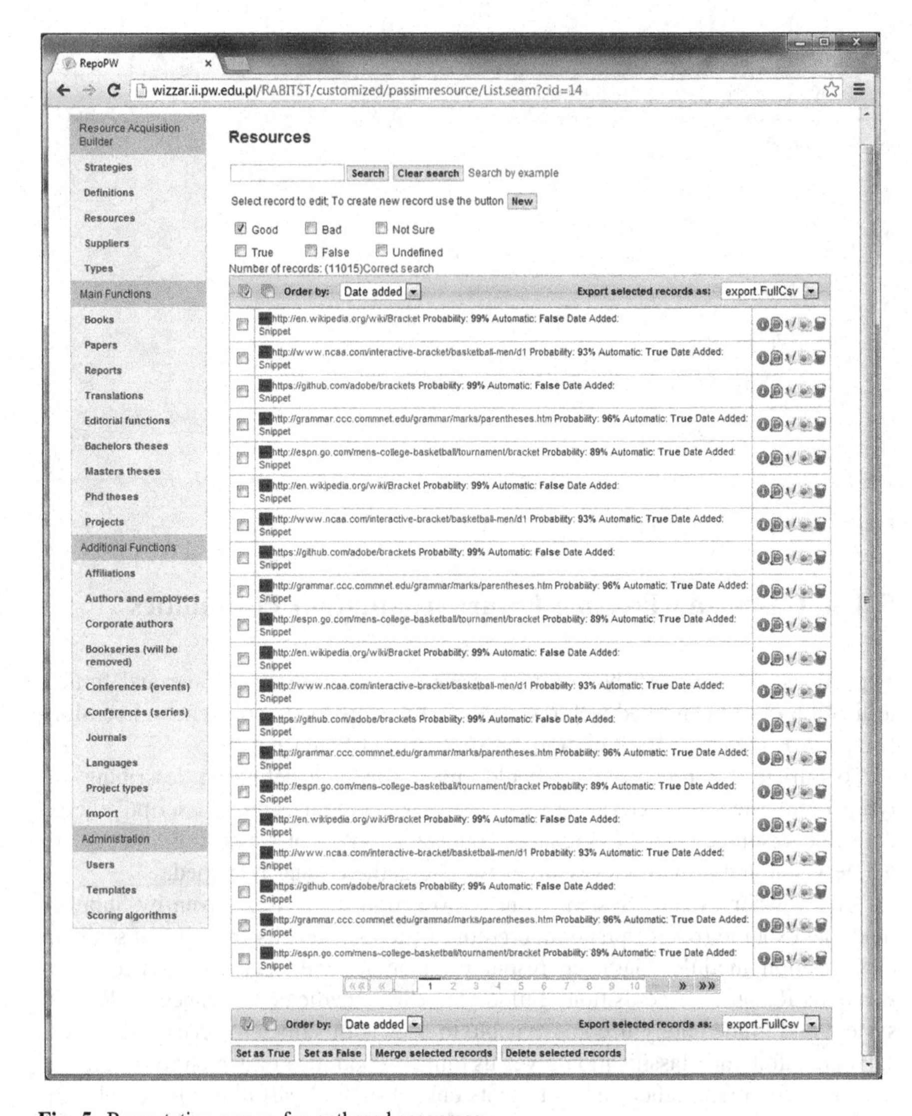

Fig. 5 Presentation screen for gathered resources

execution. At the current stage of work done, some intuition and experience in building queries for specific search engine is needed. In order to avoid unnecessary loss of query limits for gathering irrelevant content some limitations in access to this part of the system are needed. Currently only a specific, narrow group of users are allowed to browse and create new Definitions and Strategies. In the future it would be highly desirable to automate query optimization process. Some heuristic, ontology based algorithms have already been designed for this purpose. Also a semi-automatic query builder for deeper exploitation of a given category of knowledge is possible to implement. This would minimize user time spent on this stage even further.

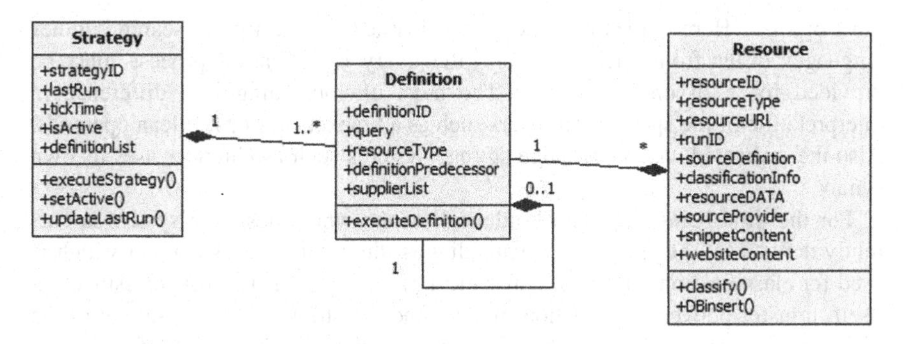

Fig. 6 Data model [2]

The next phase is completely automated. The agent system queries database for active Strategies periodically and if it encounters new one, it immediately starts its execution. Definition objects are interpreted and proper action on search engines are triggered. The gathered resources are then classified by the dedicated module and sent back to database with use of database module.

Second time when the human intervention is needed is for user assessment of already gathered resources. For the time being dedicated interface (Fig. 6) simplifies this process with the use of dedicated filters and controls. In the future, especially for well exploited domains it may be partially or even completely automated. A motivation for further automation are both better learning base for classifiers and better understanding of the acquired data. For example if for large amount of data every resource with score over 95 % is classified by a user as relevant, it is highly probable, that all other resources with such a score will be relevant as well. Other option for partial or complete replacement of this stage is to use more than one classifier on each resource.

The built system may be also used in other ways, some of them are as follows:

- an interface for semi-automatic information retrieval and usefulness or relevance classification processing,
- data source for learning systems,
- a higher level (with pre-cleaned data) interface for other information based systems,
- base of pre-classified or user-classified resources for learning, or re-learning of classifiers,
- self-learning and self-expanding Knowledge Database.

6 Lessons Learned

During first tests performed on the system over 20000 of resources have been facilitated. Its correctness and speed of facilitation are covered in other related papers [2, 5, 6]. Also the query provided as a test example possibly was not the most optimal. Here comes the next lesson learned for the future: search engines languages differ from each other and the query optimization process must be provided for each one separately. The most obvious thing is a difference in interpretation of the special characters such as a bracket char or Boolean operands. Also the dedicated web knowledge storage systems such as CiteSeer uses its own syntax.

For the time being the main bottleneck of the whole designed system is currently defined by the speed of downloading of the Internet sites content which is used for classification purposes. Other most probable are as follows: classification itself, transfer between acquisition module and the storage module, saving to the JCR, indexing, and user assessment. First one is unlikely to became relevant, as the classification of single resource takes a few orders of magnitude less than downloading of the site content.

Transfer and saving to the JCR database may became the bottleneck especially if the system would transform into a certain form of distributed one, with single frontend/storage node and multiple acquisition nodes. Loads of data transformation from xml into a java objects and finally into a form of JackRabbit records may lead to java heap space exhaustion up to the whole system crash. In order to prevent this, direct connection between facilitation system and the storage space may be provided. For example, data acquisition module may be connected directly to the database of the presentation system (e.g. JackRabbit).

Indexing and user assessment are not as crucial as the mentioned ones, because they may be processed in the background without interrupting data facilitation. Presented system is a working prototype of a platform for searching, acquiring and storage of the references to the Internet resources. It can easily became a datasource for further stages of gathering of the scientific information.

References

1. Jacobfeuerborn, B., Muraszkiewicz, M.: Media, information overload, and information science. In: Intelligent Tools for Building a Scientific Information Platform, Springer, Berlin, Heidelberg (2013)
2. Omelczuk, A., Andruszkiewicz, P.: Agent-based web resource acquisition system for scientific knowledge base. In: Studies in Computational Intelligence, Springer. (not published yet)
3. REPO system: http://repo.pw.edu.pl
4. Koperwas, J., Skonieczny, L., Rybiski, H., Struk, W.: Development of a university knowledge base. In: Intelligent Tools for Building a Scientific Information Platform, Springer, Berlin, Heidelberg (2013)
5. Omelczuk, A.: Composition and Utilization of Strategies for Scientific Information Gathering in a Multiagent System. Master's thesis, Warsaw University of Technology, Warsaw (2013)
6. Lieber, K.: The universal system for management and testing of classifiers. In: Studies in Computational Intelligence, Springer. (not published yet)

Analyzing Logs of the University Data Repository

Janusz Sosnowski, Piotr Gawkowski, Krzysztof Cabaj and Marcin Kubacki

Abstract Identification of execution anomalies is very important for the maintenance and performance refinement of computer systems. For this purpose we can use system logs. These logs contain vast amounts of data, hence there is a great demand for techniques targeted at log analysis. The paper presents our experience with monitoring event and performance logs related to data repository operation. Having collected representative data from the monitored systems we have developed original algorithms of log analysis and problem predictions, they are based on various data mining approaches. These algorithms have been included in the implemented tools: LogMiner, FEETS, ODM. Practical significance of the developed approaches has been illustrated with some examples of exploring data repository logs. To improve the accuracy of problem diagnostics we have developed supplementary log database which can be filled in by system administrators and users.

Keywords System monitoring · Event and performance logs · Dependability

J. Sosnowski (✉) · P. Gawkowski (✉) · K. Cabaj (✉) · M. Kubacki (✉)
Institute of Computer Science, Warsaw University of Technology,
Nowowiejska 15/19, 00-665 Warsaw, Poland
e-mail: jss@ii.pw.edu.pl

P. Gawkowski
e-mail: gawkowsk@ii.pw.edu.pl

K. Cabaj
e-mail: kcabaj@ii.pw.edu.pl

M. Kubacki
e-mail: M.Kubacki@ii.pw.edu.pl

R. Bembenik et al. (eds.), *Intelligent Tools for Building a Scientific Information Platform: From Research to Implementation*, Studies in Computational Intelligence 541,
DOI: 10.1007/978-3-319-04714-0_9,

1 Introduction

Complex information systems are subject to various disturbances resulting from hardware or software errors, unpredicted environmental, operational and maintenance impacts, etc. They may result in system unavailability and user dissatisfaction. Detection, diagnosis or prediction of such problems are important issues. That can be supported with extended on-line system monitoring based on registered event logs and collected system performance data, e.g. syslog, application logs, CPU and memory usage, transmission load. A lot of research has been done in this area, compare [1–7] and reference therein. Nevertheless, each system has its own specificity and needs individual treatment.

Monitoring the operation of the developed data repositories [8] targeted at scientific purposes we have found that the approaches proposed in the literature are too general. They do not fit the specificity of the collected logs and operational profiles of these systems. In particular, we have noticed the need of deeper textual analysis of event logs and advanced exploration of performance logs targeted at unknown problems. This was neglected in the literature. Moreover, we have found insufficiency of important information needed for better interpretation of collected logs. We have presented the scope of possible monitoring in [9]. The performed analysis of this research is based on dedicated developed tools and standard system mechanisms, e.g. system and application logs, performance counters. The main contribution of this paper is development of new tools: LogMiner, FEETS, and ODM which are integrated with the previous ones presented in [8]. These new tools are targeted at fine grained analysis of log features, as opposed to coarse grained approaches described in the literature. In particular, they improve classification of event logs and detection of normal or suspected operational patterns. Moreover, they facilitate adapting the analysis to the properties of different logs, e.g. system, application or access logs.

Section 2 presents an original approach to event log analysis in a large extent related to text mining methods. Section 3 outlines the implemented log management system which provides the capability of collecting remarks from users and administrators and correlates them with system standard logs. Section 4 deals with temporal analysis of performance logs integrated with the mentioned log management system. Conclusions are summarized in Sect. 5.

2 Text Analysis of Event Logs

Event logs comprise useful information to identify normal and abnormal system behavior. Hence, it is important to trace their content, frequency of appearance, correlations between different events within the same log and other logs. Detected events are registered in partially structured text records. They comprise fields related to registration time, source of event and unstructured messages written in a loose form including technical jargon. These messages may include natural language words, special technical words, specific acronyms, file names and paths, codes, etc. Taking into account the multitude of registered events an important issue is their classification based on various attributes, identified variables, their correct or incorrect ranges, etc. For this purpose we have developed some algorithms with filters specified as regular expressions and variable detection [10]. Recently, we have concentrated on enhancing them with deeper and fine grained text mining schemes. In [2], for this purpose, some text similarity measures have been used in the analysis of specific application logs. In our case this approach needs tuning due to big dispersion of messages.

Providing a set of similarity measuring algorithms and their modifications we could look for new, not similar events, or look for similar events within specific event hyper-classes, e.g. correct operation, erroneous behavior. Calculating distance of similarity we can exclude some fields e.g. date, PID, host name, etc. However, some care is needed in similarity measures e.g. two events with almost the same text message but comprising phrases: "file accessed" and "file aborted" cannot be considered similar. Here, we touch the problem of positive and negative meaning, e.g. available, unavailable, suspended, not suspended. Well derived keywords will be helpful here. So similarity distance cannot be a unique decision. In [2] higher significance has been attributed to initial words. However, weighing similarity according to word positions in the text can also be misleading. This is illustrated in the following two log messages:

- *Performance counters for the <service name> service were loaded successfully. The Record Data contains the new index values assigned to this service.*
- *Performance counters for the <service name> service were removed successfully. The Record Data contains the new values of the system Last Counter and Last Help registry entries.*

The brackets <> specify an identified variable which assumes various text contents in different messages.

Dealing with event similarity measures we could also introduce structural similarity of words, e.g. words of comparable structures. This can be applied to file paths, IP ports or Web pages, which can be defined using regular expressions. Another similarity measure may take into account word prefixes and suffixes and skip them in finding similar words. In some cases such words as *error1*, *error2* can be considered as very similar, moreover this can be conditioned with similarities on other fields or word sub sequences. We discuss this problem further below. Event classification should be supported with a deeper textual analysis of messages. Hence, we decided to support the previously developed tool EPLogAnalyser [8, 9, 11] with LogMiner which provides additional capabilities in the area of text mining, clustering and correlation analysis.

Analyzing texts comprised in event logs we base on appropriate dictionaries. In our approach we used WordNet base (http://wornet.princton.edu) which comprises a set of English words dictionaries specified in XML format. Here, we have separate dictionaries for nouns, verbs, adjectives, adverbs, etc. Words are grouped according to their meanings. Having performed a manual preliminary analysis of event log records we have found the need of specifying some additional dictionaries to facilitate text mining. In particular, we have introduced the following dictionaries: negative nouns (13), negative verbs (49), negative adjectives (2118), negative adverbs (437). In the case of negative adjectives we selected words with negative prefixes -il, -un, etc. These words have been verified and supplemented with some others selected manually. Moreover, we have added dictionaries of technical terms related to hardware, software, internet, file formats.

Basing on classical dictionaries we deal with words comprising only letters (regular words). In event logs we have also more complex words e.g. comprising numerical or special characters (file paths, process names, etc.). Hence, developing a special tool for text mining (LogMiner) we have admitted two options of word analysis: (i) related to standard lexicographic words, (ii) enhanced with partitioning complex words into elementary words. In the last case we deal with complex words comprising elementary words concatenated using special characters (e.g. /). Despite a significant extension of word definitions LogMiner does not classify all words leaving them in a large class called unknown. This group comprises various acronyms, words comprising numbers, etc. Classification of such words needs more complex algorithms and can be combined with the developed process of deriving variables and classifying event logs (compare [10]).

The developed LogMiner facilitates text mining. Using this tool we have derived some interesting properties within the collected event logs. For an illustration we give statistics (Table 1) of word classification within logs collected from Neptun server (used in Institute of Computer Science); it covers 6 months—over 100,000 events. Table 1 characterizes 11 dictionaries related to adjectives (AD), negative adjectives (NAD), adverbs (ADV), computer acronyms (ACR), hardware (HW), internet (INT), nouns (N), negative nouns (NN), software (SW), verbs (V), negative verbs (NV). All regular words not included in these dictionaries are classified as unknown words (UW). Table 1 gives the total number of detected words (accumulated frequency) belonging to specified dictionaries. This statistics includes words selected from complex words—option (ii) of LogMiner. Cancelling this extension, option (i), we have got much smaller number of identified words i.e. only 12,600 nouns as opposed to 472,519 in the first case. Here, complex words have been rejected as not conforming with regular word definition. This confirms that log messages comprise much more specialized complex words than regular lexicographic words. Such words can be analyzed with regular expressions and other variable classification algorithms (compare [8–10]). Another interesting issue is the distribution of different (unique) words (cardinality of used dictionaries). This is illustrated in Table 2 for two considered servers used for data repositories.

Here, it is also worth noting that the used vocabulary in event logs is relatively poor as compared with classical documents analyzed using text mining [12, 13]. Nevertheless, we should comment here that this analysis does not take into account so called non regular words i.e. words comprising numbers, concatenated words (without special characters). In particular in this group we can have file paths, port

Table 1 Statistics of words for Neptun server

UW	AD	NAD	ADV	ACR	HW
199664	460	129	129433	38467	1984
INT	N	NN	SW	V	NV
1013	472519	42	144050	327682	77

Table 2 Number of unique words used in Neptun and Catalina servers

UW	AD	NAD	ADV	ACR	HW	INT	N	NN	SW	V	NV
Neptun server											
548	219	14	71	51	48	20	606	4	47	301	9
Catalina server											
125	40	3	14	7	6	9	130	1	15	55	1

numbers or such specifications as: errorl, user5, severl1. These words as well as the words classified in dictionary of unknown words (e.g. specific acronyms) can be analyzed with the available classification algorithms and regular expressions.

It is worth noting that different event logs have different text properties, in particular different word statistics, different used words, etc. For comparison we give some statistics of Catalina server from NASK and access log (application level log). Here, we have got about 280 uniquely qualified words (as opposed to about 1200 in the case of Neptun). However, the event records comprised more words than in the case of Neptun (many of them non regular or not classified). The semantic diversity here was much lower. In general, different logs have different properties and the analysis has to be tuned to these properties.

An interesting issue is checking dictionary properties in time. For example, in Catalina server during the subsequent 6 months we have notified the following number of negative noun *stop*: 43, 33, 64, 36, 226, 156 and for word *failed*: 80, 124, 176, 175, 277, 334 cases, respectively. During the last 2 months the frequency of these words increased, and there were many problems with the system.

Using more advanced text mining we have studied Inverse Document Frequency (IDF) defined as follows (adapted from [12]):

$$IDF(w) = log\left(\frac{|D|}{|\{d \in D{:}w \in d\}|}\right) \quad (1)$$

where w is the considered word and D is a set of registered event records d in the analyzed log, $|S|$ denotes the cardinality of the specified set S.

For the considered Neptun server and analyzed 1184 unique words we got a logarithmic shape of this parameter (Fig. 1). The x-axis covers unique words from 11 dictionaries (ordered from the highest to the lowest IDF values). We should notice that some words appear simultaneously in different dictionaries (e.g. adjectives and negative adjectives, nouns and verbs, etc.). In classical text mining the IDF parameter is helpful in identifying interesting words (e.g. keywords), as those with high IDF value [12]. In our case we have many words with high IDF, so

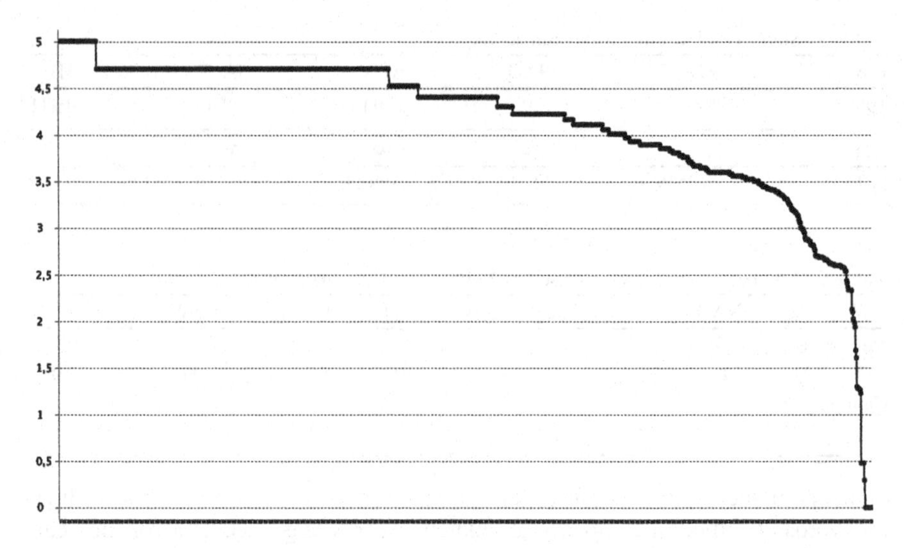

Fig. 1 Distribution of IDF parameter for Neptun server (x-axis [0,1184] words, y-axis [0–5])

it is not sufficient criteria for selecting interesting event records. Hence, some other criterion should be added. In particular, we have decided to check IDF for negative words and here we have received interesting results. Within the set of 219 used adjectives 13 have been defined as negative, for 11 of them IDF was in the range 2.62–5.01, similarly 3 negative nouns: *fault, warning, error* achieved IDF 2.96, 3.58 and 4.53, respectively. Event records comprising these words were interesting for the users. Here is a sample of events (negative words are underlined):

2013-03-02 17:59:00 neptun kernel: [2.479525] PM: Resume from disk Failed
2013-01-24 09:07:01 neptun kernel: [0.000000] Phoenix BIOS detected: BIOS may corrupt low RAM, working around it
2013-03-02 17:59:00 neptun kernel: [2.010461] sd 2:0:0:0: [sda] Cache data unavailable

Distribution of IDF for words in access logs of Catalina server was quite different (close to linear decrease) with only 40 words with IDF exceeding 2.8 (as opposed to over 1,000 in the case of Neptun).

Dealing with IDF we should be prepared for situations with bursts of replicated events within some time window. In the case of anomalous situations this may happen quite frequently. This effect decreases IDF of words appearing in bursts, so it is reasonable to filter them appropriately in calculations.

We have extended IDF analysis for word phrases, i.e. ordered sequence of words. LogMiner has several options dedicated to this purpose. In particular, we can select phrase length (2–4 words) or various filtering schemes, e.g. phrases comprising words belonging to specified dictionaries. This allowed us to select interesting log entries. We have also checked parameter IDF for phrases (treating them as a generalized word). For an illustration we give interesting phrases composed of 4 words with relatively high values of IDF: [*link, is, not, ready*]—IDF = 4.71, [*file, could, not, be*]—IDF = 4.71, [not, provided, by, any]

IDF = 4.41. In Catalina access log usually event messages are relatively long, so we searched even for longer phrases. For example phrase [*non, clean, shutdown, detected, on*] attained IDF = 2.5 and it related to interesting events. An example of such an event is given below:

> *2012-09-27 16:11:45 INFO: Non clean shutdown detected on log [/opt/apache-tomcat-7.0.16/data/passim-entity/nioneo_logical.log.1]. Recovery started...*

This event is related to incorrect closing of library Neo4j by application Passim.

Using LogMiner we have also detected events preceding (within a preprogrammed time window) events with specified properties. For example, events preceding events comprising word *error*. In the case of Neptun server we have got 5 different events satisfying this condition (for time window 4 s). They occurred 20 times (support) with probability 40 % (confidence) before the specified condition and can be considered as problem prognostic (prediction).

Another interesting capability is searching for events correlated with selected performance parameters (e.g. CPU usage, transmission load, memory usage). Here, we specify time window of searching events before a précised performance condition. For example, we can select events preceding CPU load average (the CPU queue length) threshold exceeding 3. For Neptun server we have identified 11 event types which occurred up to 3 min before CPU load average exceeded 3. The confidence of this correlation was 100 % and the support was in the range from 2 to 12. Similarly, we have identified events preceding a significant increase in network transmissions, they related to periodic backup processes.

Sometimes it is interesting to check the values of some identified variables within logs, e.g. accessed file paths in the access log. This allows us to check whether the accesses correspond to the normal operational profile. This can be derived using LogMiner. The classical log analysis tools are targeted mostly at detecting known problems (compare [14–16] and references therein). The developed LogMiner supports tracing log features to extract situations which may be symptoms of various anomalies including the unknown ones. This has been achieved with a deep textual analysis based on extended IDF measure covering also word phrases. This process is enhanced with the use of the derived supplementary dictionaries. Moreover, it is integrated with the previously developed algorithms used to identify variables [10] in the event records. Most of the identified words relate to the unclassified group, however we use some structural analysis which extracts specific parameters within this group, e.g. ports, file paths. Hence, the constructed dictionaries can be further decomposed into acceptable/unacceptable parameters, which simplify detecting anomalies. In our approach we can trace events preceding the problem appearance and even correlate them with performance measures. Tracing dynamic properties of the created dictionaries we can get some knowledge on normal or abnormal changes, e.g. newly appearing or disappearing words. In this way significant reduction of potentially interesting events has been achieved.

3 Enhancing and Managing System Logs

There are plenty of events, performance and system state tracing tools available. However, it is hard to point a single one that would be universal enough to address all the aspects related in particular to complex on-line anomaly detection, historical data availability, etc. Monitoring systems like Nagios and Nagios-based systems (e.g. op5 Monitor, opsview, Icinga) are quite simple monitoring solutions offering simple alerting in case of failure detection (based usually on exceeded thresholds detection). Both the events and other measures (sensors readouts, performance measures, environmental conditions, workloads etc.) are still not correlated with historical observations to provide administrators a comprehensive insight into the systems operation.

Basing on our experience with different sources of information about the system states we propose the Failures, Errors and Events Tracking System (FEETS). The main goal is to integrate many different sources of information about the monitored IT systems, starting from the most hardware level events, real time captured sensor data, through the operating system and virtualization hypervisors' level mechanisms of performance and event logging to the application specific logs and performance information. Moreover, FEETS collects users' and administrators' notes to complement the collected raw data from sensors and logs. This facilitates analysis and identification of problems in historical data (described later in this section). Contrary to the existing monitoring and logging systems, FEETS can be iteratively extended by sophisticated analytical on-line, as well as offline failure and anomalies detectors.

Figure 2 presents the proposed FEETS concept. One of the most important components in the FEETS would be *On-Line Analysis* module. It should implement two functionalities:

- a proxy layer for keeping the most recent data from the monitored data sources (from the data collection layer),
- some basic analysis and detection of obvious violations on the predefined thresholds and conditions (e.g. dependability, security-related, operational conditions, performance and resources bottlenecks).

Additionally, all the collected data should be stored in the corresponding transactional databases. After transformation stage it should also supply data warehouses for advanced analyses. As some correlations, complex anomalies (probably critical) and conditions could be found, they can also be used for raising the administrative alerts. FEETS should support two kinds of alerting. First of all, some critical conditions related to the on-line detection of threshold violation should be implemented. It covers some basic checking of performance, status and events at real-time. However, for the system dependability some historical analyses should

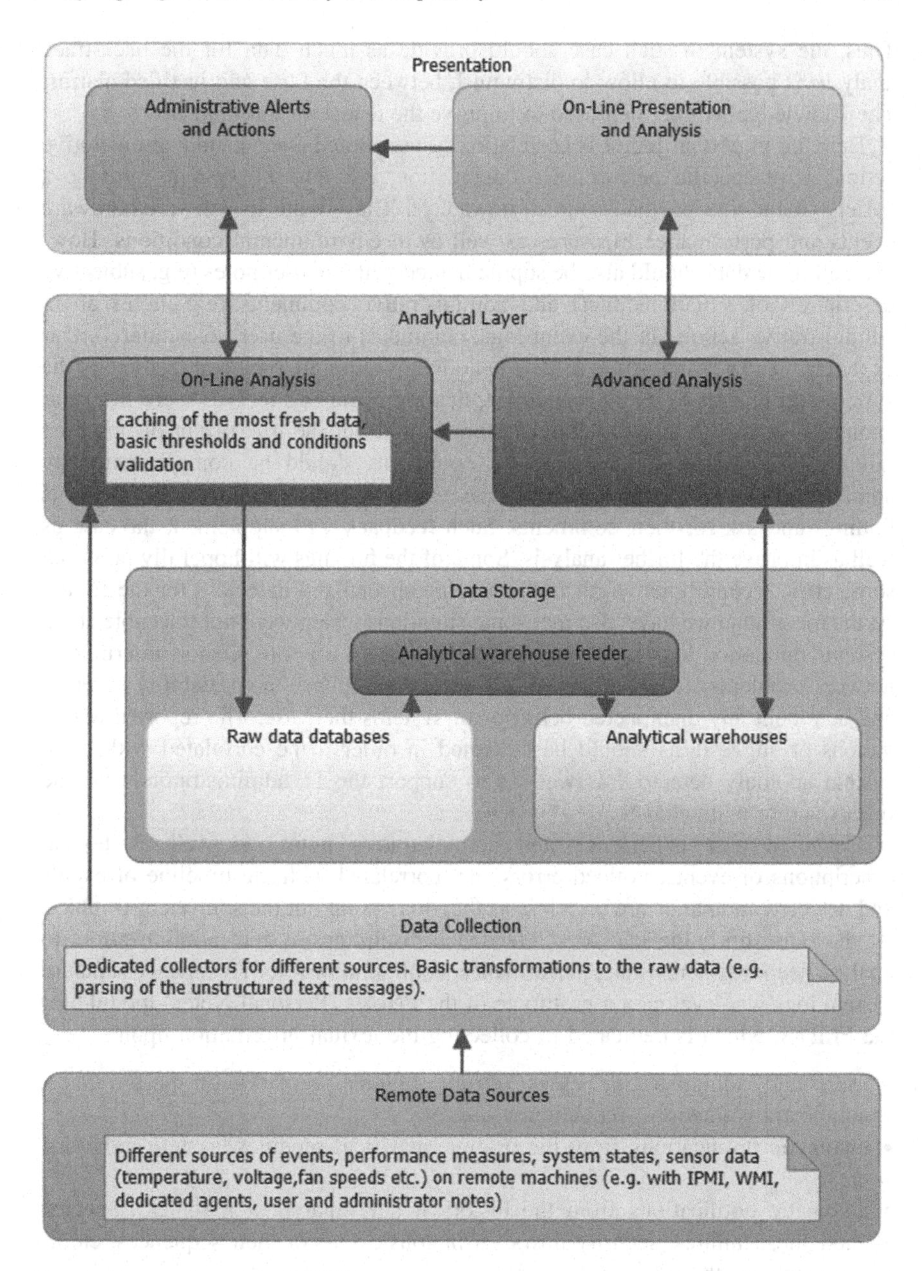

Fig. 2 An overview of the FEETS layers

also be available. Moreover, some of the raised real-time alerts could be due to the oversensitivity of the monitoring system or some specific actions made at the target systems (e.g. installation of updates, hardware replacement or reconfigurations).

Thus, the system in such case should provide as much data for the back-trace analysis as possible to allow to distinguish between the false and justified alarms. The knowledge will be then used to improve the detection mechanisms.

Here, an important factor is correlation of user-noted events and administrative actions with specific performance degradation, etc. The IT systems provide a variety of monitoring data sources nowadays. They relate to software/hardware events and performance measures as well as to environmental conditions. However, all these data should also be supplemented with the user notes (e.g. subjective description of situations met) and administrative comments (e.g. notes about administrative actions in the event logs, comments to the user notes, alerts). Our experience showed that some detected anomalies required to be commented by the system administrators as the system/application event log records were not clear enough—we had to talk to administrators to find out the true origin of observed failures. So, finding some suspicious conditions should be complemented by administrators—FEETS should notify the related systems' administrators about its findings and ask for their comments. Such feedback will supplement the case as well as improve the further analysis. Some of the findings will hopefully point out some critical conditions worth implementing specialized detectors for the future. At the other hand we have also met some situations which were not traceable at the system/application logs but were notified by system users (e.g. short intermittent network problems). Thus, the normal IT users should have the possibility to report and comment any unexpected behavior of systems their use. The reported observations or suggestions should be collected in order to be correlated with other system anomaly detectors as well as to support the IT administrators with the infrastructure maintenance.

Collected user experience and administrative notes, as well as textual descriptions of events, noticed errors, etc. correlated with the timeline of events and detected anomalies are priceless as they may point out the sources of troubles. At the same time, the end users' feedback disciplines system administrators to analyse deeper systems state, performance, configuration, etc. To enhance standard system logs we developed a prototype of the PNotes (Personal Notes) module for the FEETS, which is dedicated to collecting the textual information upon:

- users and administrative cases (errors, failures, remarks on performance, administrative actions, reconfigurations, etc.),
- automatic notifications from the on-line analytical modules, system monitors, etc.
- automatic notifications from the historical data analytical modules (detected anomalies, failures, security flows, suspicious events or their sequences, etc.),
- comments on the registered cases.

After a case creation (no matter who or what was its source) the users and administrators can share their knowledge by commenting the case. It is worth to note that several cases (e.g. from different sources) can be related or even provoked by the same reason. So, it is reasonable to allow users and administrators to

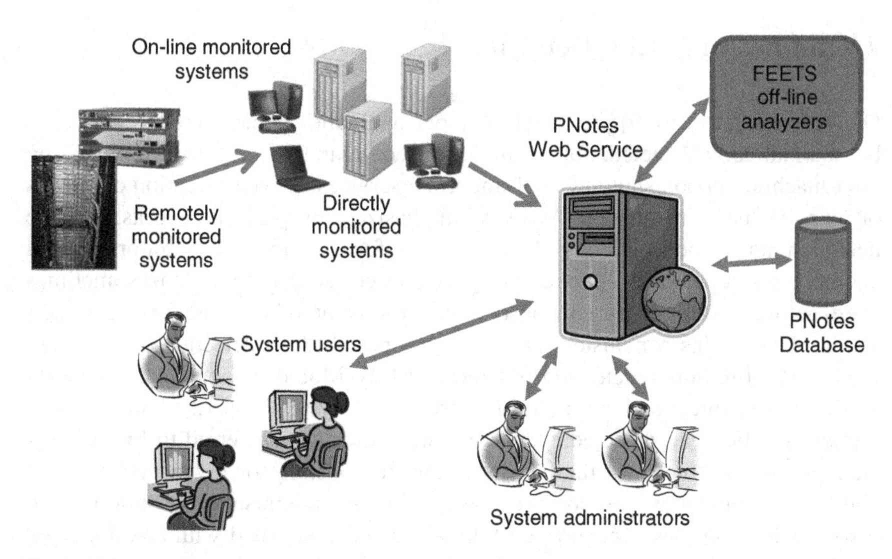

Fig. 3 PNotes architecture overview

link their comments with several cases. Similarly, a single case can be related to several systems. Finally, after information exchange between the users and administrators, a case solution comment can be marked.

The prototype implementing the most part of the PNotes operation concept described above was developed. It was successfully installed on the resources of the Division of Software Engineering and Computer Architecture at Institute of Computer Science. Moreover, the exemplary anomaly detection module was integrated with PNotes service (presented in Sect. 4) and the user-friendly client application was developed (exemplary screenshot is shown in Sect. 4). The proposed PNotes architecture is shown in Fig. 3. All the data of the PNotes module is stored in the central database. Here, the SQL Server 2008 R2 database is used. As heterogeneous sources of information have to be collected, it was decided to use web-service technology to support connectivity on many operating systems from different programming languages. The service implements API to store and retrieve the information to/from the PNotes database. Using a web-service also simplifies the installation of the info sources agents and end-user application as the firewall rules are typically allowing HTTP/HTTPS connections. Moreover, the web-service connectivity can be easily achieved from different systems, operating systems and programming languages—so, many existing systems towards e.g. anomaly detections can be easily integrated as information sources for the PNotes (see Sect. 4). Right now the PNotes service is hosted by the IIS web server (Microsoft Internet Information Server) along with the PNotes database.

4 On-line Anomalies Detection

Our previous work [8, 9] proves that various anomalies in monitored systems can be automatically detected, for example, misconfiguration of software appearing after machine reboot, shutdowns of important services or even detection of attacks or compromised machines. As was emphasized in previous works, on-line detection and reporting could be beneficial for system administrators. Rapid anomaly detection and on-line reporting enables correction of problems sometimes even before system users can observe its results or other more severe effects appear. Due to this conclusion, previously proposed [8, 9] detection methods are implemented in online detection modules (ODM). Moreover, implemented ODM software was integrated with FEETS PNotes subsystem described in previous section. In effect, all detected events are automatically transferred to FEETS and later presented directly to the administrators responsible for given system. This ability is beneficial at least for two reasons. Firstly, administrators using desktop PNotes client can observe, sort and filter all events associated with administrated systems. Secondly, all events, even irrelevant at the first glance, are archived and can be analyzed in the future, giving clues concerning reasons of future events. This ability could also be useful for preparation of various reports concerning observed events in the monitored network.

Implemented ODM software is executed at a dedicated monitoring station. Performance counters used for anomaly detection are collected from a set of monitored machines using SNMP protocol. For this purpose some custom scripts responsible for collecting data and passing it to the ODM module are used. After successful collection of the latest data, the ODM software places that data into the moving averages for various time intervals (see [8, 9]). Comparing calculated adjacent averages for the given interval and calculated values concerning various time scales, the decision concerning detection of anomaly is taken. When anomaly is detected in any monitored performance counter, appropriate information is directly reported to the FEETS PNotes to alert administrators responsible for the affected systems. For this purpose the Web Service API provided by the FEETS PNotes is used. To separate the detection and reporting functionalities in ODM software an additional program was implemented that is responsible just for reporting new cases and comments into FEETS PNotes system. The program is directly called from ODM software when new anomalies are detected. For communication with FEETS PNotes Web Service API the open source *gsoap* library in version 2.8.14 is used.

Listing 1 presents the exemplary output from the ODM software showing the real-life case of anomaly associated with the network traffic and related CPU usage rises.

```
1371481550 2013-06-17.17:05:50 202 121 50 1423073146
199860677 4271197797 126080418

Input traffic - spike 0
---------------------------
Current Hour average 49488.854772
Prev Hour average 60877.995851
Five minutes averages from current 49622.52 30350.86
34037.52 57746.76 68972.67 77721.71
Quarter averages from current 39164.737705 70296.459016
41684.442623 44503.278689
Output traffic - spike 1
---------------------------
Current Hour average 230570.506224
Prev Hour average 268532.572614
Five minutes averages from current 279913.90 24801.52
139358.86 315655.71 370018.38 451108.52
Quarter averages from current 152791.475410 391264.836066
138784.704918 228231.311475
CPU - spike 1 [initialRiseValue 3.032787]
---------------------------
Current Hour average 3.265560
Prev Hour average 1.004149
Five minutes averages from current 23.14 0.19 1.86 4.05
3.57 2.90
Quarter averages from current 8.672131 3.622951 0.327869
0.278689
```

Listing 1. Output of the ODM software showing anomaly detected in output traffic and CPU usage performance counters

The first line of the output shows the actual raw performance data passed to the ODM software. In the following lines three performance counters are analyzed: *input traffic*, *output traffic* and *CPU usage*. For each performance counter various moving averages are calculated: 2 for 1 h intervals, 6 for 5 min intervals and 4 for 15 min intervals. The flag *spike 1* denotes that the observed value is currently above the average level and anomaly is detected. In the presented listing such situation is observed for the output traffic and the CPU usage. For the first performance counter its 5 min average value rises almost eleven times from 24,000 to 279,000. At the same time, the second performance counter's value for 15 min interval rises more than two times (from 3.62 to 8.67), but for 5 min interval rises more than 120 times from 0.19 to 23.14. This particular abnormal event was provoked by requesting very detailed report concerning almost all stored records from the Passim small repository at 5 p.m. In effect the CPU utilization raised in noticeable manner. Additionally, generated report was so big, that downloading it from the web server (Web3 machine in this case) caused detection of output traffic anomaly. Figure 4 shows 5 and 15 min CPU load plots observed by implemented

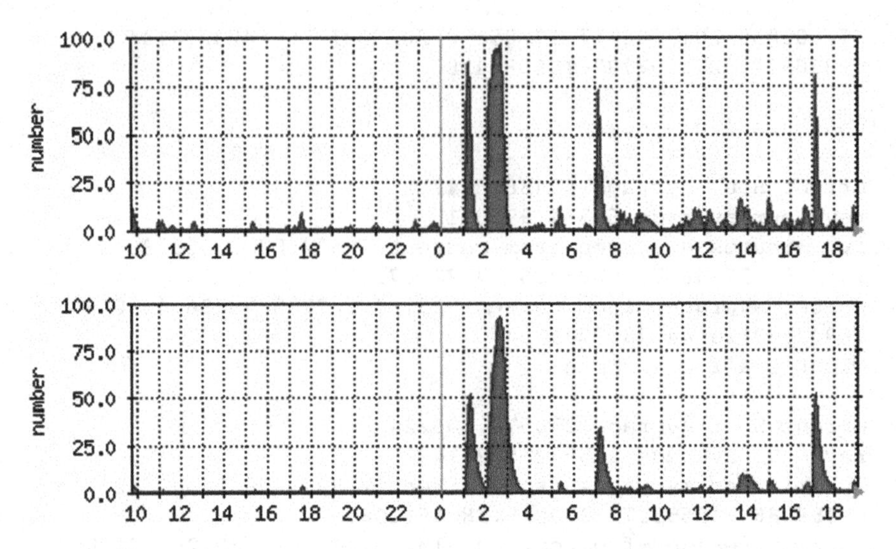

Fig. 4 CPU load in 5 (*top plot*) and 15 (*bottom plot*) min intervals at Web3 machine during experiment. Logs from ODM software during detection of this event are shown in Listing 1

MRTG monitoring system during the experiment. The sharp spike at 5 p.m was associated with the detected anomaly. One can also observe other spikes (e.g. at 1 a.m, 2 a.m and 7 a.m) which were recognized as standard backup activities (also automatically discovered by ODM).

Figure 5 presents a screenshot from PNotes FEETS desktop application with the detected anomaly described previously in the text. All anomalies which are detected by ODM software in adjacent time instants are grouped into one case. Due to this fact, this particular case consists of multiple comments which are associated with various events. In the described example anomaly was detected at two performance counters—one associated with output network traffic and the other with the CPU utilization. In effect comments appearing in this case have labels which begin with texts "Processor" and "Output traffic". Additionally, comments associated with output traffic estimate volume of data transmitted during the detected anomaly.

At the time of writing this paper, the ODM software does not implement any features that would correlate the detected anomalies in the monitored performance counters with other data sources. However, there are plans to extend the ODM functionality to automatically search for suspicious processes using data mining methods, mainly frequent sets, in case of anomaly detection. The results from such experiments performed in the off-line manner were described with details in [8].

During the last few months we have detected several types of anomalies that could be detected automatically. For instance, infection of one of the servers, normal and abnormal utilization, performance bottlenecks, failures of the backup

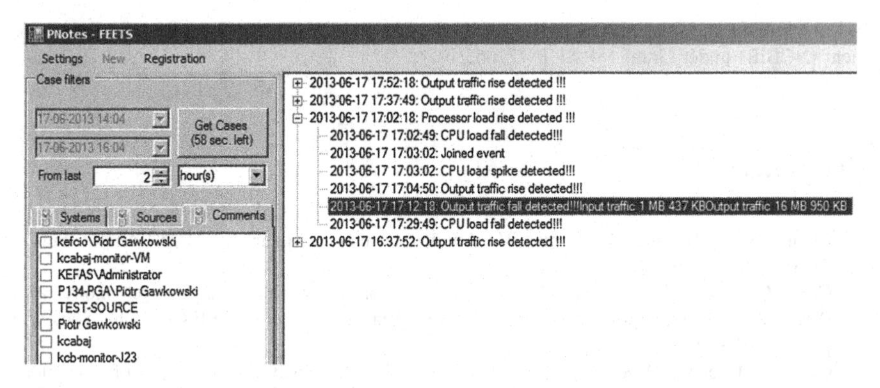

Fig. 5 Screen shot from PNotes FEETS desktop application for administrators showing the case associated with the detected anomaly. The anomaly starts with the rise in CPU usage at 5:00 p.m and after few minutes at 5:04 p.m the rise in output traffic was detected

procedures, DoS attack. The automatic detection with ODM would instantly raise the administrative alerts to limit the possible damages while in some cases the administrators reacted after several hours (in the morning).

5 Conclusion

Assessing system dependability on the basis of event and performance logs gains much research interest and practical significance. Due to the multitude and diversity of logs their analysis is not trivial and needs supplementary tools adapted to the monitored systems. Having studied the operational profiles of the developed repositories and their image in event and performance logs we have developed and tuned special tools facilitating this analysis: LogMiner, FEETS, and ODM. They involve advanced text mining, exploration of correlations and data pattern analysis which are not available in other tools. In particular, we have proposed advanced event log classification based on message text analysis. The introduced dictionaries and extended IDF measures for words and phrases facilitate extracting symptoms of correct and abnormal behavior. The gained practical experience proved the need of adding user oriented logs and the log management system called PNotes. This approach allowed us to trace log features and detect interesting situations needing attention of system administrators. As opposed to reports in the literature (compare [2, 5, 17, 18]), we faced more problems related to external attacks, configuration inconsistencies and administration flows than hardware errors.

In the further research we plan to develop more advanced log analysis algorithms, in particular correlating features of different logs, performing multi-dimensional analysis and integrating this with the developed knowledge database [11] on dependability. This will provide domain experts with suggestions to distinct between normal and anomalous system behavior.

Acknowledgment This work is supported by the National Centre for Research and Development (NCBiR) under Grant No. SP/I/1/77065/10.

References

1. Chandola, V., Baerjee, A., Kumar, V.: Anomaly detection, a survey. ACM Comput. Surv. **41**(3), 1–58 (2009)
2. Chen, C., Singh, N., Yajnik, M.: Log analytics for dependable enterprise telephony. In: Proceedings of 9th European Dependable Computing Conference, IEEE Computer Society, pp. 94–101 (2012)
3. Cinque, M., et al.: A logging approach for effective dependability evaluation of computer systems. In: Proceedings of 2nd IEEE International Conference on Dependability, pp. 105–110 (2009)
4. Naggapan, M., Vouk, M.A.: Abstracting log lines to log event types for mining software system logs. In: Proceedings of Mining Software Repositories, pp. 114–117 (2010)
5. Oliner, A., Stearley, J.: What supercomputers say: A study of five system logs. In: Proceedings of the IEEE/IFIP International Conference on Dependable Systems and Networks (2007)
6. Salfiner, F., Lenk, M., Malek, M.: A survey of failure prediction methods. ACM Comput. Surv. **42**(3), 10.1–10.42 (2010)
7. Yu, L., Zheng, Z., Lan, Z.: Practical online failure prediction for blue gene/p: period-based vs. event-driven. In: Proceedings of the IEEE/IFIP International Conference on Dependable Systems and Networks Workshops, pp. 259–264 (2011)
8. Sosnowski, J., Gawkowski, P., Cabaj, K.: Exploring the space of system monitoring. In: Bembenik, R., et al. (eds.) Intelligent Tools for Building a Scientific Information Platform: Advanced Architectures and Solutions. Studies in Computational Intelligence, vol. 467, pp. 501–517 (2013). ISBN 978-3-642-35646-9
9. Sosnowski, J., Gawkowski, P., Cabaj, K.: Event and performance logs in system management and evaluation. In: Jałowiecki, P., Orłowski, A. (eds.) Information Systems in Management XIV, Security and Effectiveness of ICT Systems, pp. 83–93. WULS Press, Warsaw (2011). ISBN 978-83-7583-371-3
10. Sosnowski, J., Kubacki, M., Krawczyk, H.: Monitoring event logs within a cluster system. In: Zamojski, W. et al. (eds.) Complex Systems and Dependability. Advances in Intelligent and Soft Computing, vol. 170, pp. 259–271. Springer, Berlin (2012)
11. Kubacki, M., Sosnowski, J.: Creating a knowledge data base on system dependability and resilience. Control Cybern. **42**(1), 287–307 (2013)
12. Berry, M.W., Kogan, J.: Text Mining Applications and Theory. Wiley, Chichester (2010)
13. Han, J., Kamber, M., Pei, J.: Data Mining Concepts and Techniques. Elsevier, Amsterdam (2012)
14. Hershey, P., Silio, C.B.: Systems engineering approach for event monitoring and analysis in high speed enterprise communication systems. In: IEEE International Systems Conference SysCon (2009)
15. Kufel, L.: Security event monitoring in a distributed systems environment. IEEE Secur. Priv. **11**(1), 36–42 (2013)
16. Cinque, M., Cotroneo, D., Pecchia, A.: Event logs for the analysis of software failures, a rule based approach. IEEE Trans. Softw. Eng. **39**(8), 806–821 (2013)
17. Fu, X., Rebn, R., Jianfeng, Z., Wei, Z., Zhen, J., Gang, L.: LogMaster: mining event correlations in logs of large-scale cluster systems. In: Proceedings of IEEE Symposium on Reliable Distributed Systems, pp. 71–80 (2012)
18. Vaarandi, R.: A data clustering algorithm for mining patterns from event logs. In: Proceedings of 3rd IEEE Workshop on IP operations and Management, pp. 119–126 (2003)

Content Analysis of Scientific Articles in Apache Hadoop Ecosystem

Piotr Jan Dendek, Artur Czeczko, Mateusz Fedoryszak, Adam Kawa, Piotr Wendykier and Łukasz Bolikowski

Abstract Content Analysis System (CoAnSys) is a research framework for mining scientific publications using Apache Hadoop. This article describes the algorithms currently implemented in CoAnSys including classification, categorization and citation matching of scientific publications. The size of the input data classifies these algorithms in the range of big data problems, which can be efficiently solved on Hadoop clusters.

Keywords Hadoop · Big data · Text mining · Citation matching · Document similarity · Document classification · CoAnSys

P. J. Dendek (✉) · A. Czeczko (✉) · M. Fedoryszak (✉) · A. Kawa (✉) ·
P. Wendykier (✉) · Ł. Bolikowski (✉)
Interdisciplinary Centre for Mathematical and Computational Modelling,
University of Warsaw, Warsaw, Poland
e-mail: p.dendek@icm.edu.pl

A. Czeczko
e-mail: a.czeczko@icm.edu.pl

M. Fedoryszak
e-mail: m.fedoryszak@icm.edu.pl

A. Kawa
e-mail: kawa.adam@gmail.com

P. Wendykier
e-mail: p.wendykier@icm.edu.pl

Ł. Bolikowski
e-mail: l.bolikowski@icm.edu.pl

R. Bembenik et al. (eds.), *Intelligent Tools for Building a Scientific Information Platform: From Research to Implementation*, Studies in Computational Intelligence 541,
DOI: 10.1007/978-3-319-04714-0_10,

1 Introduction

Growing amount of data is one of the biggest challenges both in commercial and scientific applications [1]. General intuition is that well embraced information may give additional insight into phenomena occurring in the data. To meet this expectation, in 2004, Google proposed the MapReduce paradigm. Four years later, the first version of an open-source implementation of MapReduce was released under the name of Hadoop. The ecosystem of Apache Hadoop gives a way to efficiently use hardware resources and conveniently describe data manipulations. In the Centre for Open Science (CeON) we employed that solution and produced Content Analysis System (CoAnSys) the framework for finer scientific publication mining. CoAnSys enables data engineers to easily implement any data mining algorithm and chain data transformations into workflows. In addition, during the development of CoAnSys, the set of good implementation practices and techniques has been formulated [2]. In this article we share practical knowledge in the ground of big data implementations, based on the three use cases: citation matching, document similarity and document classification.

The rest of this chapter is organized as follows. Section 2 presents an overview of CoAnSys. Section 3 describes algorithms developed at CeON, which are well suited for MapReduce paradigm. Section 4 contains conclusions and future plans.

2 CoAnSys

The main goal of CoAnSys is to provide a framework for processing a large amount of text data. Currently implemented algorithms allow for knowledge extraction from scientific publications. Similar software systems include Behemoth,[1] UIMA [3], Synat [4], OpenAIRE [5] and currently developed OpenAIREplus [6]. The difference between CoAnSys and aforementioned tools lies in the implementation of algorithms. CoAnSys is used to conduct a research in text mining and machine learning, all methods implemented in that framework have been already published or will be published in a future. An architecture overview of CoAnSys is illustrated in Fig. 1.

While designing the framework, we paid a close attention to the input/output interfaces. For this purpose CoAnSys employs Protocol Buffers[2]—a widely used method of serializing data into a compact binary format. Serialized data is then imported into the HBase using Thrift framework (REST protocol is also supported, but it is slower than Thrift). This allows for simultaneous import of data from multiple clients. On the other hand, querying a large number of records from HBase is slower than performing the same operation on a sequence file stored in the HDFS. Therefore, in the input phase of CoAnSys workflow, the data is copied

[1] https://github.com/DigitalPebble/behemoth

[2] http://code.google.com/p/protobuf/

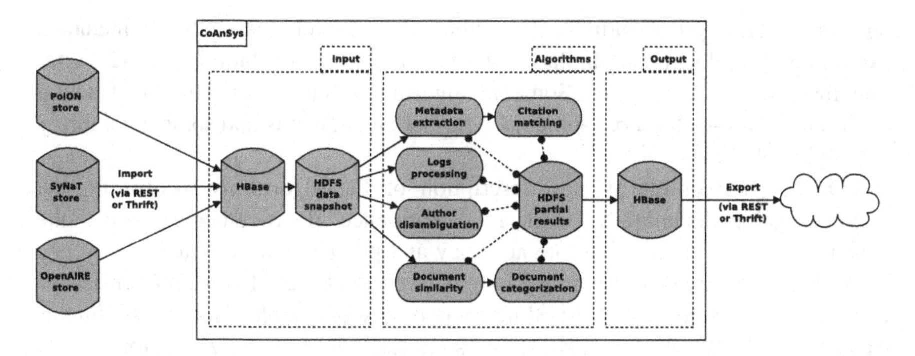

Fig. 1 A generic architecture of CoAnSys

from an HBase table to an HDFS sequence file and such format is recognized as a valid input for the algorithms.

Six modules currently implemented in CoAnSys are illustrated in the Algorithms box in Fig. 1. Each module performs a series of MapReduce jobs that are implemented in Java, Pig[3] or Scala. Apache Oozie[4] is used as a workflow scheduler system that chains modules together. Each module has well defined I/O interfaces in the form of Protocol Buffers schemas. This means, that sequence files are also used as a communication layer between modules. The output data from each workflow is first stored as an HDFS partial result (sequence file containing records serialized with Protocol Buffers) and then it is exported to the output HBase table where it can be accessed via Thrift or REST.

Even though CoAnSys is still in an active development stage, there are at least three ongoing projects that will utilize parts of CoAnSys framework. POL-on[5] is an information system about higher education in Poland. SYNAT[6] is a Polish national strategic research program to build an interdisciplinary system for interactive scientific information. OpenAIREplus[7] is the European open access data infrastructure for scholarly and scientific communication.

3 Well-Suited Algorithms

In this section a few examples of MapReduce friendly algorithms are presented. MapReduce paradigm puts a certain set of constraints, which are not acceptable for all algorithms. From the very beginning, the main effort in CoAnSys has been put

[3] http://pig.apache.org/

[4] http://oozie.apache.org/

[5] http://polon.nauka.gov.pl

[6] http://www.synat.pl/

[7] http://www.openaire.eu/

on document analysis algorithms, i.e. author name disambiguation [7–9], metadata extraction [10], document similarity and classification calculations [11, 12], citation matching [13, 14], etc. Some of algorithms can be used in the Hadoop environment out-of-the-box, some need further amendments and some are entirely not applicable [15].

For the sake of clarity, the description of the algorithms focuses on the implementation techniques (such as performance improvements), while the enhancements intended to elevate accuracy and precision are omitted.

All benchmarks were performed on our Hadoop cluster [16] which consists of four 'fat' slave nodes and a virtual machine on a separate physical box serving as NameNode, JobTracker and HBase master. Each worker node has four AMD Opteron 6174 processors (48 CPU cores in total), 192 GB of RAM and four 600 GB disks which work in RAID 5 array. Hadoop FairScheduler was used to create pools of 40 and 80 % of available computational resources. These pools were used to assess scalability of our solutions.

3.1 Citation Matching: General Description

A task almost always performed when researching scientific publications is citation resolution. It aims for matching citation strings against the documents they reference. As it consists of many similar and independent subtasks, it can greatly benefit from the use of MapReduce paradigm. A citation matching algorithm can be described in the following steps (illustrated in Fig. 2):

1. Retrieve documents from the store.
2. Map each document into its references (i.e. extract reference strings).
3. Map each reference to the corresponding document (i.e. the actual matching).
   (a) Heuristically select a possibly small set of best matching documents (map step).
   (b) Among them return the one with the biggest similarity (but not smaller than a given threshold) to the reference (reduce step).
4. Persist the results.

3.2 Citation Matching: Implementation Details

This section presents the implementation details of map and reduce steps in the citation matching algorithm. In particular, the most important design choices and used data structures are described.

Index The heuristic citation matching (map step) is implemented using an approximate author index allowing retrieval of elements with edit distance [17]

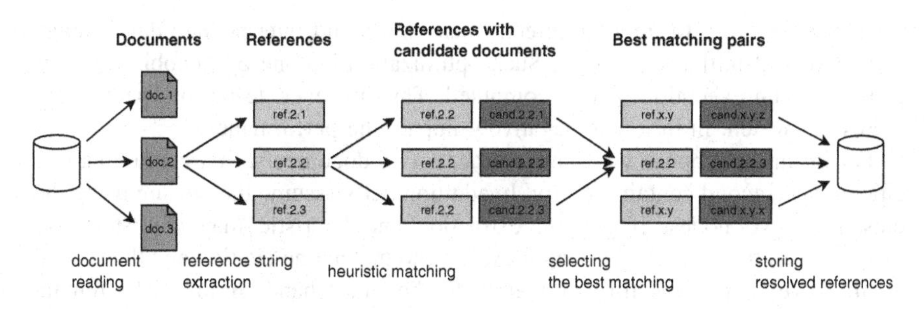

Fig. 2 Citation matching steps

less than or equal to 1. Its realisation in CoAnSys is based on the ideas presented by Manning et'al. in Chap. 3 of [18].

An efficient implementation of the map step requires an approximate author index to be stored in a data structure that provides fast retrieval and scanning of sorted entities. Hadoop MapFile turned out to be a good solution in this case. It extends capabilities of a SequenceFile (which is a basic way of storing key-value pairs in the HDFS) by adding an index to the data stored in it. The elements of a MapFile can be retrieved quickly, as the index is usually small enough to fit in a memory. The data is required to be sorted which makes changing a MapFile laborious, yet it is not a problem since the index is created from scratch for each execution of the algorithm. Hadoop exposes an API for MapFile manipulation which provides operations such as sorting, entity retrieval and data scanning.

Distributed Cache Since every worker node needs to access the whole index, storing it in the HDFS seemed to be a good idea at first. Unfortunately, this approach has a serious performance issues, because the index is queried very often. The speed of a network connection is a main bottleneck here. While seeking for a better solution, we have noticed the Hadoop Distributed Cache. It allows to distribute some data among worker nodes so that it can be accessed locally. The achieved performance boost was enormous—citation matching on the sample of 2000 documents worked four times faster.

Scala and Scoobi As MapReduce originates from a functional programming paradigm, one might expect it would fit well Scala language. Indeed, during citation matching implementation, we have exercised Scoobi[8] library which enables easy Hadoop programming in Scala by providing an API similar to Scalas native collections. In spite of great reduction of the lines of code, Scoobi does not restrict access to some low-level Hadoop features. However, when one desires complete control over a job execution, the default Hadoop API may need to be used.

Task Merging Subsequent map steps illustrated in Fig. 2 can be implemented as MapReduce jobs with zero-reducers. Sometimes it might be beneficial to merge

[8] http://nicta.github.com/scoobi/

such tasks in order to improve effectiveness and avoid intermediate data storage and additional initialization cost. Such optimization is done by Scoobi automatically when an execution plan is computed. This however, leads to a parallelism reduction, which, in turn, can negatively impact the performance.

For instance, suppose we want to process two documents, first containing one citation and second containing fifty. In addition, let's assume that we are using the cluster of two nodes. If citation extraction and heuristic matching steps are merged, then the first mapper would extract and match one citation and the second would have to process fifty of them. On the other hand, if the tasks remain independent after citation extraction and before actual matching, a load balancing will occur. As a result, a citation matching workload will be more equally distributed. Since the current version of Scoobi does not allow to disable task merging, this part of the process has been implemented using the low-level Hadoop API.

3.3 Citation Matching: Evaluation

For citation matching performance evaluation purposes a dataset containing over 748000 documents with more than 2.8 million citations was created. Those documents come from BazHum,[9] BazEkon,[10] BazTech,[11] AGRO,[12] CEJSH,[13] DML-PL[14] and PrzyrBWN[15] bibliographic databases. The phases of the citation matching process along with the amount of time spent in each of them and the performance gain when scaling from 40 to 80 % of resources are presented in Table 1. The algorithm shows high scalability in all phases but heuristic candidate selection. It might be due to the fact that this phase executes many index queries causing heavy disk usage. Because of that, any additional computing power does not improve the overall performance.

Nevertheless, one can see that the most exhaustive part, namely assessment, scales well. It is not surprising as it consists of measuring similarity of many

[9] http://bazhum.icm.edu.pl/

[10] http://bazekon.icm.edu.pl/

[11] http://baztech.icm.edu.pl/

[12] http://agro.icm.edu.pl/

[13] http://cejsh.icm.edu.pl/

[14] http://pldml.icm.edu.pl/

[15] http://przyrbwn.icm.edu.pl/

Table 1 Citation matching performance evaluation results

Phase	Time spent [hrs:mins:secs]		Ratio
	Benchmark40	Benchmark80	
Reference extraction and parsing	0:57:59	0:30:18	1.91
Heuristic candidate selection	0:39:51	0:37:25	1.07
Assessment	8:57:24	4:45:02	1.89

<citation, metadata record> pairs. As they are independent tasks, they can be easily distributed among computational nodes and benefit from additional processing power.

3.4 Document Similarity: General Description

In the previous sections we have shown how to solve the problem of citation matching efficiently on a Hadoop cluster. Here we consider the computation of a document similarity in a large collection of documents. Let's assume that the similarity between two documents is expressed as the similarity between weights of their common terms. In such approach, the computation can be divided into two consecutive steps:

1. the calculation of weights of terms for each document
2. the evaluation of a given similarity function on weights of terms related to each pair of documents

CoAnSys uses the term frequency inverse-document frequency (TFIDF) measure and the cosine similarity to produce weights for terms and calculate their similarity respectively. The process is briefly depicted in Fig. 3. At first, each document is split into terms. Then, the importance of each term is calculated using TFIDF measure (resulting in the vector of weights of terms for each document). Next, documents are grouped together into pairs, and for each pair, the similarity is calculated based on the vectors of weights of common terms associated with the documents.

Term Weighting Algorithm TFIDF is a well-known information retrieval algorithm that measures how important a word is to a document in a collection of documents. Generally speaking, the word becomes more important to a document, if it appears frequently in this document, but rarely in other documents.

Formally, TFIDF can be described by Eq. 1–3.

$$tfidf_{i,j} = tf_{i,j} * idf_i \tag{1}$$

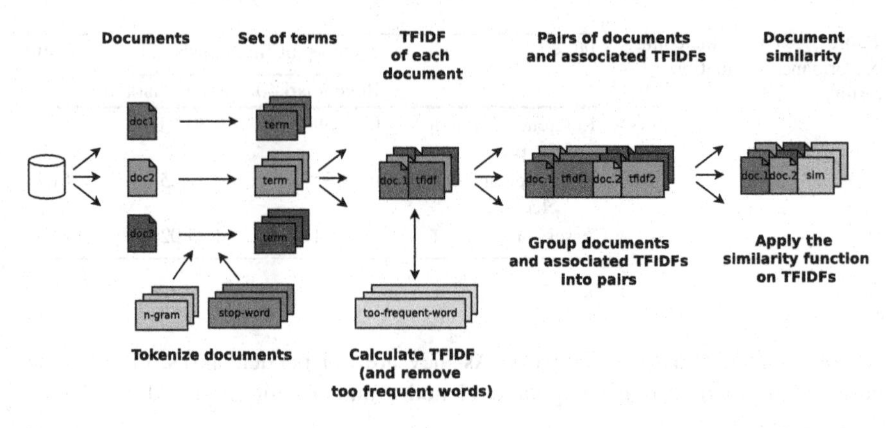

Fig. 3 Document similarity steps

$$tf_{i,j} = \frac{n_{i,j}}{\sum_k n_{k,j}} \quad (2)$$

$$idf_i = log \frac{|D|}{|d : t_i \in d|} \quad (3)$$

where

- $n_{i,j}$ is the occurrence of term t_i in document d_j
- D is the corpus of documents
- $d : t_i \in d$, document d containing term t_i

The way to use TFIDF in the document similarity calculation is presented in Eq. 4

$$cosineDocSim(d_x, d_y) = \frac{\sum_{t_i \in d_x \cap d_y} tfidf_{i,x} * tfidf_{i,y}}{\sqrt{\sum_{t_i \in d_x} tfidf_{i,x}^2} * \sqrt{\sum_{t_j \in d_y} tfidf_{j,y}^2}}, for\ x<y<|D| \quad (4)$$

For each document d, this algorithm produces the vector W_d of term weights $w_{t,d}$ which indicates the importance of each term t to the document. Since it consists of separate aggregation and multiplication steps, it can be nicely expressed in the MapReduce model with several map and reduce phases [19–21]. In addition, the following optimization techniques have been used in CoAnSys:

1. stop words filtering (based on a predefined stop-word list)
2. stemming (the Porter stemming algorithm [22])

3. applying n-grams to extract phrases (considering a statistically frequent n-gram as a phrase)
4. removal of the terms with the highest frequencies in a corpus (automatically, but with a parametrized threshold)
5. weights tuning (based on the sections where a given term appears).

Similarity Function Having each document represented by the vector W_d of term weights $w_{t,d}$, one can use many well known functions (e.g. cosine similarity) to measure similarity between the pair of vectors. Our implementation is based on ideas from [23], but provide more generic mechanism to deploy any similarity function that implements one-method interface i.e.

$$similarity\left(id(d_i), id\left(d_j\right), sort\left(w_{t_i,d_i}\right), sort\left(w_{t_j,d_j}\right)\right)$$

(where $id(d_i)$ denotes document id, and $sort\left(w_{t_i,d_i}\right)$ is a list of weights of common terms ordered by terms lexicographically). Sorting is necessary in order to use merge join mechanism present in Apache Pig. The similarity function receives only terms that have non-zero weights in both vectors, thus, the final score is calculated faster. This assumption remains valid only for the pairs of documents that have at least one common term.

3.5 Document Similarity: Implementation Details

Apache Pig Similarly to citation matching, the module of document similarity requires multiple map and reduce passes that lead to a long and verbose code. To make the code easier to maintain and less time-consuming to implement, CeON team uses Apache Pig (enhanced by UDFs, User Defined Functions written in Java). Apache Pig provides a high-level language (called PigLatin) for expressing data analysis programs. PigLatin supports many traditional data operations (e.g. group by, join, sort, filter, union, distinct). These operations are highly beneficial in multiple places such as stop words filtering, self-joining TFIDFs output relations or grouping relations with a given condition. Algorithm 1 contains a Pig script that is used in CoAnSys to calculate TFIDF.

Input Dataset Document similarity module takes advantage of rich metadata information associated with each document. Keywords needed to compute the similarity are extracted from the title, the abstract and the content of a publication and then, they are combined with the keyword list stored in the metadata. The information in which sections a given keyword appears is taken into account during the computation of the final weights wt;d in the TFIDF algorithm. A user may configure how important a given section is in the final score.

Algorithm 1 Pig script for TFIDF calculation.

```
DEFINE calculate_tfidf(in_relation,id_field,token_field,tfidfMinValue)
RETURNS tfidf_values {
   -- Calculate the term count per document
   doc_word_group = group $in_relation BY ($id_field,$token_field);
   doc_word_totals = foreach doc_word_group generate
   FLATTEN(group) AS ($id_field, token),
   COUNT($in_relation) AS doc_total;
   -- Calculate the document size
   pre_term_group = group doc_word_totals BY $id_field;
   pre_term_counts = foreach pre_term_group generate group AS $id_field,
   FLATTEN(doc_word_totals.(token,doc_total)) AS (token,doc_total),
   SUM(doc_word_totals.doc_total) AS doc_size;
   -- Calculate the TF
   term_freqs = foreach pre_term_counts generate $id_field AS $id_field,
   token AS token,
   ((double)doc_total / (double)doc_size) AS term_freq;
   -- Get count of documents using each token, for idf
   token_usages_group = group term_freqs BY token;
   token_usages = foreach token_usages_group generate
   FLATTEN(term_freqs) AS ($id_field,token,term_freq),
   COUNT(term_freqs) AS num_docs_with_token;
    -- Get document count
   just_ids = foreach $in_relation generate $id_field;
   just_unique_ids = distinct just_ids;
   ndocs_group = group just_unique_ids all;
   ndocs = foreach ndocs_group generate
   COUNT(just_unique_ids) AS total_docs;
   -- Calculate idf
   tfidf_all = foreach token_usages {
       idf = LOG((double)ndocs.total_docs/(double)num_docs_with_token);
       tf_idf = (double)term_freq * idf;
       generate $id_field AS $id_field,token AS $token_field,
       tf_idf AS tfidf,idf,ndocs.total_docs,num_docs_with_token;
   };
   -- Get only important terms
   $tfidf_values = FILTER tfidf_all BY tfidf >= $tfidfMinValue;
};
```

Additional Knowledge The main output of document similarity module is the set of triples in the form of $\langle d_i, d_j, sim_{i,j} \rangle$, where d_i and d_j are documents and $sim_{i,j}$ denotes the similarity between them. However, during the execution of this module, an additional output is generated. It contains potentially useful information such as:

- top N terms with the highest frequencies that might be considered as additional stop words,
- top N terms with the highest importance to a given document,
- top N articles with the lowest and highest number of distinct words.

3.6 Document Similarity: Evaluation

While evaluating document similarity, the Springer Verlag collection of about 1.8 million document's metadata was used. The length of each phase of the document similarity and performance gain between using 40 and 80 % of resources are presented in the Table 2.

Table 2 Document similarity performance evaluation results

Phase	Time spent [hrs:mins:secs]		Ratio
	Benchmark40	Benchmark80	
Data load and filtering	0:13:57	0:25:36	0.54
Similarity calculations	0:40:36	0:38:33	1.05
Final results calculations	0:22:41	0:15:16	1.49

The results show poor scalability of the solution on the investigated dataset. It can be explained by two facts. First of all, the usage of merge join in Apache Pig v.0.9.2 forces many data store and load operations. The self-join on a single dataset requires loading the data, ordering and storing it in two separate locations, and loading both copies. Only then one can perform memory efficient self-join, which is one of the operations used in the document similarity. The second bottleneck is our current cluster installation, where each slave is connected to the same hard drive matrix via an appropriate controller. This means that I/O operations represented in MapReduce paradigm cannot benefit from parallelism on our cluster.

MapReduce paradigm was designed to use multiple weak machines, each of which can be easily and cheaply replaced. Each Hadoop installation has its own strengths and weaknesses, which have to be known and exploited in the most efficient way. Nevertheless, the presented solution is invaluable when large datasets have to be processed and memory capacities are limited (compare this approach with a naïve implementation evaluated in Sect. 3.9).

3.7 Document Classification: General Description

The most common application of document similarity is an unpersonalized recommendation system. However, in CoAnSys, the document similarity module is used as a first step in a document classification algorithm based on the k-nearest neighbors.

In the context of document classification, it is important to distinguish two topics—a model creation (MC) and a classification code assignment (CCA), each of which starts in the same way, as depicted in Fig. 4. In the first step, documents are split into two groups—classified and unclassified (in case of MC both groups contain the same documents). Then, TFIDF is calculated for both of these groups. Finally, document similarity between groups is calculated (excluding self-similarity) and for each document from unclassified group, n closest neighbors are retained. After this initial phase, the subsequent step is different for MC and CCA.

For MC, the classification codes from neighbors of a document are extracted and counted. Then, for each classification code, the best threshold is selected against given criteria, e.g. an accuracy or a precision to fit "unclassified" documents classification codes. Finally, the pairs ⟨classification code, threshold⟩ are persisted.

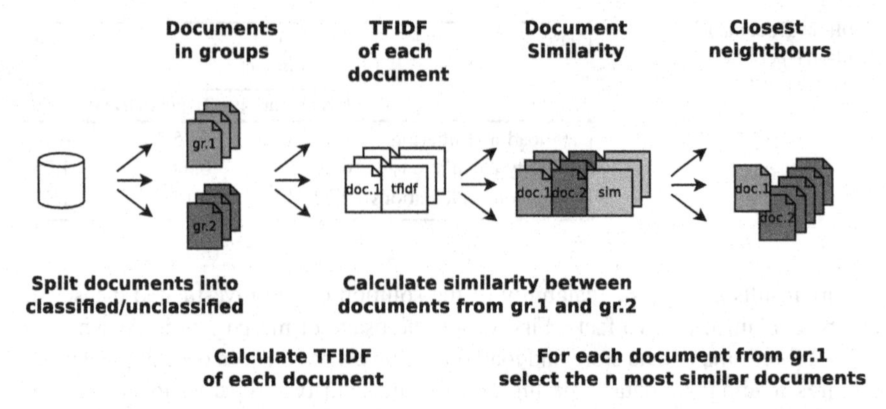

Fig. 4 The initial phase for model Creation (MC) and classification code assignment (CCA)

In case of CCA, after extraction of the classification code, the number of classification code occurrences is compared with a classification code threshold from a model and retained only if it is greater or equal to the threshold value.

3.8 Document Classification: Implementation Details

Optimizations The amount of data to be transferred between Hadoop nodes has a great influence on the performance of the whole workflow. One can perform certain optimizations that will reduce the number of comparisons in the document similarity step. First of all, since model creation uses only document's metadata containing classification codes, the other records may be filtered out. One of the best techniques to avoid the over-fitting of a model is to employ an n-fold cross-validation. During that step, the data is split into n groups, from which, training and testing sets for each fold are created. This again results in the reduction of comparisons by $\frac{n-1}{n}$ for training sets and $\frac{1}{n}$ for testing sets.

Similarly to MC, during CCA only similarity between "labeled" and "unlabeled" data is vital, hence the calculations within the groups can be omitted.

Language Choice For data scientists dedicated to implement and enhance algorithms in MapReduce, it is crucial to take advantage of programming languages created specifically for MapReduce. Again, Apache Pig looms as the natural candidate for document classification. Besides its strengths (predefined functions, UDFs), it should be noted that Pig (as a MapReduce paradigm) lacks some general purpose instructions like loops or conditional statements. However, it is easy to encapsulate Pig scripts into workflow management tools such as Apache Oozie or simply use the Bash shell which offers such operations. Moreover, due to the presence of *macro* and *import* statements, one can abbreviate the size of description by extracting popular transformations into macros and inserting them into separate files. In this approach, a variant of an operation (e.g. a way of calculating document similarity) can be passed to a general script as a parameter used in the import statement.

In order to optimize performance and memory utilization, it is important to use specialized types of a general operation. In case of *join* operation, there are dedicated types for joining small data with a large one ("replicated join"), joining data with a mixed, undetermined size ("skewed join") and joining sorted data ("merge join").

Data Storage The most utilized ways of storing data in the Apache Hadoop ecosystem are Hadoop database—HBase and Hadoop file system—HDFS. When massive data calculations are considered, then the better choice is the HDFS. When many calculating units are trying to connect to the HBase, then not all of them may be served before timeout expires. That results in a chain of failures (tasks assigned to calculation units are passed from failed to working ones, which, in turn, become more and more overwhelmed by the amount of data to process). On the other hand, such failure cannot happen when the HDFS is used.

Using HDFS in MapReduce jobs requires pre-packing data into SequenceFiles (already discussed in Sect. 3.2). To obtain the most generic form, it is recommended to collect the key and value objects as a BytesWritable class, where a value object contains data serialized via Protocol Buffers. This approach makes it easy to store and extend schema of any kind of data. Our experiments have shown that reading and writing ⟨BytesWritable, BytesWritable⟩ pairs in Apache Pig v.0.9.2. may results in some complications. In that case, one may consider to encapsulate BytesWritable into NullableTuple class or use ⟨Text, BytesWritable⟩ pairs.

Workflow Management As mentioned previously, one of the best ways to build a chain of data transformations is to employ a workflow manager or a general purpose language. Our experiences with using Apache Oozie and Bash were strongly in favour of the former one. Apache Oozie is a mature solution, strongly established in the Apache Hadoop ecosystem, aimed for defining and executing workflows (that can be triggered by a user, time event or data arrival). In fact, using Bash or Python would require a burden of implementing an Oozielike tool to support useful functionalities such as the persistence of an execution history.

3.9 Document Classification: Evaluation

The Springer Verlag collection was used again to evaluate document classification algorithms. Timings for each phase of the model creation and performance gain between using 40 and 80 % of resources are presented in Table 3. These numbers are averages for five models created during the evaluation.

After applying optimizations from Sect. 3.8, only about 53 k documents were accepted for further calculations. In the second stage data was split into training and testing sets containing about 42 and 11 k items respectively. Knowing that document similarity calculations with reduced number of data are far below

Table 3 Document classification performance evaluation results

Phase	Time spent [hrs:mins:secs]		Ratio
	Benchmark40	Benchmark80	
1. Data extraction and filtering	0:01:16	0:00:50	0.66
2. Folds splitting	0:00:38	0:00:39	1.03
3. Feature vector calc.	0:10:32	0:08:30	0.81
4. Similarity calc. and neighbours selection	4:52:25	2:52:31	0.59
5. Categories addition	0:02:34	0:02:23	0.93
6. Model building	0:05:39	0:03:56	0.70
7. Model testing	0:05:57	0:04:25	0.74
Total	5:19:01	3:13:14	0.61

computational capacities of the cluster, we decided to employ crude implementation of the module, with no sophisticated technical enhancements. That solution, although scalable (5 h on benchmark40 and 3 h on benchmark80, phases 3–4 in Table 3), takes much longer than an optimized implementation evaluated in Sect. 3.6. It should be noted that the naïve implementation can be used only to relatively small input data. The problem of document classification on larger datasets requires more advanced techniques for computing similarity, such as the ones used in the document similarity module.

4 Summary and Future Work

In this article we have described the experience gained in the implementation of CoAnSys framework. Decisions we took in the development process required about half a year of tries and failures. It is hard to find coherent studies of different algorithms' implementations and therefore we hope that this contribution can save time anyone who is preparing to embrace MapReduce paradigm and especially Apache Hadoop ecosystem into data mining systems.

This description is the snapshot of an on-going work, hence many more improvements and observations are expected to be done in a future.

References

1. Manyika, J., Chui, M., Brown, B., Bughin, J., Dobbs, R., Roxburgh, C., Byers, A.H.: Big data: the next frontier for innovation, competition, and productivity. Technical report, Mc Kinsey (2011)
2. Dendek, P.J., Czeczko, A., Fedoryszak, M., Kawa, A., Wendykier, P., Bolikowski, Ł.: How to perform research in Hadoop environment not losing mental equilibrium—case study. arXiv:1303.5234 [cs.SE] (2013)
3. Ferrucci, D., Lally, A.: UIMA: an architectural approach to unstructured information processing in the corporate research environment. Nat. Lang. Eng. **10**(3–4), 327–348 (2004)

4. Bembenik, R., Skonieczny, L., Rybinski, H., Niezgodka, M.: Intelligent Tools for Building a Scientific Information Platform Studies in Computational Intelligence. Springer, Berlin (2012)
5. Manghi, P., Manola, N., Horstmann, W., Peters, D.: An infrastructure for managing EC funded research output—the OpenAIRE project. Grey J: Int. J. Grey Lit. **6**, 31–40 (2010)
6. Manghi, P., Bolikowski, Ł., Manola, N., Schirrwagen, J., Smith, T.: OpenAIREplus: the European scholarly communication data infrastructure. In: D-Lib Magazine, vol. 18(9/10) (2012)
7. Dendek, P.J., Bolikowski, Ł., Lukasik, M.: Evaluation of features for author name disambiguation using linear support vector machines. In: Proceedings of the 10th IAPR International Workshop on Document Analysis Systems, pp. 440–444 (2012)
8. Dendek, P.J., Wojewodzki, M., Bolikowski, Ł.: Author disambiguation in the YADDA2 software platform. In: Bembenik, R., Skonieczny, L., Rybinski, H., Kryszkiewicz, M., Niezgodka, M. (eds.) Intelligent Tools for Building a Scientific Information Platform. Studies in Computational Intelligence, vol. 467, pp. 131–143. Springer, Berlin Heidelberg (2013)
9. Bolikowski, Ł., Dendek, P.J.: Towards a flexible author name disambiguation framework. In: Sojka, P., Bouche, T., (eds.): Towards a Digital Mathematics Library, pp. 27–37. Masaryk University Press (2011)
10. Tkaczyk, D., Bolikowski, Ł., Czeczko, A., Rusek, K.: A modular metadata extraction system for born-digital articles. In: 2012 10th IAPR International Workshop on Document Analysis Systems (DAS), pp. 11-16. (2012)
11. Lukasik, M., Kusmierczyk, T., Bolikowski, Ł., Nguyen, H.: Hierarchical, multilabel classification of scholarly publications: modifications of ML-KNN algorithm. In: Bembenik, R., Skonieczny, L., Rybinski, H., Kryszkiewicz, M., Niez- godka, M., (eds.): Intelligent Tools for Building a Scientific Information Platform. Studies in Computational Intelligence, vol. 467 pp. 343–363. Springer, Heidelberg (2013)
12. Kusmierczyk, T.: Reconstruction of MSC classification tree. Master's Thesis, The University of Warsaw (2012)
13. Fedoryszak, M., Bolikowski, Ł., Tkaczyk, D., Wojciechowski, K.: Methodology for evaluating citation parsing and matching. In: Bembenik, R., Skonieczny, L., Rybinski, H., Kryszkiewicz, M., Niezgodka, M. (eds.) Intelligent Tools for Building a Scientific Information Platform. Studies in Computational Intelligence, vol. 467, pp. 145–154. Springer, Heidelberg (2013)
14. Fedoryszak, M., Tkaczyk, D., Bolikowski, Ł.: Large scale citation matching using apache hadoop. In: Aalberg, T., Papatheodorou, C., Dobreva, M., Tsakonas, G., Farrugia, C. (eds.) Research and Advanced Technology for Digital Libraries. Lecture Notes in Computer Science, vol. 8092, pp. 362–365. Springer, Heidelberg (2013)
15. Lin, J.: MapReduce is Good Enough? If All You Have is a Hammer, Throw Away Everything That's Not a Nail! Sept 2012
16. Kawa, A., Bolikowski, A., Czeczko, A., Dendek, P., Tkaczyk, D.: Data model for analysis of scholarly documents in the mapreduce paradigm. In: Bembenik, R., Skonieczny, L., Rybinski, H., Kryszkiewicz, M., Niezgodka, M. (eds.) Intelligent Tools for Building a Scientific Information Platform. Studies in Computational Intelligence, vol. 467, pp. 155–169. Springer, Heidelberg (2013)
17. Elmagarmid, A., Ipeirotis, P., Verykios, V.: Duplicate record detection: a survey. IEEE Trans. Knowl. Data Eng. **19**(1), 1–16 (2007)
18. Manning, C.D., Raghavan, P., Schtze, H.: Introduction to Information Retrieval. Cambridge University Press, New York (2008)
19. Cloudera: Mapreduce algorithms. http://blog.cloudera.com/wp-content/uploads/2010/01/5-MapReduceAlgorithms.pdf (2009)
20. Lee, H., Her, J., Kim, S.R.: Implementation of a large-scalable social data analysis system based on mapreduce. In: 2011 First ACIS/JNU International Conference on Computers, Networks, Systems and Industrial Engineering (CNSI), pp. 228–233 (2011)

21. Wan, J., Yu, W., Xu, X.: Design and implement of distributed document clustering based on mapreduce. In: Proceedings of the 2nd symposium international computer science and computational technology (ISCSCT), pp. 278–280 (2009)
22. Porter, M.F.: Readings in information retrieval, pp. 313–316. Morgan Kaufmann Publishers, San Francisco (1997)
23. Elsayed, T., Lin, J., Oard, D.W.: Pairwise document similarity in large collections with mapreduce. In: Proceedings of the 46th Annual Meeting of the Association for Computational Linguistics on Human Language Technologies: Short Papers. HLT-Short '08, pp. 265–268. Association for Computational Linguistics, Stroudsburg, PA, USA (2008)

Implementation of a System for Fast Text Search and Document Comparison

Maciej Wielgosz, Marcin Janiszewski, Pawel Russek, Marcin Pietron, Ernest Jamro and Kazimierz Wiatr

Abstract This chapter presents an architecture of the system for fast text search and documents comparison with main focus on N-gram-based algorithm and its parallel implementation. The algorithm which is one of several computational procedures implemented in the system is used to generate a fingerprint of analyzed documents as a set of hashes which represent the file. This work examines the performance of the system, both in terms of a file comparison quality and a fingerprint generation. Several tests were conducted of N-gram-based algorithm for Intel Xeon E5645, 2.40 GHz which show approximately 8x speedup of multi over single core implementation.

Keywords N-gram-based model · Document comparison · Parallel implementation · Information retrieval · Pattern recognition

---

M. Wielgosz (✉) · P. Russek · E. Jamro · K. Wiatr
AGH University of Science and Technology, Al. Mickiewicza 30,
30-059 Krakow, Poland
e-mail: wielgosz@agh.edu.pl

P. Russek
e-mail: russek@agh.edu.pl

E. Jamro
e-mail: jamro@agh.edu.pl

K. Wiatr
e-mail: wiatr@agh.edu.pl

M. Janiszewski · M. Pietron
ACK Cyfronet AGH, Ul. Nawojki 11, 30-950 Krakow, Poland
e-mail: Marcin.Janiszewski@cyfronet.krakow.pl

M. Pietron
e-mail: pietron@agh.edu.pl

R. Bembenik et al. (eds.), *Intelligent Tools for Building a Scientific Information Platform: From Research to Implementation*, Studies in Computational Intelligence 541,
DOI: 10.1007/978-3-319-04714-0_11,

1 Introduction

With the growth of online information, searching for and comparing documents using different measures of similarity becomes important. There is no single definition of document similarity and it can be expressed differently depending on the comparison goal. Suppose that one wants to find the latest version of a research paper and journal it was published in. Provided that all versions of the paper have a substantial part of the text in common, text-related document search is able to trace them and, in particular, the most up-to-date one.

Copyright violation and plagiarism can be given as another application example of a document comparison system. In the Internet era it is very easy to get access to documents, modify them, and create a new document based on existing ones. Such cases lead to the question of authorship of a paper. More generally, the Internet raises major plagiarism and copyright problems because of the availability and ease with which papers may be copied and modified.

There are different plagiarism detection strategies. Watermarks or digital signatures may be embedded in a file or attached to it [1]. Those may be file checksums or other fingerprints. Those components are well suited for detection of partial modifications but they can be easily removed from the files. The other approach involves using a database and different file comparison techniques. In this case, however, the following issues arise: how to provide an appropriate storage for all the files, how to meet security restrictions—in other words how to protect the files from being stolen and how to convert files to a common text format with a little or no information loss. Finally, the most demanding task is a reliable text matching. This is an open issue which may be addressed differently depending on information content taken into account in a comparison.

The main focus of the authors' research is how to build a system capable of comparing millions of files in a reasonable time. The system is to be used for information extraction and fast text searches and file comparison. This chapter examines the feasibility of using N-gram-based method [2, 3] to address those research issues.

2 N-gram Based Algorithm

Authors decided to use N-grams for document comparison because they are reported to be well-suited for pattern matching and text content filtering [2, 3]. An N-gram is a continuous substring of n-letter length which is chosen from the original document. The length of N-gram may vary from few letters to longer sequences which reflects a scope of text which is examined [2]. The number of N-grams generated from the document is large, almost equal to a number of letters in the document (Eq. 1).

$$N = (n - k + 1) \quad (1)$$

where; N—total number of N-grams, n—number of letters in a document, k—size of gram.

Using all the N-grams (Eq. 1) for a file comparison would result in an overwhelming number of computations. Therefore only a fraction of N-grams denoted as a fingerprint is considered. This raises the question of fingerprint selection i.e. what N-grams should be selected. The choice of selection patterns significantly impact the result of a file comparison. Several approaches were proposed in [4] such as 0 mod p or a minimum [2] or maximum value in a given range. Fixing the number of N-grams per document makes the system more scalable, but does not guarantee that documents which significantly differ in size can be compared meaningfully.

It is important to provide equal document coverage, i.e. the fingerprint must be composed of N-grams equally distributed over the whole document. It makes the algorithm immune to simple modifications such as words and sentences swapping or text relocations within a file. Therefore the winnowing [2] procedure was employed to choose a fingerprint, the procedure is presented in Fig. 1. A core idea of the winnowing which makes it different to other N-gram-based algorithms is window filtering of hashes. In the first step N-grams are generated as in all N-gram-based algorithms then hashes are computed which makes it possible to apply the window filter. In the third step a single hash per window is chosen according to the algorithm expressed in python and described in [2].

Winnowing procedure—Python code

```
b = []
FingerPrint = []
tmp = []
for i in range(0, (len(HashedNgrams) - WindowSize + 1)):
     b = HashedNgrams[i:(WindowSize+i)]
     if (min(b) != tmp) or (b[WindowSize - 1] == tmp):
         tmp = min(b)
         FingerPrint.append(tmp)
```

In the preprocessing stage input text is compressed i.e. all the white spaces are removed. Thereafter N-grams are generated and passed on to the third stage in which hashes are calculated. In the final stage a set of hashes (denoted as a fingerprint) is chosen according to the winnowing. The authors use 32 bit simple hash function, in the future more advanced one will be implemented. The mechanics of the winnowing procedure and its mathematical background is provided in [2].

File comparison is performed using similarity measures based on the inclusion formulas: 2, 3.

$$similarity(A, B) = |Ha \cap Hb| / |Ha| \quad (2)$$

$$similarity(B, A) = |Ha \cap Hb| / |Hb| \quad (3)$$

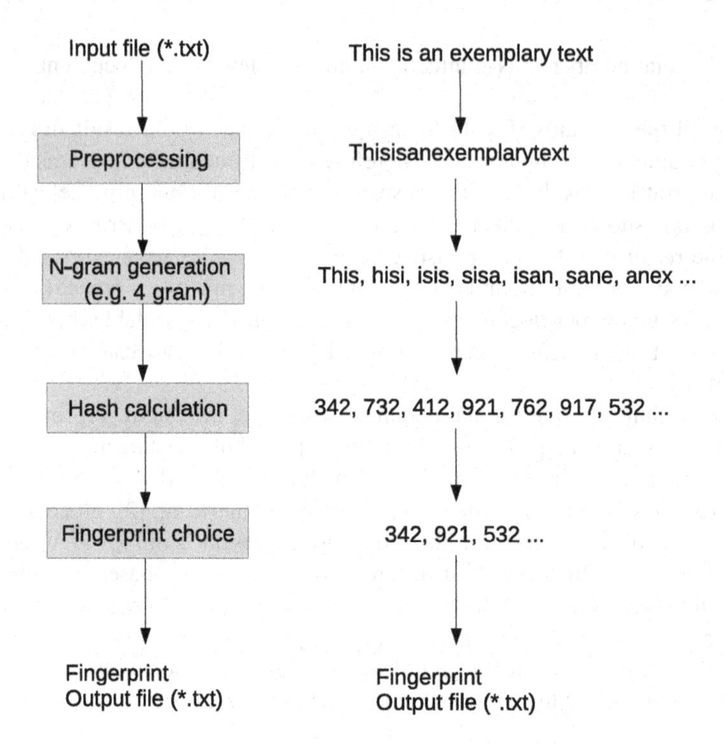

Fig. 1 N-gram generation scheme (example of 4gram generation)

where Ha and Hb are fingerprints. Consequently, examined files are considered more similar if they have more hashes in common.

3 System Architecture

The architecture of the system is modular and easily expendable due to its flexible structure.

The system is composed of the following components:

- *algorithm bundle* which contains a complete processing path of the algorithm
- *local daemon* is a process which handles all the operations related to modules interactions within each algorithm bundle
- *global daemon* is a process which coordinates algorithm bundles and merging their results
- *retrieval phase* is a first processing stage in the plagiarism detection system
- *alignment phase* is a second processing stage in the plagiarism detection system.

The files uploaded for a comparison are temporally stored in the local directory on the server. Once the system daemon notices that new files are available it notifies all the algorithm bundles and feeds them with the incoming file. Before this happens, the files get processed by a pdf2txt standard tool that converts them to txt format. All the files are uploaded in a raw format (txt) to the database so they can be used for further processing when the system gets extended with more algorithm bundles.

The files are processed independently by all the algorithm bundles (Figs. 2 and 3) and the results are stored in the database. Algorithm bundles prompt the report generator module as their tasks are finished so it can extract statistical information from the database in order to create a final report. The format of the report is chosen in the website menu (front-end interface) by the user. It can be either brief or a long report depending on a user's choice.

An algorithm quality test wrapper is used to evaluate the efficiency (detection quality) of the algorithms (Figs. 3 and 4) from all the bundles. It is an evaluation framework based on the one proposed in [5]. It comprises four detection quality measures: precision, recall, granularity and overall.

All the users of the system are required to have an account created in order to use the service. Files uploaded to the system are associated with the user and tagged accordingly. The authorization module (Fig. 2) is an autonomous part of the system within the service functionality.

Each algorithm bundle is a separate processing unit within the system which operates on the same input data, but generates different fingerprints. Therefore, all the bundles have access to a separate space in the database (separate table) or use their individual storage (e.g. store data in files). Local daemon receives a prompt message from the global coordination daemon once a new input file is ready to be processed by a bundle. The daemon also coordinates the flow of data and processing sequence within each bundle.

In the first step, input data is fed to the preprocessing unit in which such operations as stemmization and lemmatization are carried out. Stoplist or a wordnet [6] is considered to be used in the future in the preprocessing stage. At the source retrieval stage it is decided how many files the input file is to be compared to in the text alignment stage (Fig. 3).

Text alignment is performed by the second module of the algorithm bundle. This is the most tedious and computationally demanding stage of the file comparison process because it involves pattern matching. Sufficient data is produced to generate a complete report as the result of this processing stage.

The source retrieval unit (Fig. 4) is composed of three sections each of which extensively uses a local Database to temporarily store data. Since the fingerprint is unique for each algorithm it does not make sense to share it among the bundles. Therefore only the results of the alignment module are passed further to the report generation unit.

A fingerprint is generated in the first stage within the source retrieval unit which represents the file. The bigger the fingerprint is, the more costly the fingerprint

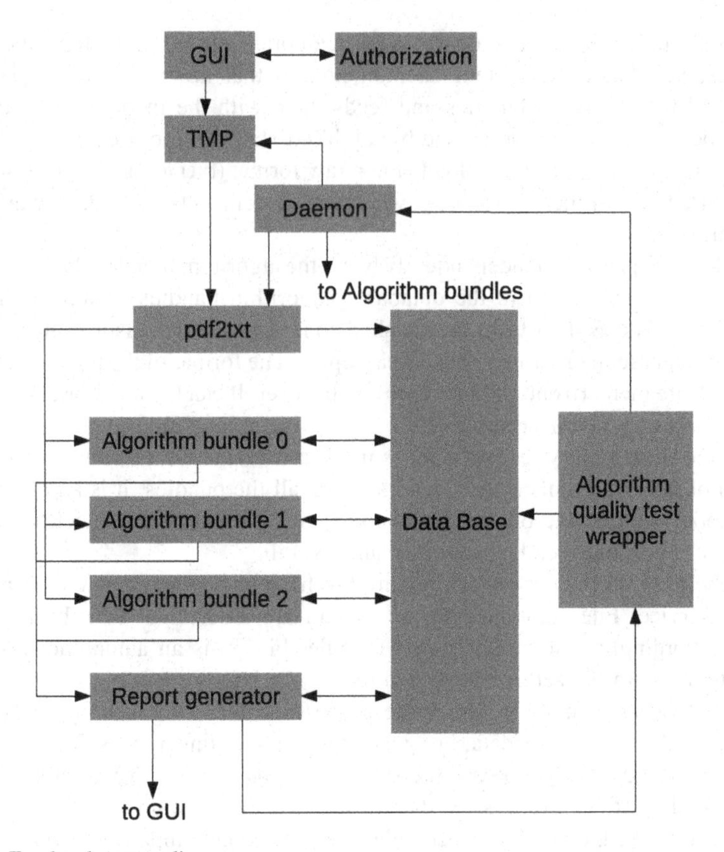

Fig. 2 Top level system diagram

comparison is. Thus the fingerprint should be small and contain as much information as possible.

Fingerprint comparison is performed by a specialized engine which is optimized for performance. This stage is intended to be implemented in hardware in order to accelerate computations.

The final stage of processing involves a decision regarding whether a given file should be qualified for further processing and passed on the alignment module. The choice of the threshold is discretionary and should be done carefully to achieve the best overall performance of the system (Fig. 4). The choice of the threshold depends mostly on a precision and recall of a given stage (retrieval or alignment). The definitions of those measures are available in [7]. Recall is at the premium in the retrieval stage as opposed to the alignment stage where the precision is more important. This is so because in the retrieval stage the biggest possible number of plagiarized documents should detected. There can be even some more detected which does not affect recall. Thus the threshold should be approximately 0.2 (the typical value) which means that if 20 % of hashes within a

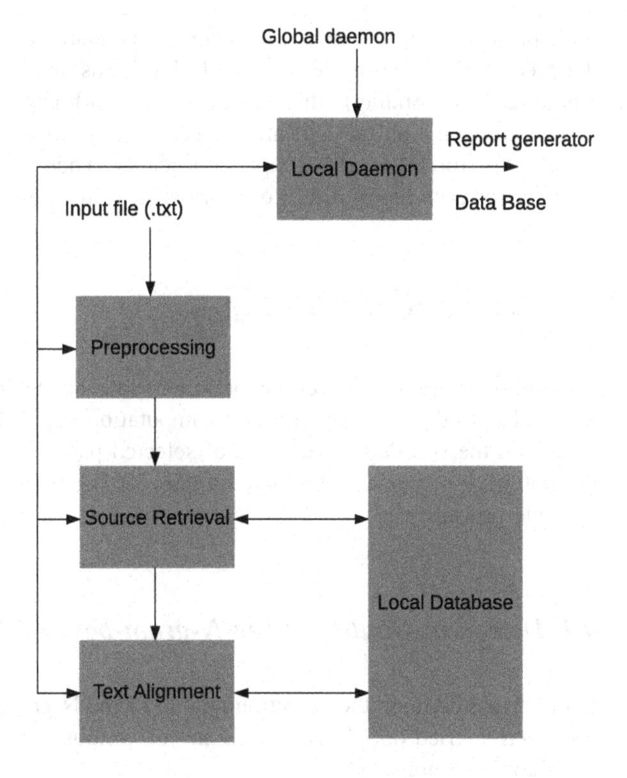

Fig. 3 Block diagram of the algorithm bundle

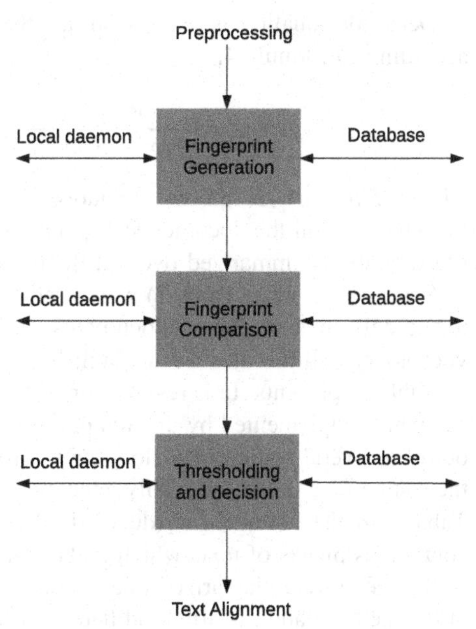

Fig. 4 Block diagram of the source retrieval section of the system

fingerprint print are the same the document is qualified to the alignment phase. In the second phase the more advanced algorithms are employed (and more computationally demanding) which work with much higher precision.

It is worth noting that all three stages (Fig. 4) contribute independently to the system's performance and comparison quality. Thus they should be chosen individually for each algorithm in order to achieve the best results.

4 Experiments and Discussion

The authors carried out several experiments to evaluate the system performance in terms of both detection quality and computation speed. Furthermore, the authors examined the speedup capabilities of selected parts of the algorithm (fingerprint generation) by implementing them on Intel Xeon E5645, 2.40 GHz as a single and multiple process.

4.1 Detection Quality of the N-gram-based Algorithm

In order to determine the detection quality of the N-gram-based algorithm several tests were carried out. They were done for various gram sizes and preprocessing stage configurations.

Detection quality is expressed as the relative error which is calculated according to formula 4.

$$Error = \frac{|Summary - Expected|}{max(|Summary - Expected|)} \times 100\,\% \tag{4}$$

where: *Summary* is an accumulation of the measured values according to formula 2 for all the documents; *Expected* is the similarity value generated by the file scrambler summarized over all the documents under analysis.

Six different files (Table 1) were used for the tests, which ranged from 40 KB to 1.5 MB in size. These documents cover different topics and use different vocabulary. All the files are in English.

Table 1 presents test results for the files generated artificially using file scrambler implemented by the authors in Python. The scrambler builds a new file out of a several source files along with information regarding the contribution of the source files to the artificially generated one. Those values are presented in the Table 1 in the bottom row (denoted as *real*). The scrambler copies and pastes continuous blocks of texts which makes plagiarism detection easier. Contribution of the files to the plagiarized one is denoted as *con** an in Table 1 Data was not obfuscated nor additionally modified. It is worth noting that (Table 1) the smaller N-gram is used the larger error is generated. This results from a way scrambler

Table 1 Detection quality of the unstemmed artificially generated file, where *con* represent contributions of the source files, *summary* is a sum of all the contributions, *expected* is an expected contribution, *error* is a relative quality measure

N-gram	con0	con1	con2	con3	con4	con5	Summary	Expected	Error [%]
3	0.107	0.338	0.337	0.286	0.215	0.32	1.603	0.422	100
5	0.037	0.246	0.146	0.134	0.103	0.19	0.856	0.422	36.775
7	0.017	0.196	0.082	0.081	0.06	0.13	0.566	0.422	12.23
9	0.013	0.181	0.066	0.063	0.046	0.1	0.469	0.422	4.054
11	0.012	0.175	0.061	0.055	0.04	0.097	0.44	0.422	1.557
13	0.011	0.175	0.06	0.052	0.038	0.093	0.429	0.422	0.66
15	0.011	0.175	0.06	0.051	0.037	0.091	0.425	0.422	0.296
17	0.011	0.175	0.061	0.05	0.036	0.089	0.422	0.422	0.034
19	0.011	0.175	0.061	0.05	0.036	0.089	0.422	0.422	0.034
21	0.011	0.175	0.062	0.05	0.036	0.089	0.423	0.422	0.11
23	0.011	0.175	0.062	0.05	0.036	0.089	0.423	0.422	0.11
25	0.011	0.175	0.063	0.049	0.036	0.089	0.423	0.422	0.11
27	0.011	0.175	0.063	0.049	0.036	0.089	0.423	0.422	0.11
29	0.011	0.175	0.063	0.049	0.036	0.089	0.423	0.422	0.11
31	0.011	0.175	0.063	0.048	0.036	0.089	0.422	0.422	0.017
33	0.011	0.175	0.063	0.047	0.036	0.089	0.421	0.422	0.068
real	0.011	0.152	0.089	0.046	0.038	0.086	0.422		

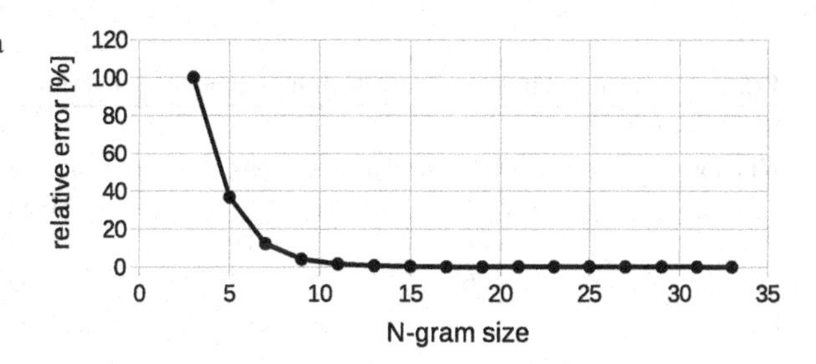

Fig. 5 Detection quality as a function of N-gram size for unstemmed input text

works. It chunks data into blocks which form a plagiarized file. Thus longer N-grams work better.

Figure 5 presents the detection quality of the algorithm understood as the change of relative error as a function of N-gram size. It is worth noting that for N-gram sizes larger than 15 the curve flattens out in Fig. 5.

4.2 Implementation

As it is presented in Tables 2 and 3, N-gram fingerprint generation can be efficiently calculated using Intel Xeon E5645, 2.40 GHz with approx. 8x speedup over single process implementation. There were 10 cores used for fork-based C implementation of the N-gram-based algorithm of fingerprint generation.

Table 2 Implementation results of N-gram winnowing algorithm for different gram size

W = 4, time [s]				
N-gram	core cnt = 1	core cnt = 4	core cnt = 8	core cnt = 10
4	2.41	0.7	0.37	0.37
8	3.03	0.86	0.45	0.4
12	3.66	1.04	0.54	0.48
16	4.53	1.2	0.63	0.55
W = 8, time [s]				
N-gram	core cnt = 1	core cnt = 4	core cnt = 8	core cnt = 10
4	3.78	1.05	0.55	0.51
8	4.4	1.21	0.64	0.54
12	5.07	1.39	0.73	0.62
16	5.89	1.55	0.81	0.69
W = 12, time [s]				
N-gram	core cnt = 1	core cnt = 4	core cnt = 8	core cnt = 10
4	4.55	1.25	0.65	0.59
8	5.19	1.41	0.74	0.63
12	5.85	1.58	0.83	0.7
16	6.68	1.75	0.91	0.77

Table 3 Implementation results of N-gram winnowing algorithm for different window size

N = 4, time [s]				
Window	core cnt = 1	core cnt = 4	core cnt = 8	core cnt = 10
4	2.41	0.7	0.37	0.37
8	3.81	1.05	0.56	0.51
12	4.06	1.24	0.65	0.59
16	5.21	1.4	0.73	0.66
N = 8, time [s]				
Window	core cnt = 1	core cnt = 4	core cnt = 8	core cnt = 10
4	3.05	0.87	0.46	0.41
8	4.43	1.22	0.64	0.55
12	5.23	1.41	0.74	0.63
16	5.85	1.57	0.82	0.7
N = 12, time [s]				
Window	core cnt = 1	core cnt = 4	core cnt = 8	core cnt = 10
4	3.69	1.04	0.55	0.48
8	5.07	1.39	0.73	0.62
12	5.89	1.58	0.83	0.7
16	6.48	1.74	0.91	0.77

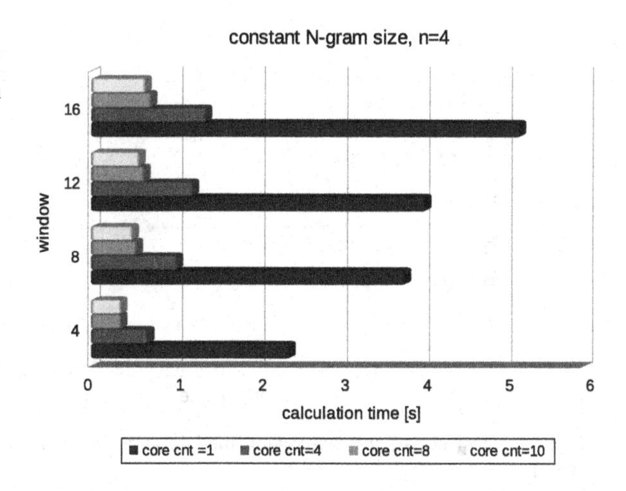

Fig. 6 Intel Xeon E5645, 2.40 GHz implementation of N-gram winnowing algorithm for N = 4 (constant N-gram size)

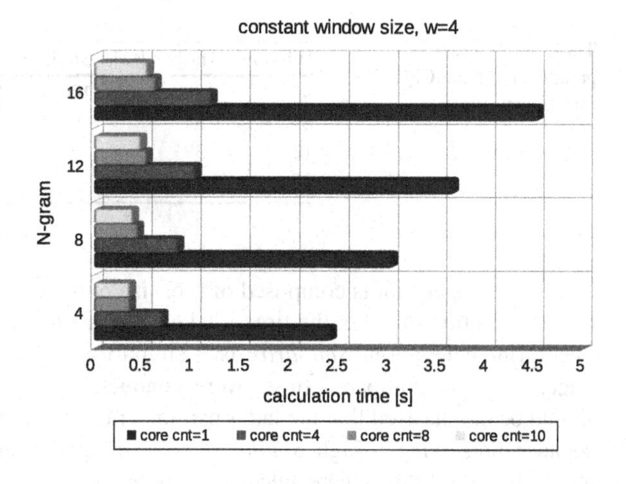

Fig. 7 Intel Xeon E5645, 2.40 GHz implementation of N-gram winnowing algorithm for W = 4 (constant window size)

It contributes significantly to the performance of a system that works on a huge volume of data (e.g. file comparison/plagiarism detection), especially when a database is recompiled in order to compute a new set of fingerprints. In that case, all the fingerprints are calculated again according to a new comparison profile. It may be essential when the database is extended with a new set of files that are very different than the ones already stored in the system.

Fast fingerprint calculation makes the system flexible. Furthermore, in streaming data analysis applications (e.g. malware detection) it is a must, since it directly affects the performance of the system.

As presented in Figs. 6 and 7 the speed-up rises with a growing number of cores and implementation results (Tables 2 and 3) are obtained for an exemplary file of 38 MB and integer (32 bit) hash function.

Figure 8 presents performance of the software implementation of the N-gram-based algorithm characterized by three parameters: *Hashes* denoting a number of

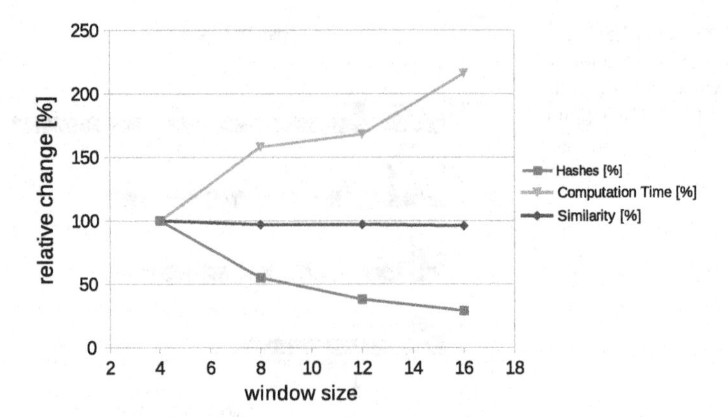

Fig. 8 Performance of the winnowing module as a function of a window size (N = 4, core cnt = 1)

Table 4 Different parameters of the CPU implementation

Window	Hashes [%]	Similarity [%]	Computation time [%]
4	100	100	100
8	55	97	158
12	38	97	168
16	29	96	216

hashes the fingerprint is composed of (i.e. size of fingerprint), *Comp. time* which is the computation time i.e. the time it takes to perform a complete procedure of the fingerprint generation, *Similarity* is a similarity measure of the files for which fingerprints are compared (tests were conducted for a set of different files). It should be emphasized that the latter parameter reflects the quality of the algorithm, because preserving a high *Similarity* value despite a drop of a fingerprint size means that the hashes were selected properly.

According to the conducted tests, a change of the window size affects comparison results only to a small percentage (marked in red Fig. 8) but significantly reduces the number of hashes constituting the fingerprint. This results in a large drop in the computational effort at a comparison stage—there is a lower number of hashes to process. The comparison stage of the algorithm has not been discussed herein but is considered to be highly demanding with its computational complexity of $O(n^2)$ if the fingerprint is not presorted, where n is a number of hashes within a fingerprint. The sorting operation is also computationally demanding with its lowest computational complexity of $O(n * logn)$. Therefore it is important to reduce the cardinality of the fingerprint (Table 4).

It is worth noting that reduction of the fingerprint size computed by CPU is achieved at the expense of the processing time which grows substantially with the size of the window (Fig. 8). The bigger the window, the longer it takes to compute a fingerprint. Consequently, the decrease of the fingerprint size and, subsequently, its comparison time is traded for the growth of the fingerprint computation time.

5 Conclusions and Future Work

The chapter presents the analysis of the N-gram-based algorithm performance used in the file comparison system being developed at ACK Cyfronet.

The system is currently in its early development stage and has been partially implemented. Nevertheless, the authors have described its architecture along with the results of the preliminary detection quality measurements and its parallel implementation on Intel Xeon E5645, 2.40 GHz.

Since the alignment stage is missing in the system, it was not possible to take advantage of the quality measures implemented as the system component (Fig. 2). Therefore the results presented herein should be considered as preliminary ones. In their future work, the authors are going to finish implementation of the system and build an ensemble of algorithms in order to increase the detection quality as well as implement the most computationally demanding tasks on hardware platforms (i.e. FPGA and GPU).

Acknowledgments The work presented in this chapter was financed through the research program—Synat [8].

References

1. Jalil, Z., Mirza, A.M., Iqbal, T.: A zero-watermarking algorithm for text documents based on structural components. In: International Conference on Information and Emerging Technologies (ICIET). vol. 1(5), pp. 14–16, June 2010
2. Schleimer, S., Wilkerson, D.S., Aiken, A.: Winnowing: local algorithms for document fingerprinting. In: Proceedings of the 2003 ACM SIGMOD International Conference on Management of Data (SIGMOD'03). pp. 76–85
3. Miller, E., Shen, D., Liu, J., Nicholas, Ch., Chen, T.: Techniques for gigabyte-scale N-gram based information retrieval on personal computers. In: Proceedings of the 1999 International Conference on Parallel and Distributed Processing Techniques and applications—PDPTA'99
4. Heintze, N.: Scalable document fingerprinting. In: Proceedings USENIX Workshop on Electronic Commerce. 1996
5. Forner, P., Karlgren, J., Womser-hacker, Ch., Potthast, M., Gollub, T., Hagen, M., Gra*β*egger, J., Kiesel, J., Michel, M., Oberländer, A., Barrón-cedeño, A., Gupta, P., Rosso, P., Stein, B.: Overview of the 4th international competition on plagiarism detection. In: Forner, P., Karlgren, J., Womser-Hacke, C. (eds.) Notebook Papers of CLEF 2012 LABs and Workshops, CLEF-2012 17–20 September. Rome, Italy
6. Chin, O.S., Kulathuramaiyer, N., Yeo, A.W.: Automatic discovery of concepts from text. In: IEEE/WIC/ACM International Conference on Web Intelligence. pp. 1046–1049. 18–22 Dec 2006

7. Amine, A., Elberrichi, Z., Simonet, M., Malki, M.: WordNet-based and N-grams-based document clustering: a comparative study, broadband communications. In: 3rd International Conference on Information Technology and Biomedical Applications. pp. 394–401. 23–26 Nov 2008
8. Synat NCBiR Project: SP/I/1/77065/10. http://synat.pl
9. Stevenson, M., Greenwood, M.A.: Learning information extraction patterns using WordNet. In: Proceedings of the 5th International Conference on Language Resources and Evaluations, LREC 2006. pp. 95–102. 22–28 May 2006
10. Natural Language Toolkit. http://nltk.org/

Chrum: The Tool for Convenient Generation of Apache Oozie Workflows

Piotr Jan Dendek, Artur Czeczko, Mateusz Fedoryszak, Adam Kawa, Piotr Wendykier and Łukasz Bolikowski

Abstract Conducting a research in an efficient, repetitive, evaluable, but also convenient (in terms of development) way has always been a challenge. To satisfy those requirements in a long term and simultaneously minimize costs of the software engineering process, one has to follow a certain set of guidelines. This article describes such guidelines based on the research environment called Content Analysis System (CoAnSys) created in the Center for Open Science (CeON). In addition to best practices for working in the Apache Hadoop environment, the tool for convenient generation of Apache Oozie workflows is presented.

Keywords Hadoop · Research environment · Big data · CoAnSys · Text mining

P. J. Dendek (✉) · A. Czeczko (✉) · M. Fedoryszak (✉) · A. Kawa (✉) ·
P. Wendykier (✉) · Ł. Bolikowski (✉)
Interdisciplinary Centre for Mathematical and Computational Modelling,
University of Warsaw, Warszawa, Poland
e-mail: p.dendek@icm.edu.pl

A. Czeczko
e-mail: a.czeczko@icm.edu.pl

M. Fedoryszak
e-mail: m.fedoryszak@icm.edu.pl

A. Kawa
e-mail: kawa.adam@gmail.com

P. Wendykier
e-mail: p.wendykier@icm.edu.pl

Ł. Bolikowski
e-mail: l.bolikowski@icm.edu.pl

R. Bembenik et al. (eds.), *Intelligent Tools for Building a Scientific Information Platform: From Research to Implementation*, Studies in Computational Intelligence 541,
DOI: 10.1007/978-3-319-04714-0_12,

1 Introduction

1.1 Distributed Computing

Currently, no single machine can be employed to consume, in a convenient way and in a reasonable time, the massive amounts of data. To tackle this issue, several approaches to GRID computing have been introduced.

In 2004, Google proposed MapReduce paradigm [4, 5], which gives an opportunity to maximize computer resources utilization and enables better parallelization trade-off between the number of machines used and a computational time. Many articles discussed the topic of embracing MapReduce for data analysis purposes, covering also its optimization and adaptation [2, 3, 10–12, 14, 15]. MapReduce is a great candidate to face challenges brought by the universe of big data.

Since 2008, the second name of MapReduce is its open-source, Java based implementation of Apache Hadoop. It has proved its adaptability for business purposes in companies like Amazon, Last.fm, LinkedIn, Twitter, Mendeley, and Yahoo!, e.g. for rapid storing and retrieving huge amounts of data, machine learning and information retrieval mechanisms.

Authors assume that all readers of this article have a basic knowledge about MapReduce and Apache Hadoop environment. In case of doubts we encourage to familiarize with adequate literature [8, 9, 13].

1.2 Research Environment

Thanks to services like Amazon Elastic Compute Cloud, use of the powerful environment of Apache Hadoop is not accompanied with a burden of purchasing appropriate machines, followed by onerous administrative tasks. Nevertheless, there are many other difficulties, especially in conducting research, which should be anticipated and resolved before they arise in major impediments.

To omit such occasions, researchers should put on the first place a certain dojo,[1] which can be condensed as follows:

1. Separate input data, working data and output data and keep them for a later comparison (e.g. for other researchers or in case of detecting malformation in a code). Save everything to disk frequently [**comparable**, **convenient**].
2. Separate options from parameters. Provide easy way of swapping options [**convenient**].
3. Modify input data with care [**comparable**].
4. Record parameters and options used to generate each run of the algorithm [**comparable**].

[1] http://arkitus.com/PRML/

5. Make it easy to execute only portions of the code [**convenient**].
6. Support automatic way of triggering all workflows [**convenient**, **efficient**].

Naturally, obtaining and shipping research results is the top priority goal—paraphrasing a popular quote "The technology you use impresses no one. The experience you create with it is everything". To do so, a suitable background and a daily routine are indispensable. Intuitively, following all those rules is equivalent to a time overhead in a research process, especially at the beginning, but later on, it is rewarded by results acceleration and convenience. Moreover, some desperate rescue actions in the final stage of writing articles, typically close to the submission deadline, are no longer needed. Those requirements are put not only on particular persons, but also on the whole research system.

The rest of this paper is organized as follows. Section 2 introduces CoAnSys—a framework for mining scientific publications using Apache Hadoop, Sect. 3 describes tools and practices used for the development of CoAnSys. Section 4 describes Chrum—a tool developed by the authors to simplify the description of a workflow. Finally, Sect. 5 derives conclusions about developing data processing environment and conducting research in Apache Hadoop.

2 CoAnSys

The Center for Open Science (CeON, part of Interdisciplinary Center for Mathematical and Computational Modeling, University of Warsaw) has been developing, the Content Analysis System (CoAnSys)[2] embracing Apache Hadoop ecosystem. This framework is suitable for research on big data, e.g. document metadata extraction and further analysis. An architecture overview of CoAnSys is illustrated in Fig. 1. It consists of the following parts:

1. Data import mechanisms.
2. Data processing mechanisms by a sequential or parallel set of algorithms.
3. Data export mechanisms.

To provide unified input data, CoAnSys utilizes Protocol Buffers,[3] a method of serializing data into a compact binary format. We have chosen Protocol Buffers over similar Apache Thrift[4] or Apache Avro[5] due to the fact that it had already established position and was widely used at Google, Twitter and Mendeley. Once the data is serialized, it can be imported into the HBase using REST protocol

[2] http://github.com/CeON/CoAnSys

[3] http://code.google.com/p/protobuf/

[4] http://thrift.apache.org/

[5] http://avro.apache.org/

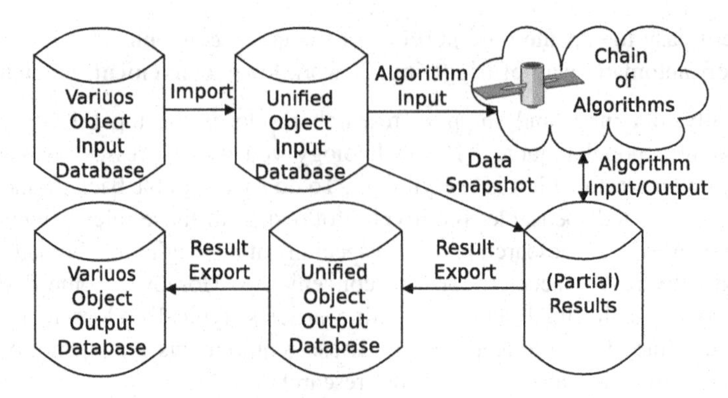

Fig. 1 A generic framework for performing unified, repetitive experiments on data with eventual algorithm comparison and evaluation

(HBase has a build-in REST server called Stargate). REST allows for simultaneous import of data from multiple clients, however, to perform complex, offline operations on the data, it is beneficial to use HDFS instead of HBase.

For that reasons, in the next step of the CoAnSys workflow, the data is copied from an HBase table to an HDFS sequence file. Data in that form (Unified Object Input Database in Fig. 1) is then used as an input for the algorithms. MapReduce jobs are implemented in Java, Apache Pig [8] or Scala (depends on the preference of a developer) and chained together using Apache Oozie[6]—a workflow scheduler system for Hadoop. To maintain the consistency of the framework, each module has well defined I/O interfaces in the form of Protocol Buffers schemas. This means, that each module reads input data from an HDFS sequence file (that contains records serialized with Protocol Buffers) and stores the results in the same way (output records can have different schemas). In the last part of the workflow, the output data from each module is stored in an HBase table and can be accessed by the clients via REST protocol. From the high-level architecture perspective, CoAnSys workflow is quite similar to UIMA's [7] collection processing engine, where user defined functions play a role of nodes in the chain of execution.

There are three ongoing projects that use the elements of CoAnSys framework. OpenAIREplus[7] is the European open access data infrastructure for scholarly and scientific communication. PBN[8] is a portal of the Polish Ministry of Science and Higher Education, collecting information on publications of Polish scientists and on Polish and foreign scholarly journals. Infona[9] is a user portal for SYNAT [1]—Polish national strategic research program to build an interdisciplinary system for interactive scientific information.

[6] http://oozie.apache.org/

[7] http://www.openaire.eu/

[8] http://pbn.nauka.gov.pl/

[9] http://www.infona.pl/

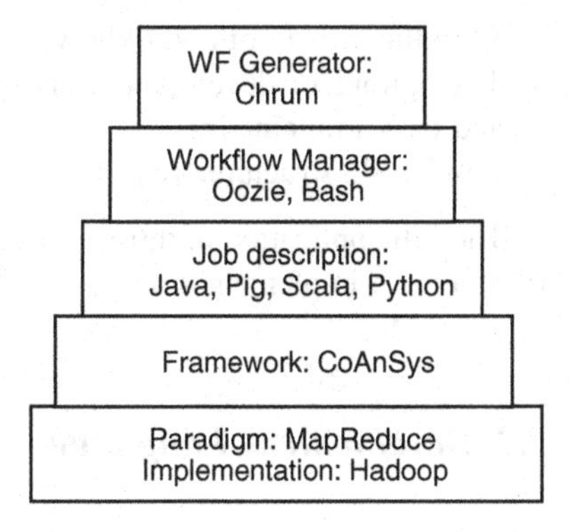

Fig. 2 CeON analytic stack

Research conducted in CeON may be analyzed in five layers depicted in Fig. 2. Since its very beginning, the development of CoAnSys has been a great lesson to CeON team. We investigated both technology and ours limitations, applied solutions crystallized, both vertically (e.g. use of Bash scripts to chain Apache Pig scripts) and horizontally (e.g. shift from Bash usage to Apache Oozie). Taking into consideration all the efforts, we decided to disseminate our experiences in order to recommend CoAnSys, highlight tools available in the Apache Hadoop environment and familiarize readers with the development of research framework.

3 Tools and Practices

This section presents some important aspects of a development process and workflow creation tools that are being used in CoAnSys.

3.1 Development Aspects

Workflows implemented in CoAnSys algorithms (e.g. [6]) are quite straightforward—no traps, no mazes, very transparent ideas. Crafting them against guidelines mentioned in Sect. 1.2 had impact not only on practices used, but also on the choice of tools. CeON team treated with a particular care the following development aspects:

1. Workflows parts have to be loosely coupled.
2. Workflows need to be executed in parallel.
3. Each module should be parametrized.
4. Experiment's results have to be maintained by an automatic procedure.
5. Experiment's options should be separated from execution variables.
6. Code repetitions should be kept at the lowest possible level.

7. With the growth of a workflow, a comprehensibility should remain unchanged.
8. The options of development and production environment should be separated and easy to maintain.
9. Basic code structures (e.g. loops) have to be provided.

The full application of those guidelines results in a developer experience remote from a nave implementation.

3.2 Development Language

Java is a native programming language for Apache Hadoop. However, contrasting the term frequency–inverse document frequency (TF-IDF) calculation code written in Java (418 lines of code) with Apache Pig (28 lines of code) leads to the conclusion that development in Apache Pig may be more productive and transparent. Additionally, incorporating into Pig scripts mechanisms such as the definition of parameters or macros increases their flexibility (point 3). However, due to the lack of loop blocks in Pig, scripts are strongly coupled (points 1 and 9) and weakly reusable. This stresses the need of using gluing scripts. The most natural choice in a Linux-based development environment is the Bash shell.

Sequences of data transformations described in Pig scripts can be divided into the sets of sub-procedures activated by Bash scripts (point 1). Limitations of this approach become apparent only when one starts to use it. First of all, some amendments are needed in Bash code to enable fully parallel execution (point 2). Naturally, those mechanisms are not managed by any Hadoop environment load manager. Furthermore, with each aspect of an experiment, the number of nested loops will be continuously growing, leading to a long code that is hard to develop or even comprehend (point 7). Finally, the parameters of an experiment (e.g. the number of neighbour documents) are mixed together with auxiliary variables (e.g. fold number) (point 5) and results' management (e.g. choosing localization) have to be done by hand or by additions in bash scripts (point 4).

Development team at CeON was aware that at some point, a general workflow embracing not only document classification module, but also all other algorithms has to be prepared and it was quite obvious at that point that Bash is not a convenient solution in this scale.

3.3 Apache Oozie

A workflow scheduler system called Apache Oozie has become our tool of choice. That framework outperforms Bash in a few aspects: it is easy to separate parameters from options (point 5) and it allows creation and management of sets or parameters for development or production environment (point 8).

On the other hand, the workflow management system from Apache rises some obstacles: it requires writing long XML files (point 8) (see Listing 1.1), it does not support for loop statements (point 9; simulation via tedious copy-pasting), the output and subsequent executions of one code or its continuously improved versions, have to be manually managed by a developer, and finally, there is no automatic mechanism for creating a link between code versions and the results.

Listing 1.1: The example of an Apache Oozie action block. Each action has (a) its own distinct name, (b) a subsequent nodes for successful (ok) and unsuccessful (error) script finalization, (c) a set of parameters with a path to JAR libraries among others and (d) a script to be executed with.

```
<action name='tfidf'>
   <pig>
      <job-tracker>${jobTracker}</job-tracker>
      <name-node>${nameNode}</name-node>
      <prepare>
         <delete path="${ds_similarityOutputPath}"/>
      </prepare>
      <configuration>
         <property>
            <name>mapred.job.queue.name</name>
            <value>${queueName}</value>
         </property>
      </configuration>
      <script>${pigScriptsDir}/document-similarity.pig</script>
      <param>inputPath=${ds_bwndataMetadataInputPath}</param>
      <param>outputPath=${ds_similarityOutputPath}</param>
      <param>commonJarPath=${ds_commonJarPath}</param>
      <file>${pigScriptsDir}/macros.pig#macros.pig</file>
   </pig>
   <ok to='end'/>
   <error to='kill'/>
</action>
```

4 Chrum

Chrum (eng. "oink")[10] is a Python-based tool implemented by the CeON team to alleviate the drawbacks of Apache Oozie mentioned in Sect. 3.3. There are two main goals of Chrum:

1. The simplification of a workflow description (no repetitive code).
2. Better experiment management (separate spaces for results, a code used in a given time and for each combination of parameters).

[10] http://github.com/CeON/chrum

4.1 Workflow Generation

Chrum introduces additional layer over Apache Oozie configuration and workflow files. In a workflow file, the following new blocks are present: REPLACE, ACTION and FORK_MERGE. With these three simple operations, a workflow development process is much more convenient. A scope of a Chrum block is limited by BEG and END keywords, as presented in Listing 1.2.

Listing 1.2: The definition of the ACTION BLOCK docs2neigh_01. Chrum will substitute REPLACE BLOCKS VALUES present inside of the ACTION BLOCK (e.g. WF-1 shown in Listing 1.3 and subsequently expand it to a final form similar to the code presented in Listing 1.1)

```
# BEG:ACTION name=docs2neigh_01 ok=createDocClassif_02 error=kill
   @PIG_START@
       @PR-1@
       @CONFIG-1@
       @WF-1@
   @PIG_END@
# END:ACTION
```

The ACTION block, presented in Listing 1.2, gathers all important Oozie's action node information, i.e. a node name and a final node after success or failure. This block may contain other blocks, e.g. replace block variables.

The REPLACE block, shown in Listing 1.3 allows extraction and placing in the same part of a workflow file the essence of a node description i.e. parameters, files needed and a script used. In addition, the REPLACE block enables abbreviation of other longer portion of the code by referencing to it with a replace block variable.

Listing 1.3: The definition of the REPLACE block WF-1, which contains regular Apache Oozie XML code. A definition of one block may include another Chrum block. The definitions of blocks cannot create a substitution cycle.

```
# BEG:REPLACE @WF-1@
   <script>${pigScriptsDir}/1_MODEL_CREATE_01_docs2neig.pig</script>
   <param>dc_m_double_sample=${dc_m_double_sample}</param>
   <param>dc_m_hbase_inputDocsData=${dc_m_hbase_inputDocsData}</param>
   <param>dc_m_hdfs_neighs=${dc_m_hdfs_neighs}</param>
   <param>dc_m_int_folds=${dc_m_int_folds}</param>
   @AUXIL@
# END:REPLACE
```

Finally, the FORK_MERGE block, presented in Listing 1.4, enables a user to execute a script in a way similar to a (nested) "for" loop. In the first line of the FORK_MERGE block one declares the name of a block, accompanied with the names of subsequent nodes in case of success and failure.

In the next line(s) of FORK_MERGE block, parameters with their values separated by spaces are defined, e.g. "@src@/tmp/1/tmp/2/tmp/3". It is possible to dynamically generate the sequence of numeric values using the idiom "seq", e.g.

"@fold@ seq(0,$dc_m_int_folds,1)", where 0 is the initial number of a sequence, $dc_m_int_folds is the maximum number of that sequence (defined in the Oozie's configuration file) and 1 is the default step between consecutive values. Idioms like "seq" can be easily added to Chrum by a developer. Eventually, the code for substitution begins at the first line that does not describe parameters.

Having described information, Chrum translates FORK_MERGE block into Apache Oozie fork and join nodes in the resulting workflow.xml file. Between them, the set of *n* parallel Apache Oozie action nodes is created (*n* is the number of combinations of parameters' values). In each new action node, parameters (denoted by @param_name@) are substituted by corresponding values. As in regular usage of Apache Oozie fork-join, calculations in all precedent nodes have to be successfully completed before the execution of a subsequent node begins.

Listing1.4: The definition of the FORK_MERGE block. FORK_MERGE block contains the name of a fork node ("split_03"), the next node after merge ("enrich_04") and an error node, where control goes in case of any failure.

```
# BEG:FORK_MERGE name=split_03 node_after_join=enrich_04 error=kill
@src@ ${dc_m_hdfs_neighs} ${dc_m_hdfs_docClassifMapping}
@fold@ seq(0,${dc_m_int_folds},1)
   @PIG_START@
      @PR-3@
      @CONFIG-1@
      <script>${pigScriptsDir}/1_MODEL_CREATE_03_split.pig</script>
      <param>dc_m_hdfs_src=@src@</param>
      <param>dc_m_int_concreteInvestigatedFold=@fold@</param>
      @AUXIL@
   @PIG_END@
# END:FORK_MERGE
```

These three simple commands allow to significantly shorten the number of lines of code. For example, Chrum shortened an XML workflow file describing the document classification model from 1.5 KLOC to 0.1 KLOC.

4.2 Experiments Management

Frequent inconvenience for researchers and developers is that, even if one works with a version control system, it is hard to follow a bind between the code version and the results. Similarly, when several combinations of parameters have to be tested against a given code, a control system has to be written or tedious manual actions have to be taken.

Chrum Experiment Manager (CEM) is a remedy for these problems. CEM, as an input, takes Oozie's workflow and property files, accompanied with a Chrum configuration file. The following information is specified in the Chrum configuration file:

1. Key information
    (a) project name
    (b) path to the input data in HDFS
    (c) local path to Chrum's trigger scripts
    (d) address and port of Oozie's server.
2. Folders to be created in the $PROJECT directory with a content indicated (e.g. "lib ←/usr/lib/pig/pig-0.9.2-cdh4.0.1.jar").

Execution of CEM triggers the extraction of key information from Chrum's configuration file. For each combination of multivalued properties in Oozie's properties file, a modified copy of that file is created on a local file system, each of which carries CEM execution time (in the property "COMPILATION_TIME") and the combination of parameters ("PARAMETER_COMBINATION"). Multivalued properties (denoted as "@var@ val1 val2 val3") are flattened to one variable (e.g. "var = val2"), where each combination appears only once. Property files composed in such a manner, accompanied with generated workflow files and a workflow submission script (execute-in-oozie.py), are stored in a local file system ($LOCAL/$PROJECT/$COMPILATION_TIME/$PARAMETER_COMBINATION), while libraries, scripts, workflow files and other data mentioned in the Chrum configuration file are sent to the HDFS folders $HDFS/$PROJECT/$COMPILATION_TIME/$PARAMETER_COMBINATION. Finally, when the execute-in-oozie.py script is started, the current time is stored in a property file ("EXECUTION_TIME"), and the workflow is submitted to the Oozie Server.

The above procedure fulfills the following goals:

1. A user, in an easy way, can send to HDFS all data needed for further consideration (e.g. source code, personal notes).
2. Each call to CEM and each combination of parameters are stored in a different location ($HDFS/$PROJECT/$COMPILATION_TIME/$PARAMETER_COMBINATION).
3. Each workflow's execution output may be stored in a separate location for subsequent investigation ($HDFS/$PROJECT/$COMPILATION_TIME/$PARAMETER_COMBINATION/results/$EXECUTION_TIME).

4.3 Different Approaches to Workflow Creation and Management

Contributing to CoAnSys as well as seeking for the most clear and coherent code has been a path from using only Apache Pig scripts, mixture of Apache Pig and Bash, Apache Oozie and finally Chrum. Workflows described with those approaches may be analyzed in terms of the following aspects:

Table 1 Subjective assessment of different approaches to a workflow creation

Approach	Strong code coupling	Loops support	Cost of code maintenance in small scale	Cost of code maintenance in large scale	Extensibility	Maturity	Option/parameters separation
Pig	High	None	High	Very high	Low	High	Low
Bash	Low	Present	Moderate	High	High	High	Moderate
Oozie	Low	None	Low	High	Low	Moderate	Moderate
Chrum	Low	Present	Low	Moderate	High	Low	High

1. Strong code coupling.
2. Loop statement support.
3. Separation of options and parameters.
4. Cost of code maintenance—both in micro and macro scale.
5. Extensibility.
6. Solution maturity.

According to our subjective opinion, reflected in Table 1, the mixture of Apache Pig and Chrum provides the best solution for developing sophisticated algorithms in the Hadoop environment. However, it has to be stressed that this solution is both immature and implicitly dependent on Apache Oozie.

5 Summary and Future Work

Chrum is still far from its maturity and as so, there is plenty of room for improvements. Proposed transformations are the most frequent, but there are still a few Apache Oozie's operations which could be transformed into standard procedural language statements such as "while", "if", etc. As Chrum enables an easy way to store subsequent versions of the code and its output, it would be beneficial to develop a convenient way of comparing results in terms of a given metric.

In this article we have described an effective approach to conducting research in the Apache Hadoop environment, based on presented Content Analysis Framework and the analytic stack used in CeON. We have shown obstacles emerging in a development and a research process and provided suitable solutions which may have positive effect on both of them.

We hope that the provided description of our experience gained from the usage of Apache Hadoop will lower the entry barrier for users and researchers who want to benefit from MapReduce paradigm and it will reduce inconveniences arising in the process of learning a new technology.

References

1. Bembenik, R., Skonieczny, L., Rybinski, H., Niezgodka, M.: Intelligent Tools for Building a Scientific Information Platform. Studies in Computational Intelligence. Springer, Berlin (2012)
2. Chu, C.T., Kim, S.K., Lin, Y.A., Ng, A.Y.: Map-reduce for machine learning on multicore. Architecture **19**(23), 281 (2007). http://www.cs.stanford.edu/people/ang/papers/nips06-mapreducemulticore.pdf
3. Dean, B.Y.J., Ghemawat, S.: MapReduce: a flexible data processing tool. Commun. ACM **53**(1), 72–77 (2010). http://dl.acm.org/citation.cfm?id=1629198
4. Dean, J., Ghemawat, S.: MapReduce: Simplified Data Processing on Large Clusters. Commun. ACM **51**(1), 1–,,13 (2004). http://dl.acm.org/citation.cfm?id=1251254.1251264
5. Dean, J., Ghemawat, S.: System and Method for Efficient Large-scale Data Processing (2010). http://patft.uspto.gov/netacgi/nph-Parser?Sect1=PTO2&Sect2=HITOFF&p=1&u=%2Fnetahtml%2FPTO%2Fsearch-bool.html&r=1&f=G&l=50&co1=AND&d=PTXT&s1=7,526,461&OS=7,526,461&RS=7,526,461
6. Fedoryszak, M., Tkaczyk, D., Bolikowski, Ł.: Large scale citation matching using apache hadoop. In: Aalberg, T., Papatheodorou, C., Dobreva, M., Tsakonas, G., Farrugia, C. (eds.) Research and Advanced Technology for Digital Libraries, Lecture Notes in Computer Science, vol. 8092, pp. 362–365. Springer, Heidelberg (2013). http://dx.doi.org/10.1007/978-3-642-40501-3_37
7. Ferrucci, D., Lally, A.: UIMA: an architectural approach to unstructured information processing in the corporate research environment. Nat. Lang. Eng. **10**(3–4), 327–348 (2004). http://www.journals.cambridge.org/abstract_S1351324904003523
8. Gates, A.: Programming Pig. O'Reilly Media, Sebastopol (2011)
9. George, L.: HBase: The Definitive Guide, 1 edn. O'Reilly Media, Sebastopol (2011)
10. Kawa, A., Bolikowski, Ł., Czeczko, A., Dendek, P., Tkaczyk, D.: Data model for analysis of scholarly documents in the mapreduce paradigm. In: Bembenik, R., Skonieczny, L., Rybinski, H., Kryszkiewicz, M., Niezgodka, M. (eds.) Intelligent Tools for Building a Scientific Information Platform, Studies in Computational Intelligence, vol. 467, pp. 155–169. Springer, Heidelberg (2013). http://dx.doi.org/10.1007/978-3-642-35647-6_12
11. McKenna, A., Hanna, M., Banks, E., Sivachenko, A., Cibulskis, K., Kernytsky, A., Garimella, K., Altshuler, D., Gabriel, S., Daly, M., DePristo, M.A.: The Genome Analysis Toolkit: A MapReduce framework for analyzing next-generation DNA sequencing data. Genome Res. **20**(9), 1297–1303 (2010). http://genome.cshlp.org/cgi/doi/10.1101/gr.107524.110
12. Ranger, C., Raghuraman, R., Penmetsa, A., Bradski, G., Kozyrakis, C.: Evaluating MapReduce for multi-core and multiprocessor systems. In: Proceeding of the 2007 IEEE 13th International Symposium on High Performance Computer Architecture, vol. 0, 13–24 Oct 2007. http://ieeexplore.ieee.org/lpdocs/epic03/wrapper.htm?arnumber=4147644
13. White, T.: Hadoop: The Definitive Guide, 1st edn. O'Reilly Media Inc., Sebastopol (2009)
14. Yang, H.c., Dasdan, A., Hsiao, R.l., Parker, D.S.: Map-reduce-merge: simplified relational data processing on large clusters. Rain pages, 1029–1040 (2007), http://portal.acm.org/citation.cfm?id=1247480.1247602
15. Zaharia, M., Konwinski, A., Joseph, A.D., Katz, R., Stoica, I.: Improving MapReduce performance in heterogeneous environments. Symp. Q. J. Mod. Foreign Lit. **57**(4), 29–42 (2008). http://www.usenix.org/event/osdi08/tech/full_papers/zaharia/zaharia_html/

PrOnto: A Local Search Engine for Digital Libraries

Janusz Granat, Edward Klimasara, Anna Mościcka, Sylwia Paczuska and Andrzej P. Wierzbicki

Abstract This chapter describes system PrOnto version 2.0 and results of work on this system in the SYNAT project. After the introduction, the chapter presents shortly the functionality of PrOnto that is a system of personalized search for information and knowledge in large text repositories. Further the chapter presents elements of the personalized ontological profile of the user, the problem of finding similar concepts in many such profiles, and the issue of finding interesting documents in large text repositories, together with tests of the system and conclusions.

Keywords Search engine · Local search · Digital library

1 Introduction

Even if there are very efficient global search engines for entire Internet, a local digital library might need own local search engine, with a user-friendly and transparent method of ranking documents. Such a local search engine was developed in the SYNAT project as a version 2.0 of the system PrOnto—originally

J. Granat · E. Klimasara · A. Mościcka · S. Paczuska · A. P. Wierzbicki (✉)
National Institute of Telecommunications, Szachowa Str. 1 04-894 Warsaw, Poland
e-mail: A.Wierzbicki@itl.waw.pl

J. Granat
e-mail: J.Granat@itl.waw.pl

E. Klimasara
e-mail: E.Klimasara@itl.waw.pl

A. Mościcka
e-mail: A.Moscicka@itl.waw.pl

S. Paczuska
e-mail: S.Paczuska@itl.waw.pl

R. Bembenik et al. (eds.), *Intelligent Tools for Building a Scientific Information Platform: From Research to Implementation*, Studies in Computational Intelligence 541,
DOI: 10.1007/978-3-319-04714-0_13,

oriented on supporting the work of a local research group, but specialized in this version as a local search engine for digital repositories of documents.

System PrOnto supports the user in response to her/his individual, personalized ontological model, called hermeneutic profile, composed of three layers:

- A lower layer of keywords and key phrases that is subject to semantic and logical analysis with the use of tools of ontological engineering;
- An upper layer of intuitively defined concepts that have not far reaching logical interpretations because they might be highly intuitive;
- A middle layer of relations between concepts and key phrases containing weighting coefficients defined subjectively by the user.

The definition of such hermeneutic profile is not highly automated, it is fully controlled by the user. The response of the system to the user consists of a ranking list of documents interesting for him. The ranking is also controlled by the user, who can choose one of four multiple criteria ranking methods (applied to ontological measures of importance of keywords aggregated by weighting coefficients into measures of importance of concepts for a given document):

1. Compensable multiple criteria—such that high measure for one criterion might compensate lower measures for other criteria;
2. Essential multiple criteria—such that all criteria should have reasonably high measures;
3. Fuzzy logical concepts with fuzzy “or” aggregation;
4. Fuzzy logical concepts with fuzzy “and” aggregation.

Extensive testing of the PrOnto system have shown that the best results for the user are obtained when using the method 2, while the method 4 is less elastic.

Originally, PrOnto system was designed for English language, but recently the functioning of the system was extended also to Polish texts. Polish text analysis was made possible due to using a tagger TaKIPI, developed by the Group of Language Technology of Technical University of Wrocław [1]. TaKIPI defines a morpho-syntactic description of each word in a Polish text and a correct interpretation of words depending on their context. The output of TaKIPI are words in their fundamental form; these are used by PrOnto to define frequencies of key phrases occurrence in analysed texts.

The system PrOnto was subject to diverse tests, including a test on Internet commerce by a cooperating Chinese doctoral student, a test on internal library of the National Institute of Telecommunications, etc. The results confirmed that PrOnto system is more user-friendly than known global search engines (who usually do not make their rankings transparent because of commercial reasons), and that the most efficient ranking method is 2. The tests on digital libraries resulted in a demand for changing the user interface of PrOnto (which was designed originally to support research group cooperation) to support typical enquiries of a library user.

Currently, system PrOnto is being integrated with the platform Infona of SYNAT project. An integration includes user authorization and initial search for documents containing key phrases, together with modification of some components of the system. The chapter presents results of this integration and conclusions concerning the use of the system as a local search engine.

2 PrOnto System

PrOnto system will be a module enriching the Infona platform developed in the SYNAT project. Originally, the main function of PrOnto was to support the work of a group of researchers (Virtual Research Community, VCR) using a radically personalized user interface. The radical personalization of the interface results from an assumption that the research preferences of a user cannot be fully formalized logically or probabilistically (at most 0.01 % of neurons in our brain is concerned with rational, logical reasoning, see [2, 3, 4, 5]). Therefore, the interface should preserve and stress the intuitive character of user decisions, but nevertheless support the user in cooperation with the tools of ontological engineering. The latter choice results from the conviction (see [6–24]) that the organization of knowledge in an ontological structure performs practically much better as a basis for knowledge representation than using only keywords. Ontological concepts and relations between them enable an intuitive knowledge systematization that expresses the interpretation of a problem by a person or a group of them. The highly personalized ontological profiles can be thus called hermeneutic profiles, since they express hermeneutic perspective (horizontal assumptions, see [2, 25–27]) of the user. Such hermeneutic profiles provide much better tools both for sharing knowledge and for individual search in a digital library.

The model of PrOnto assumed originally [9–12] the support for a group of researchers (VRC) with functionalities serving an individual user or the group collaboration. In further work it turned out that this model can also serve as an individualized and user friendly local search engine for digital libraries (that for diverse reasons might not necessarily use the powerful global search engines accessible on the Internet.)[1]

A hermeneutic profile, or a radically personalized ontological model of the user of PrOnto system consists of three layers defined earlier:

- A layer of intuitive, beyond-logical concepts C; radical personalization means precisely that we treat these concepts as intuitive compositions, personally

[1] Some of such reasons might be the diversification of access rights and the issues of intellectual property. It is telling that Google search engine never publishes their methods of ranking (even in Google Scholar, where only additional logical constraints are used to select, e.g., publications of a given researcher) but recently came to the conclusion that the used methods of ranking should be modified; this indicates the importance of research on methods of ranking for search engines.

belonging to the user, and we do not attach to them too far-reaching logical interpretations (a concept represents what the user means by the concept, not what is popularly meant by the concept); we refrain from too much automation of these concepts, even if we accept an intuitive definition of relations between them.

- A layer of classical keywords or key phrases K (that are subject to semantic and logical analysis with the use of tools of ontological engineering).
- A layer of relations between concepts and key phrases f: CxK→R (in original version they are expressed by weighting or importance coefficients defined intuitively by the user, but precisely in this relation layer we can give diverse interpretations and extensions of these relations).

3 Functionality of PrOnto System v.2

System PrOnto supports the user in search for information contained in electronic documents, while the search is based on the hermeneutic profile of the user, modelled as a personalized ontological profile. Therefore, each user creates her/his own ontology that consists of a set of concepts and their relations to a set of keywords of key phrases. Keywords and key phrases characterize diverse documents, each concept is related to some keywords or key phrases, importance coefficients in such relations are used in ranking the documents most interesting for the user. Such representation of personalized ontology helps in an intuitive systematization of knowledge corresponding to the hermeneutical perspective of the user; the resulting ranking of documents should express this perspective.

System PrOnto v. 2.0 consists of following modules:

- A graphical module supporting the creation of an hermeneutic profile;
- A module for defining importance coefficients in relations of concepts with keywords and key phrases;
- A module for search of documents corresponding to the hermeneutic profile of the user;
- A module computing rankings of documents according to several methods, with user selecting the method;
- A module for supporting knowledge sharing based on combining hermeneutic profiles—personalized ontological profiles of several users;
- A module for loading documents selected by the user into a data base of PrOnto (both in Polish and in English), or an aggregate loading of documents from a given repository.

Some selected functions of the system are described below.

3.1 Supporting Knowledge Sharing

The organization of information in most social networks is based on tags, that is, keywords or key phrases that are attached by the users to given resources in a top-down manner. This is in a sense opposite to the bottom-up search for keywords in indexing electronic documents by most search engines. Hermeneutic profiles or individualized ontological profiles are similar to but much richer than tags since they rely on sets of keywords determined in a bottom-up search but enrich them by concepts defined in a top-down manner and by the indication of relations between concepts and keywords. An individual hermeneutic profile might be further enriched by comparing it to hermeneutic profiles of other persons. A simple example is following the perspective of an expert in a given field and using the hermeneutic profile of an expert to enrich our own profile.

However, the hermeneutic profile of an expert might be considered her/his intellectual property. Therefore, sharing knowledge in a research group must be unanimously agreed upon. System PrOnto is an environment that supports sharing knowledge in creative processes, if the members of a research group agree on sharing hermeneutic perspectives. A user that creates a model of her/his research interests in the form of an hermeneutic profile not only systematizes knowledge for own use, but also supports possibility of its sharing (see Fig. 1).

Having the access to hermeneutic profiles of other members of a research group, we can use tools of ontological engineering to find similarities of concepts, similar relations of groups of key phrases to a concept, etc. The user can get a display of an ordered list of documents according to the frequency of occurrence of a given key phrase and browse interesting documents.

The tools of ontological engineering used for this purpose are typical tools of automated analysis of a repository of documents, indexing the documents etc. They might support the user in creating new concepts and defining relations between concepts and key phases or keywords. However, the final decision is left to the user. Because of large amount of information, the user might overlook some interesting relations, and the system supports the user in finding knowledge in a repository of documents—but the user can accept or disregard automated suggestions, if they do not enrich his hermeneutic perspective (see Fig. 2).

PrOnto v.2 contains also a module of aggregating individual ontological profiles. Each user has the possibility to aggregate his hermeneutic profile with selected other such profiles made accessible to the research group, together with finding similar profiles and accounting for similarity of concepts and key phrases.

3.2 Handling of Documents in PrOnto System

In the prototype PrOnto system we implemented addition and handling of documents in English language. Later we added handling of documents in Polish language. Because of inflexion character of the language, a morph-syntactic

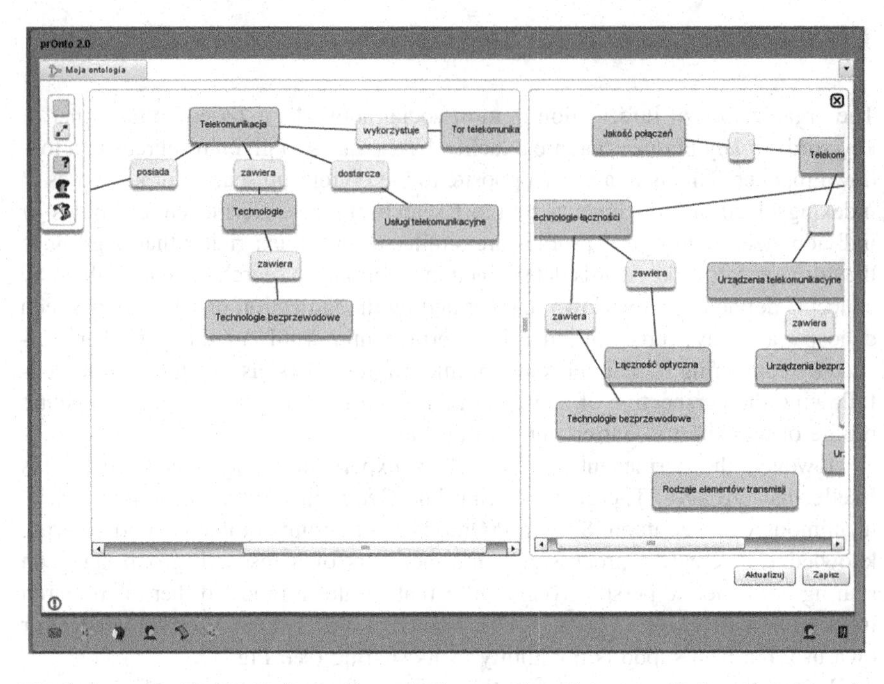

Fig. 1 Access to hermeneutic profiles of other members in system PrOnto v.2 from the perspective of a hermeneutic profile (here in Polish language) of the user

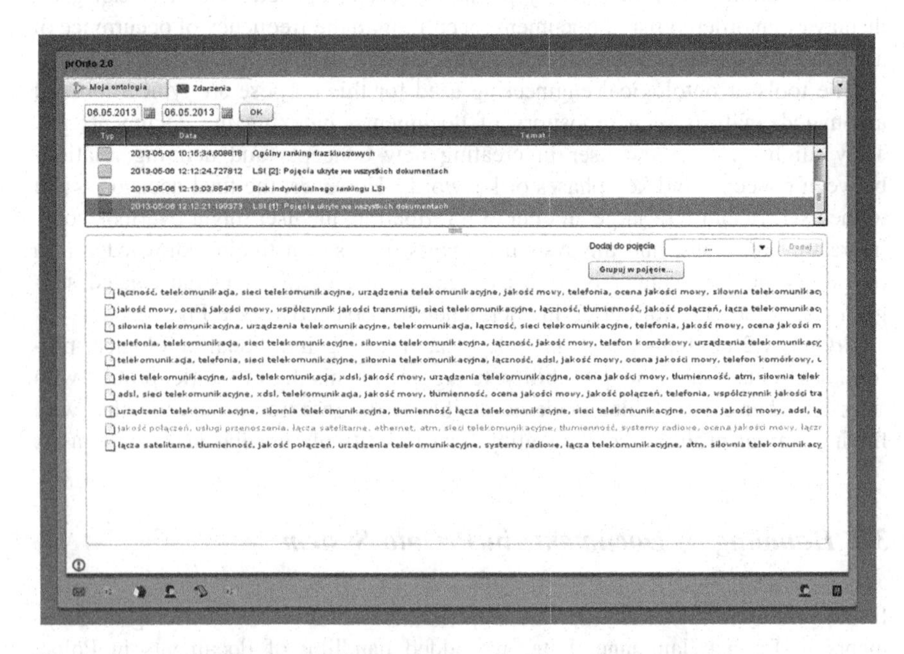

Fig. 2 Access to modifications of hermeneutic profile when preparing a request

indexing was necessary that consists of adding to each word a tag containing a grammatical class and an attribute (a basic form of the word). For performing such indexing, we implemented the tool TaKIPI [1] developed by the Group of Language Technology G4.19 of Wrocław University of Technology. TaKIPI is a tagger of Polish language that attaches to each word in a text its morph-syntactic index and determines the correct interpretation of the word depending on context. In the PrOnto system we use the basic forms given by the tagger, both for words in a document and for keywords or key phrases in the system. This results in an efficient search for the information interesting to the user (see Fig. 3).

Documents that are not a direct electronic text (scans) are converted by using the tool Tesseract-OCR [28]. This tool was selected as most efficient between open access OCR open source tools. Therefore, the repository of PrOnto system can be enlarged also by scanned documents that after conversion can be indexed, searched for key phrases, ranked etc. The system has also a function of automatic loading documents from a given outside repository, on the command of but without bothering the user.

3.3 Search of Documents

Ranking analysis is used for searching such documents in repository that contain most of key phrases or, more precisely, contain them in best accordance with the hermeneutic profile of the user. The order of documents presented depends on the selected method of ranking. The result of search presents not only the names (titles) of documents but also the frequencies of key phrases occurrence converted to the measure TF-IDF (Term Frequency—Inverse Document Frequency) [29]. The user has also an easy access to the text of selected documents together with the possibility of display of key phrases found in the document. Because of the possibility of using requests of diverse levels of complexity, of selecting the method of ranking documents, of obtaining diverse information about documents, the user has an elastic support for finding documents interesting for her/him—even in repositories of considerable size.

Ranking of documents might be determined on the basis of entire hermeneutic profile of the user, but the user has also the possibility of temporarily modifying her/his profile and constructing a request using only some selected key phrases, or a selected concept or a group of them, together with a temporal change of importance coefficients between concepts and key phrases. Thus, the use of hermeneutic profile is fully elastic.

Four separate methods of ranking were implemented in the PrOnto system v.2 (for a free selection by the user, who is informed only that the best results in testing the system were achieved when using method 2):

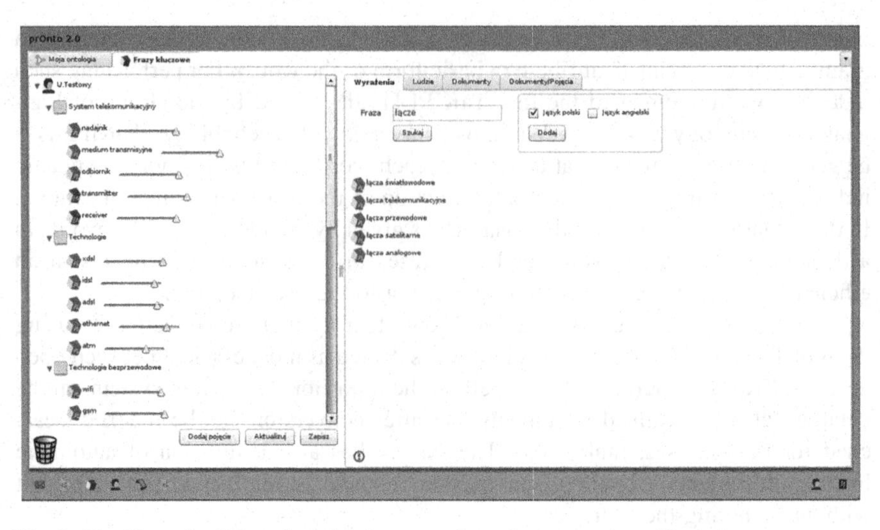

Fig. 3 Handling of polish and english language from the perspective of key phrases

1. Many compensable criteria of choice (a large value of one criterion might compensate a small value of another one);
2. Many essential criteria of choice (all essential criteria should have reasonably large values);
3. Fuzzy logical terms related by fuzzy logical operation "or";
4. Fuzzy logical terms related by fuzzy logical operation "and".

These ranking methods use the following formulae; for method (1):

$$h_1(d_i, C) = \sum\nolimits_{j \in C} \left(h(d_i, c_j) - h_{av}(c_j)\right), \tag{1}$$

where $h_1(d_i, C)$ is the importance index of document d_i with respect to all concepts C, computed for the method (1) and based on importance indexes with respect to specific concepts c_j:

$$h(d_i,\ c_j) = \sum\nolimits_{k \in K} f(c_j,\ k_k) g(d_i,\ k_k), \tag{1a}$$

while $f(c_j, k_k)$ is user-determined importance coefficient of key phrase k_k for the concept c_j and $g(d_i, k_k)$ is the (converted to the measure TF-IDF) frequency of occurrence of key phrase k_k in the document d_i. The average importance index $h_{av}(c_j)$ is computed by averaging over documents as:

$$h_{av}(c_j) = \sum\nolimits_{i \in D} h(d_i, c_j) / |D|, \tag{1b}$$

where |D| is the number of documents. If we would omit the average importance index $h_{av}(c_j)$ in Eq. (1), the importance index would be a simple sum of partial importance indexes; but some concepts might have small, other—large average

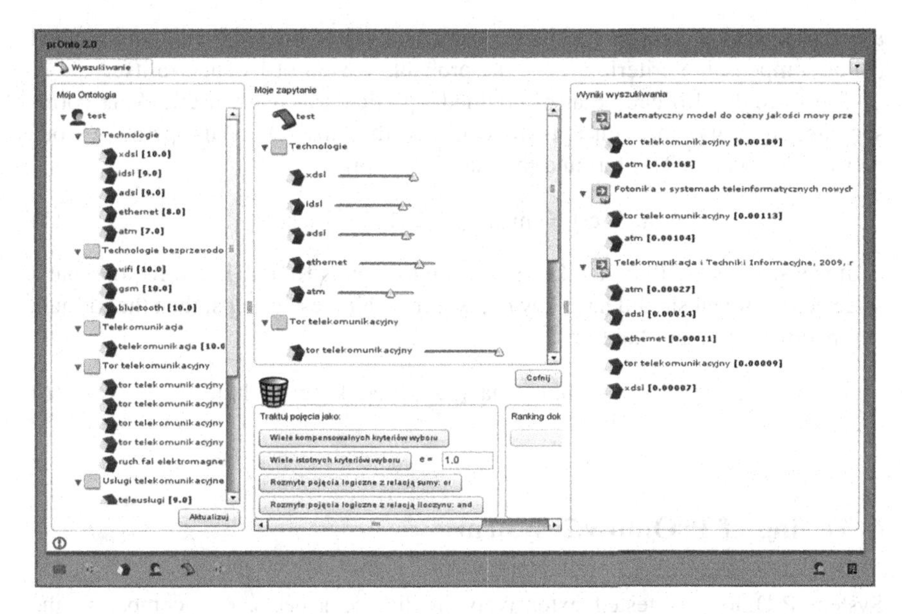

Fig. 4 Ranking of documents for the method (4) of fuzzy logical "and" operation

importance between all documents, thus it is useful to compare the deviations from the average.

For method (2) we use the following formula:

$$h_2(d_i, C) = \min_{j \in C}\left(h(d_i, c_j) - h_{av}(c_j)\right) + \varepsilon \sum_{j \in C} \left(h(d_i, c_j) - h_{av}(c_j)\right), \quad (2)$$

where ε is a small positive coefficient; this corresponds to so-called objective ranking, see [2].

For method (3), fuzzy logical "or", we use the formula:

$$h_3(d_i, C) = \max_{j \in C} h(d_i, c_j) \quad (3)$$

and for method (4), fuzzy logical "and":

$$h_4(d_i, C) = \min_{j \in C} h(d_i, c_j). \quad (4)$$

If we take $\varepsilon = 0$ and omit the comparisons to $h_{av}(c_j)$, then the case (2) is equivalent to (4). However, extensive testing of the four methods of ranking has shown that the method (2) is usually the most efficient and user-friendly, while method (4) gives less elastic results: its application gives at the top of a ranking list only those documents for which all concepts have key phrases contained in the document (which rarely happens), and all other documents have the resulting importance indexes equal zero, even if they might have many phrases corresponding to some if not all concepts (see Fig. 4). The method (2) with $\varepsilon \neq 0$ (e.g.,

$\varepsilon = 0.04$) gives a compromise between method (4) of logical “and” and method (1) of compensable criteria; this is the probable reason of its superiority.

The formula (1a) can be also modified—particularly if we use ranking corresponding to only one concept—to consider the fuzzy logical operation “or” between keywords. The formula then takes the form:

$$h(d_i, c_j) = \max_{k \in K} f(c_j, k_k) g(d_i, k_k). \tag{5}$$

It is also possible to divide key phrases into groups K(l), l∈L and require that a selected document should have key phrases from all these groups; then the formula (5) modifies to fuzzy “and-or”:

$$h(d_i, c_j) = \min_{l \in L} \max_{k \in K(l)} f(c_j, k_k) g(d_i, k_k). \tag{6}$$

4 Testing of PrOnto v.2 System

System PrOnto was tested extensively in diverse aspects, e.g. comparing the effectiveness of diverse methods of ranking documents, see [26]. As a whole system, it was tested internally by the group of co-workers of the Division of Advanced Information Technology (Z6) of the National Institute of Telecommunications, by representatives of digital libraries being developed in the Institute and in the consortium of private universities Futurus collaborating with the Institute. Moreover, the system was also tested by a Chinese doctoral student from Japan who used it for supporting commercial Internet application of selling traditional folklore products.

Generally, testing of the system included:

- Tests of functionality of the system;
- Tests of algorithms implemented in the system.

These testing was based on specially developed testing scenarios and on a questionnaire of system evaluation. The main conclusions from the testing are following:

- The interface of the user should be consolidated, e. g. icons, buttons;
- The windows of the interface should be automatically refreshed, e.g. after add new key phrase in bookmark “Key phrase”, user have to manually refresh window with bookmark “Searching” to see this key phrase in his ontology;
- Help suggestions should be included into the system;
- A clear instruction for an average user should be prepared.

Table 1 Rating of functionality of the system PrOnto

Functionality system PrOnto	Average
Login to the system	4.3
Bookmark "Editing data"	3.2
Bookmark "My ontological profile"	4
Bookmark "Event"	5
Bookmark "Key phrase"	3.8
Bookmark "Other people ontologies"	5
Bookmark "Searching"	4.1
Bookmark "Profiles aggregation"	3.9
General appraisal	4

The table below (see Table 1) represents average scores[2] of each functionality of the system PrOnto based on test results. Testers evaluated the system subjectively and intuitively responding to the questions in the survey tests.

The method of ranking which was the most frequently chosen by testers was "many essential criteria of choice". The method "many compensable criteria of choice" was often selected too. It is understandable because both of this methods are similar. However the method "many essential criteria of choice" is more flexible and allows the user to match ranking results to his actual needs.

A Chinese doctoral student Jin [30] used the methods of ranking which occur in PrOnto system to choose a product of art for a client of electronic commerce. He determinated which method of the ranking is the most satisfying for a customer (see Table 2).

Based on the average response of users, J. Jin ascertained that methods of ranking: "Many essential criteria of choice" and "Many compensable criteria of choice" provide the highest customer average satisfactions. The method "Many essential criteria of choice" has the smallest value a variance of responses and thus this method is the most stable.

The results of testing have confirmed that PrOnto system is more user-friendly than known global search engines. Another conclusion of the testing is that the most efficient method of ranking is method (2) (many essential criteria of choice), what confirms earlier internal results of testing and theoretical considerations.

Since PrOnto was originally designed to support cooperation in a research group, its application as a local search engine for digital libraries requires a change of the user interface to respond to a local digital library standards.

The current stage of PrOnto development consists of its integration with the platform and portal Infona of SYNAT project in such a way that the system will search for documents in entire repository of that platform.

[2] The final assessment is the arithmetic average rating given by the testers in the range 1–5.

Table 2 Comparison of different methods of ranking for electronic commerce of products of art [30]

Method of the ranking	Average satisfaction	Variance
Many essential criteria of choice	0.8051	13.8355
Many compensable criteria of choice	0.7867	25.1769
Fuzzy logical terms related by fuzzy logical operation "or"	0.7153	30.9358
Fuzzy logical terms related by fuzzy logical operation "and"	0.7738	27.1458

5 Conclusions

PrOnto system v.2 presented in this chapter has considerable possibility of diverse applications, starting with supporting the work of a research group and sharing knowledge in such group. The system is innovative and adaptable to individual needs of the user, helps in searching for digital documents in response to the hermeneutic profile of the user modelled as highly personalized ontological profile. The system uses algorithms of search and methods of ranking that are legible for the user—as opposed to known global search engines in Internet that are concentrated on commercial use. The user has the possibility of choosing a method of ranking and to define her/his requests more precisely (by defining concepts and importance coefficients attached to keywords) in order to obtain more adequate answers. After the integration with SYNAT (Infona) platform, the system will support large depositories of documents.

The results of the research on PrOnto system might have importance for the general problem of intelligent knowledge search in large text repositories, in relation to the research, e.g., on Future Internet in the aspect of context-aware networks, or in relation to future development of digital libraries.

Acknowledgments This work has been supported by the National Centre for Research and Development (NCBiR) under research grant no. SP/I/1/77065/10 SYNAT: "Establishment of the universal, open, hosting and communicational, repository platform for network resources of knowledge to be used by science, education and open knowledge society" as a part of a strategic programme for Research and Development: "The interdisciplinary system of scientific and technological information".

References

1. Grupa Technologii Językowych G4.19 Politechniki Wrocławskiej http://nlp.pwr.wroc.pl/takipi/
2. Wierzbicki, A.P., Nakamori, Y. (eds.): Creative environments issues of creativity support for the knowledge civilization age. Studies in computational intelligence, Vol. 59. Springer, Heidelberg (2007)
3. Wierzbicki, A.P.: The problem of objective ranking: foundations, approaches and applications. J. Telecommun. Inf. Technol. **3**, 15–23 (2008)
4. Wierzbicki, A.P.: On the role of intuition in decision making and some ways of multicriteria aid of intuition. Multiple Criteria Decis. Making **6**, 65–78 (1997)

5. Wierzbicki, A.P., Nakamori, Y.: Creative space: Models of Creative Processes for the Knowledge Civilization Age. Studies in Computational Intelligence, Vol. 10. Springer, Heidelberg (2006)
6. Antoniou, G., Van Harmelen, F.: A Semantic Web Prime. The MIT Press, Cambridge (2004)
7. Bishop, C.M.: Pattern Recognition and Machine Learning. Springer, Signapore (2006)
8. Broekstra, J., Ehrig, M., Haase, P., Van Harmelen, F., Menken, M., Mika, P., Schnizler, B., Siebes, R.: Bibster—a semantics-based bibliographic Peer-to-Peer system. In: Proceedings of the Third International Semantic Web Conference, Hiroshima, Japan, pp. 122–136 (2004)
9. Chudzian, C.: Ontology creation process in knowledge management support system for a research institute. Int. J. Telecommun. Inf. Technol. **4**, 47–53 (2008)
10. Chudzian, C., Klimasara, E., Paterek, A., Sobieszek, J., Wierzbicki, A.P.: Personalized search using knowledge collected and made accessible s model of ontological profile of the user and group in PrOnto system. SYNAT Workshop, Warsaw (2011)
11. Chudzian C., Granat J., Klimasara E., Sobieszek J., Wierzbicki A.P.: Personalized knowledge mining in large text sets. J. Telecommun. Inf. Technol. Nat. Inst. Telecommun. **3**, 123–130 (2011)
12. Chudzian, C., Klimasara, E., Sobieszek, J., Wierzbicki, A.P.: Wykrywanie wiedzy w dużych zbiorach danych: Analiza tekstu i inżynieria ontologiczna. Sprawozdanie prac PBZ Usługi i sieci teleinformatyczne następnej generacji—aspekty techniczne, aplikacyjne i rynkowe, grupa tematyczna i: Systemy wspomagania decyzji regulacyjnych: Wykrywanie wiedzy w dużych zbiorach danych telekomunikacyjnych. Instytut Łączności (2009)
13. Davies, J., Fensel, D., van Harmelen, F. (eds.): Towards the Semantic Web-Ontology-Driven Knowledge Management. John Wiley & Sons Ltd, England (2003)
14. Davies, J., Duke, A., Sure, Y.: Ontoshare—an ontology-based knowledge sharing system for virtual communities of practice. J. Univers. Comput. Sci. **10**(3), 262–283 (2004)
15. Dieng, R., Corby, O.: Knowledge Engineering and Knowledge Management: Methods, Models and Tools. Springer, Berlin (2000)
16. Ding, Y., Foo, S.: Ontology research and development. Part I—a review of ontology generation. J. Inf. Sci. **28**(2), 123–136 (2002)
17. Ehrig, M., Haase, P., Schnizler, B., Staab, S., Tempich, C., Siebes, R., Stuckenschmidt, H.: SWAP: Semantic Web and Peerto-to-Peer Project Deliverable 3.6 Refined Methods. (2003)
18. Ehrig, M., Tempich, C., Aleksovski, Z.: SWAP: Semantic Web and Peer-to-Peer Project Deliverable 4.7 Final Tools. (2004)
19. Ehrig, M.: Ontology Alignment: Bridging the Semantic Gap (Semantic Web and Beyond). Springer, New York (2006)
20. Ren, H., Tian, J., Wierzbicki, A.P., Nakamori, Y.,Klimasara, E.: Ontology construction and its applications in local research communities. In: Dolk, D., Granat, J. (eds.) Modeling for Decision Support for Network-Based Services, Springer, Heidelberg, Germany pp. 279–317, Lectures Notes in Business Processing, Vol. 42 (2012)
21. Sobieszek, J.: Towards a unified architecture of knowledge management system for a research institute. Int. J. Telecommun. Inf. Technol. **4**, 54–59 (2008)
22. Tian, J., Wierzbicki, A.P., Ren, H., Nakamori, Y.: A study of knowledge creation support in a Japanese research institute. Int. J. Knowl. Syst. Sci. **3**(1), 7–17 (2006)
23. W3C Semantic Web Best Practices and Deployment Working Group.: Ontology driven architectures and potential uses of the semantic web in systems and software engineering. http://www.w3.org/2001/sw/BestPractices/SE/ODA/ (2006)
24. Web-Ontology Working Group: OWL Web Ontology Language Guide, http://www.w3.org/TR/owl-guide/ (2004)
25. Gadamer, H.-G.: Warheit und Methode. Grundzüge einer philosophishen Hermeneutik. J.B.C. Mohr (Siebeck), Tübingen (1960)

26. Granat, J., Klimasara, E., Mościcka, A., Paczuska, S., Pajer, M., Wierzbicki, A.P.: Hermeneutic cognitive process and its computerized support. In: Bembenik, R., Skonieczny, Ł., Rybiński, H., Kryszkiewicz, M., Niezgódka, M. (eds.), Intelligent Tools for Building a Scientific Information Platform), Studies in Computational Intelligence, vol. 467, pp. 281–304. Springer, Berlin (2013)
27. Król Z.: The emergence of new concepts in science. Creative environments: Issues of creativity support for the knowledge civilization age. In: Wierzbicki, A.P., Nakamori, Y. (eds.) Chapter 17, Springer, Heidelberg (2007)
28. Tesseract-OCR http://code.google.com/p/tesseract-ocr/
29. TF-IDF http://en.wikipedia.org/wiki/Tf%E2%80%93idf
30. Jin, J., Nakamori, Y., Wierzbicki, A.P.: A study a multiatrribute aggregation approaches to product recommendation. J. Adv. Fuzzy Syst. **2013,** (2013)

Part IV
Implemented Systems

Further Developments of the Online Sound Restoration System for Digital Library Applications

Janusz Cichowski, Adam Kupryjanow and Andrzej Czyżewski

Abstract New signal processing algorithms were introduced to the online service for audio restoration available at the web address: www.youarchive.net. Missing or distorted audio samples are estimated using a specific implementation of the Jannsen interpolation method. The algorithm is based on the autoregressive model (AR) combined with the iterative complementation of signal samples. Since the interpolation algorithm is computationally complex, an implementation which uses parallel computing has been proposed. Many archival and homemade recordings are at the same time clipped and contain wideband noise. To restore those recordings, the algorithm based on the concatenation of signal clipping reduction and spectral expansion was proposed. The clipping reduction algorithm uses interpolation to replace distorted samples with the estimated ones. Next, spectral expansion is performed in order to reduce the overall level of noise. The online service has been extended also with some copyright protection mechanisms. Certain issues related to the audio copyright problem are discussed with regards to low-level music feature vectors embedded as watermarks. Then, algorithmic issues pertaining watermarking techniques are briefly recalled. The architecture of the designed system along with the employed workflow for embedding and extracting the watermark are described. The implementation phase is presented and the experimental results are reported. The chapter is concluded with a presentation of experimental results of application of described algorithmic extensions to the online sound restoration service.

J. Cichowski (✉) · A. Kupryjanow (✉) · A. Czyżewski (✉)
Multimedia Systems Department, Gdansk University of Technology,
Narutowicza 11/12, 80-233 Gdansk, Poland
e-mail: jay@sound.eti.pg.gda.pl

A. Kupryjanow
e-mail: adamq@sound.eti.pg.gda.pl

A. Czyżewski
e-mail: andcz@sound.eti.pg.gda.pl; ac@pg.gda.pl

R. Bembenik et al. (eds.), *Intelligent Tools for Building a Scientific Information Platform: From Research to Implementation*, Studies in Computational Intelligence 541,
DOI: 10.1007/978-3-319-04714-0_14,

Keywords Automatic audio restoration · Noise reduction · Impulsive distortions · Spectral signal expansion · Audio clipping

1 Introduction

In the chapter several improvements introduced recently to the online audio restoration service [1] were described. Those could be divided into two groups: modifications related to the restoration process and new directions of the watermarking applications.

Signal restoration especially for some type of distortions like impulsive noise or clipping is computationally demanding. If audio signal is stereo and has high sampling rate (e.g. 96 kS/s) then the amount of data that should be processed by the algorithm is large. Fortunately, nowadays a typical PC or a server CPU is equipped with more than one core. Therefore, the restoration procedure can be parallelized through dispersing of the signal operations into many threads. Such a solution was added to the engineered audio restoration web service (www.youarchive.net). In the experimental part of this chapter we have investigated the influence of the proposed parallel restoration schema on the algorithm efficacy.

The subject of data protection is very important in the digital era, the most popular approach for protection digital audio data involves watermarking. Various audio watermarking approaches have been proposed in the past and the significant companies related with computer technology and digital entertainment also are interested in watermarking algorithm applications. In particular, Sony, Apple and Microsoft developed commonly known mechanisms for digital rights management (DRM). Software package called Extended Copy Protection (XCP) [2] developed by Sony was the first DRM scheme for CDs used on a large scale. The software has been installed automatically on computers, therefore audio data genuity was checked at beginning of the audio material playback. Security researchers characterized the XCP software as both a "trojan horse" and a "rootkit", what led to civil lawsuit and criminal investigation which enforced Sony to discontinue use of their DRM system. Apple inserts watermarks in protected files as an indicator of legitimate content. Apple's QuickTime, iTunes and software built into iDevices looks for these watermarks in order to verify the authenticity of the digital content. This mechanism called Fairplay [3] is applicable to Apple devices only and to the derivative software. Above limitation partially led Apple to withdraw from this technology. On the other hand, Microsoft develops Windows Media DRM (WM-DRM) [4] services. The Janus DRM [5] algorithm is oriented towards portable devices and its modification called Cardea DRM is designed for network/streaming devices. Both algorithms protect audio visual content using Windows Media Player and some algorithms implemented in specific devices. WM-DRM is scheme which protects multimedia content stored as Windows Media Audio or Video (WMA and WMV). The software which enables striping DRM from files protected with WMDRM diminished popularity of this system. There are several standalone implementations of DRM mechanisms which provide licensed software for copyright protection. The commercial systems such as Varance [6] or Cinavia [7] or Audio Watermarking Toolkit (AWT) [8] provide out of the box DRM schemes. The co-existence of multiple closed and not unified DRM systems do not allow for applying a universal approach for data protection. The Coral Consortium [9] founded by Hewlett-Packard, InterTrust, Philips, Panasonic, Samsung, Sony and Twenty Century Fox works on an architecture whereby devices using different DRM technologies will be able to join a domain that allows them to exchange protected content securely.

The lack of universal DRM solution and the experience gained by the authors during previous researches founded the motivation to develop an autonomous DRM solution focused on the audio copyright protection in multimedia repository. An algorithmic background, system applications and the results obtained from the experiments performed are described in the following sections following the subject of new signal restoration algorithms introduced to the online system launched previously [1].

2 Restoration Algorithms

In this section two algorithms, which were recently added to the online audio restoration system, were presented. These are: clicks reduction algorithm and clipping reduction algorithm. Both methods are computationally demanding, therefore we have proposed an implementation supported by the parallel signal processing. This approach reduces the time duration of audio restoration procedures.

2.1 Impulsive Noise (Clicks) Reduction

Clicks in audio signal are related to the short, local discontinuity of the signal. Their duration should not be longer than 1 ms. It is assumed that in the recording at most 10 % of samples could be affected by the clicks [10]. This type of distortion is typical for gramophone discs and for optical soundtracks recorded on film tapes. The origin of the distortion could be related with dust, dirt and scratches. Clicks distortion is additive and could be described by the formula (1):

$$y(n) = x(n) + i(n) \cdot q(n) \quad (1)$$

where $x(n)$ is the undistorted signal, $y(n)$ is the distorted signal, $q(n)$ is the noise signal and $i(n)$ is the binary vector containing positive values in the clicks positions. In our implementation we have used fully automatic algorithm which operation is divided into three steps: (a) distorted samples detection, (b) distorted samples removal, (c) replacement of removed samples by interpolated ones.

Clicks detection. The simplest solution for the clicks detection is to perform a threshold analysis of the high-pass filtered signal. Unfortunately this method is not capable to detect the distortion when the input signal contains frequency components belonging to the upper frequency band. Therefore in the implementation we have used algorithm based on the AR (autoregressive) model. In this approach a sound signal is modeled as an autoregressive process according to the Eq. (2):

$$x(n) = \sum_{k=1}^{P} a_k x(n-k) + e(n) \tag{2}$$

where $\{a_k, k = 1 \ldots P\}$ are the coefficients of the all-pole filter, P is the order of the filter and $e(n)$ is the excitation signal (also called residual). For the distortion detection, the excitation signal is used. When the signal is undistorted, $e(n)$ has characteristic similar to the white noise. To get the residual signal from the input signal an inverse filter, defined by the formula (3), should be used:

$$A(z) = 1 - \sum_{k=1}^{P} a_k z^{-k} \tag{3}$$

Clicks are detected using the threshold analysis of the residual signal. Samples exceeding the threshold are marked as distorted ones. Unfortunately, such a type of algorithm is not much efficient. Therefore, Esquef et al. [11] have proposed the iterative double threshold analysis of the excitation signal. Their method provides a higher accuracy of the detection and consequently it improves the restoration quality. In the proposed restoration framework we have implemented the Esquef's algorithm. In Fig. 1 a block diagram of the iterative clicks detection algorithm was presented. In every step of the procedure an excitation signal is calculated and the detection procedure is performed. Then values of the samples detected as clicks are interpolated (by the interpolation algorithm which is presented in the next section) and distorted samples are replaced by the recovered ones.

In every step a double threshold analysis of the excitation signal is executed. First, the threshold Th_d is responsible for the clicks position extraction and then Th_w is used to determine its width. The detection procedure is as follows: in the excitation signal samples exceeding the value Th_d are found; in the neighborhood of the detected clicks samples greater than Th_w are marked as a part of that click distortion. Values of the thresholds (in every iteration) are being decreased accordingly to the routines described by the following Eqs. (3–5):

$$Th_d(i) = K \cdot \sigma(e_i(n)) \tag{4}$$

$$Th_w(i) = b_i \cdot Th_d(i) \tag{5}$$

$$b_i \triangleq r^{\left(\frac{i}{f}\right) \cdot b} \tag{6}$$

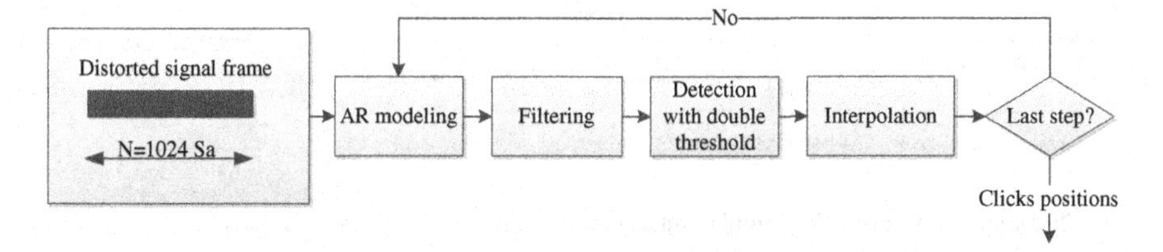

Fig. 1 Block diagram of the implemented clicks detection algorithm

where $Th_d(i)$ is a value of threshold in i-th step, $Th_w(i)$ is the threshold for the clicks width detection in the i-th step, b_i is the value calculated according to the formula (6), K, r, f are constant values selected experimentally and $\sigma(\cdot)$ is the standard deviation.

Clicks reduction. In the reduction step information about clicks positions, provided by the clicks detection algorithm, is utilized. The restoration procedure is executed only for the samples which were recognized as distorted ones. These sample values are treated as missing and replaced by the values obtained from the interpolation. In this implementation, for the sample values interpolation, the algorithm proposed by Janssen [12] was used illustrated in Fig. 2. This algorithm is based on the AR modeling and on an iterative samples estimation. The signal processing is performed in the time-domain. Originally authors of that method have proposed to use 100 iterations during the restoration. In every step a new filter, based on the AR modeling, is calculated, signal frame is filtered and the missing samples are replaced with the interpolated ones. Then the AR model is calculated for the signal obtained in the previous step and interpolated values are reestimated. As the result the estimated sample values converge to the original ones. Since for every signal frame, 100 times AR model is estimated, the computational complexity of the interpolation is high. Additionally it depends also on the order of the AR model. Originally it was proposed to use $p = 3\ m + 2$ where m is the number of missing samples in the signal frame. Therefore, the computational complexity is also related to the number of distorted signal samples. More details related to the interpolation procedure could be found in the Janssen's chapter [12].

2.2 Clipping Reduction

Clipping is one of the most often found inconveniences in the amateur digital recordings, since many of them are made in the noisy environments (e.g. concerts, public speeches) or with the microphone located close to the speaker's lips. In both situations these circumstances cause the overdrive of the recording device. The signal clipping could also occur in the old archival recordings, but in this case the characteristic of the distortion is usually different.

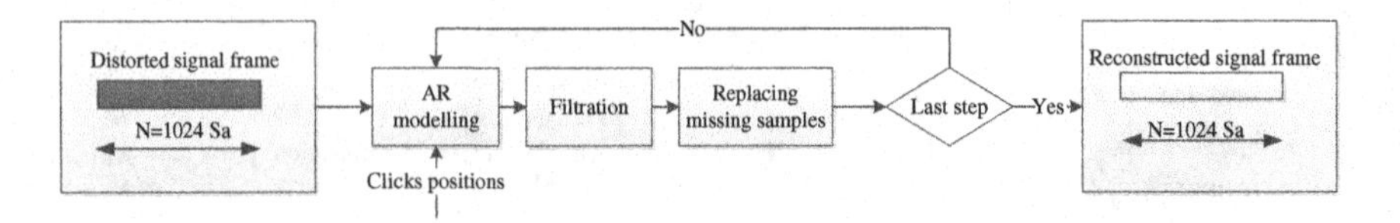

Fig. 2 Block diagram of the samples interpolation algorithm

The proposed algorithm operates in the time-domain. It is based on the threshold analysis of a distorted signal. As a clipped sample the algorithm considers a sample amplitude value which simultaneously:

- equals the maximum sample value in the distorted recording,
- differs between the preceding sample by less than a defined threshold and it fits to a predefined range of maximum–minimum $\pm\delta$.

The first assumption enables to detect single values of amplitudes which have exceeded the allowed signal range. The second one helps to detect a typical digital overdrive and also analog clipping where the distorted signal fluctuates in the peak location (i.e. the peak is not flat). An example of the analog clipping was presented in Fig. 3 (the ellipse marks the fluctuating peak).

In the restoration step all distorted samples are treated the in the same way as in the clicks reduction algorithm i.e. they are being replaced with the interpolated values. The signal restoration is then performed using the Janssen's algorithm.

2.3 Parallel Signal Restoration Schema

As it was mentioned, the interpolation algorithm used for the signal restoration (for clicks and clipping reduction) is computationally demanding. Moreover, the detection algorithm presented in Sect. 2.1 also uses a interpolation algorithm (several times, iteratively, for each frame of the signal). Therefore, we have proposed to use here parallel restoration schema which should be more efficient. Since for both algorithms (clicks and clipping reduction) the whole restoration process could be performed locally i.e. for every frame separately without using the knowledge on the other signal frames content, thus our concept of the parallel restoration can be based on the frame-by-frame processing scheme. Owing to that issues related to the signal division and concatenation may be omitted. In Fig. 4 a typical (serial) restoration layout was depicted. In such a framework the input frame of the distorted audio signal is analyzed, in terms of the distortion presence, and then it is restored based on the detection results. This approach is not optimal in the efficacy sense, since every signal frame is autonomous here.

In Fig. 5 the proposed parallel restoration schema was presented. A signal frame which is k-times longer than in the typical framework makes an input for the algorithm. Dependently on the analysis frame hop length, k is an integer (hop

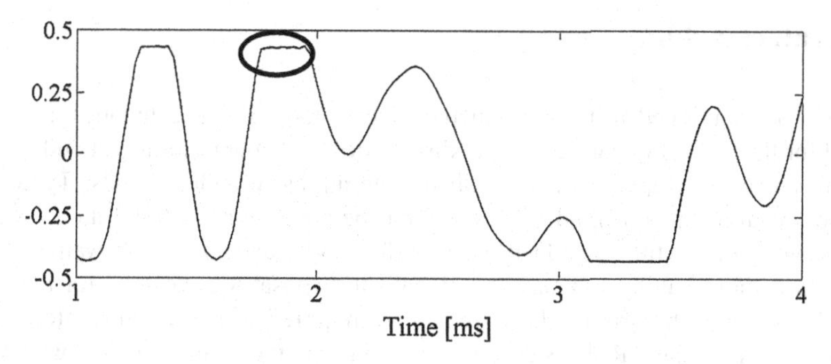

Fig. 3 Waveform of the signal distorted in the analog domain

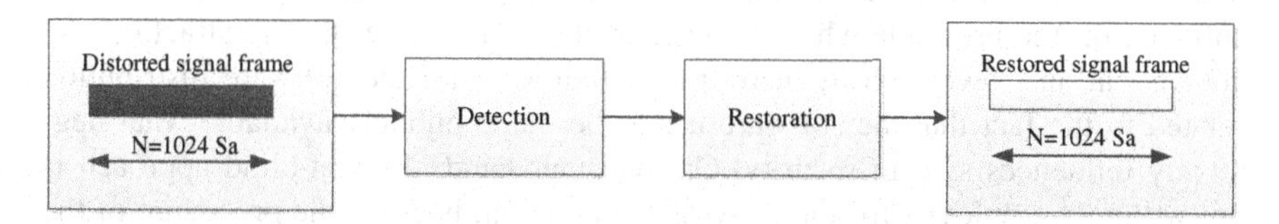

Fig. 4 Block diagram of typical restoration schema

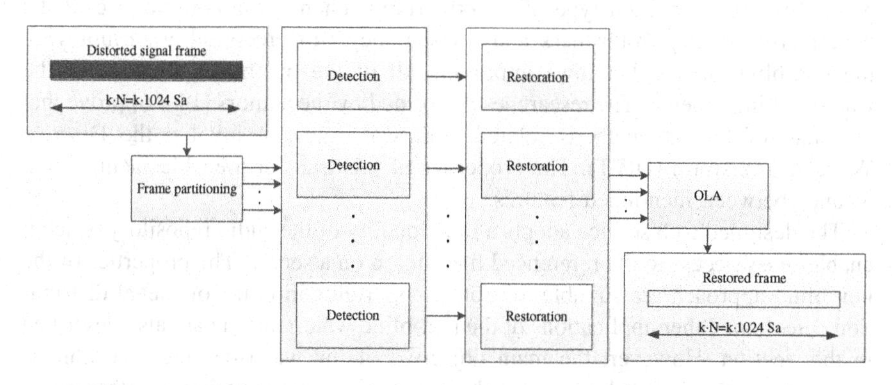

Fig. 5 Block diagram of parallel restoration schema

length equals to frame length) or a real-type value (for the hop length lower than the frame length). In the first step of the processing, a signal frame of the length equal *kN* is partitioned into *h* subframes of the length *N*, where *h* is the number of threads that can be used during the restoration and *N* is the length of the frame used by the restoration algorithms. In the second step a detection procedure is performed in parallel for *h* threads. Then distortion positions and subframes are processed by the restoration algorithms. At the end subframes are concatenated with the OLA (OverLap and Add) procedure [13]. It should be mentioned that if the hop or the frame length used by the detection algorithm is different from these values for the restoration algorithm, then above two steps should be performed separately for the whole signal.

3 Watermarking

As it was mentioned in the Introduction, the watermarking techniques are often used for the protection against illegal distribution and counterfeiting of audio data. There are several approaches to embed watermarks in audio signals. Typically, those are divided into blind and non-blind methods. For the blind audio watermarking methods, the embedded watermarks can be extracted from watermarked signals without having original host signals at disposal. Conversely, for the non-blind approach, the original host signals are required to extract the watermarks. Present commercial DRM systems generally base on the blind approach, where the watermarked signal is analyzed and the watermark is extracted. If system cannot identify the marker, the audio playback is being locked. This approach requires software on the user side which is responsible for the watermarks extraction and authentication. Nevertheless, there are difficulties with the software distribution related to the fact that the software has to be made publicly available what negatively influences system security. On the other hand, the non-blind approach is not willingly employed for the copyright protection, because the procedure of the watermark extraction requires possessing of the reference original host signal. It should be noted that audio watermarking techniques were developed for various goals. In particular, each type of watermarking scheme is a compromise of the watermark capacity, robustness and transparency (imperceptibility). Employing the non-blind approach allows improving all of the mentioned features of the watermarking scheme. The researches performed by the authors [14–16] prove that the state of the art methods related with watermarking involve the Discrete Wavelet Transform (DWT). The properties of this transform enables achieving a balance between mentioned features.

The designed web service adopts a functionality of the audio repository, since it enables easy access to the referenced files stored on a server. The properties of the non-blind approach are suitable to protect copyrights in terms of illegal distribution detection. Other applications of the non-blind watermarking are also described in this section. However, the main objective of the watermarking algorithm is detection of illegal distribution, thereby the audio signal containing watermark is transmitted to the user. The audio signals quality is crucial, since the main function of the designed web service is sound tracks restoration. Consequently, it would be unacceptable that after a signal quality improvement, obtained during the restoration, the audio quality is reduced by the watermarking algorithm operation. The research was oriented on the implementation of the watermarking algorithm focused on the intellectual property protection in the audio domain. It was also important to evaluate its efficiency in terms of preserving the audio signal quality.

3.1 Principles of Embedding Algorithm

The watermarking embedding takes place in the DWT domain. The original audio signal is divided into non-overlapping frames containing 1024 samples each. The frame can also have 256, 512 or 2048 samples. A longer frame would lead to less number of artifacts generating during the watermark embedding, but at a cost of the smaller watermark bitrate achieved. Each frame contains a single bit of the watermark. The watermark is embedded in the low frequency of the second level of DWT transformation calculated for the signal frame. The sample values in the frame are modified according to the proposed formula (7):

$$s^2_{LP_{wat}}[n] = \begin{cases} 0.9 \cdot s^2_{LP_{org}}[n] + \alpha \cdot \left| s^2_{LP_{org}}[n] \right| & \Rightarrow bit = 1 \\ 0.9 \cdot s^2_{LP_{org}}[n] - \alpha \cdot \left| s^2_{LP_{org}}[n] \right| & \Rightarrow bit = 0 \end{cases} \tag{7}$$

where $s^2_{LPwat}[n]$ is the modified value of the second level low-pass DWT component, $s^2_{LPorg}[n]$ is the original value of the second level low-pass DWT component, α is the watermarking strength and *bit* is the bit value of the embedded information.

The parameter α represents watermarking strength. It is inversely proportional to the fidelity of the watermarked file. This parameter is understood as the similarity of the content of the watermarked file to the source file. α values are in the range from 0.01 to 0.1. The procedure of watermark embedding is the following; current audio sample amplitude after the second level of DWT decomposition is multiplied by the factor 0.9 to avoid overdriving of the audio signal track caused by watermarking. The audio signal is modified using the result of multiplying α (strength parameter) by the current decomposed audio sample amplitudes, these result should fall into the range of (0.0–0.1). In case of embedding, the current bit of the watermark binary sequence equals to 1, and the result of multiplication is added to the rescaled current sample value, contrarily if the bit of the watermark binary sequence is equal to 0, the result of multiplication is subtracted from the rescaled current sample value. The high-pass components $s^{level}_{HP}[n]$ computed in the first and in the second level of DWT are not modified during embedding procedure, thus they both are used to transform signal back from the wavelet domain to the time domain.

After the embedding process a frame is transformed back into the time domain using Inverse Discrete Wavelet Transform (IDWT) and is added into the output stream as subsequent frame. A general scheme of the watermark embedding procedure is shown in Fig. 6 [14].

The presented approach uses stereo or mono input signals. The algorithm embeds the same watermark information in both channels of a stereo signal, which entails a redundancy. The mutual dependence of the signal frame length and the watermark bitrate is shown in Table 1.

The frame size of 1024 samples for 44.1 kS/s sample rate is a compromise between the high bitrate and the quality degradation. The length of the watermark

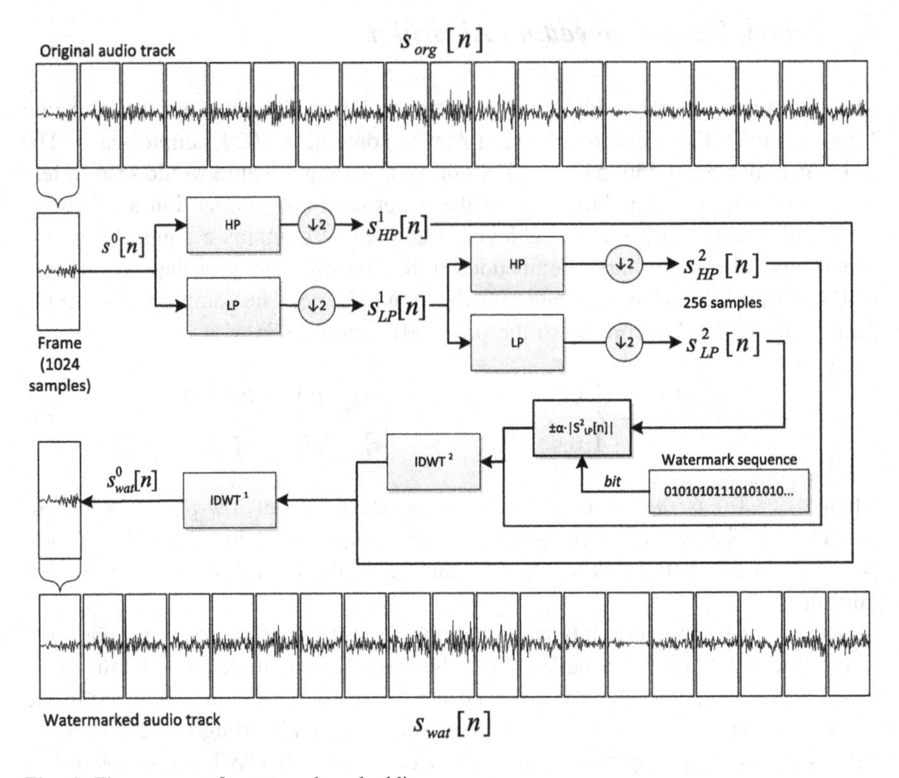

Fig. 6 The process of watermark embedding

Table 1 Dependence of frame length, watermark bitrate and time capacity

Frame length [samples]	256	512	1024	2048
Watermark bitrate [bit/s]	172	86	43	21

inside the audio signal depends on sampling rate and on the assumed processing frame size. If the audio signal capacity allows for the storing of the watermark sequence more than once, than watermark is embedded repeatedly while audio signal analysis is finished.

3.2 Principles of Extraction Algorithm

The non-blind extraction mechanism requires availability of the original and the watermarked signals, to extract the hidden information. The disadvantage of the need of providing the original signal for watermark extraction is recompensed by a high imperceptibility and a high capacity of the watermark. These properties are

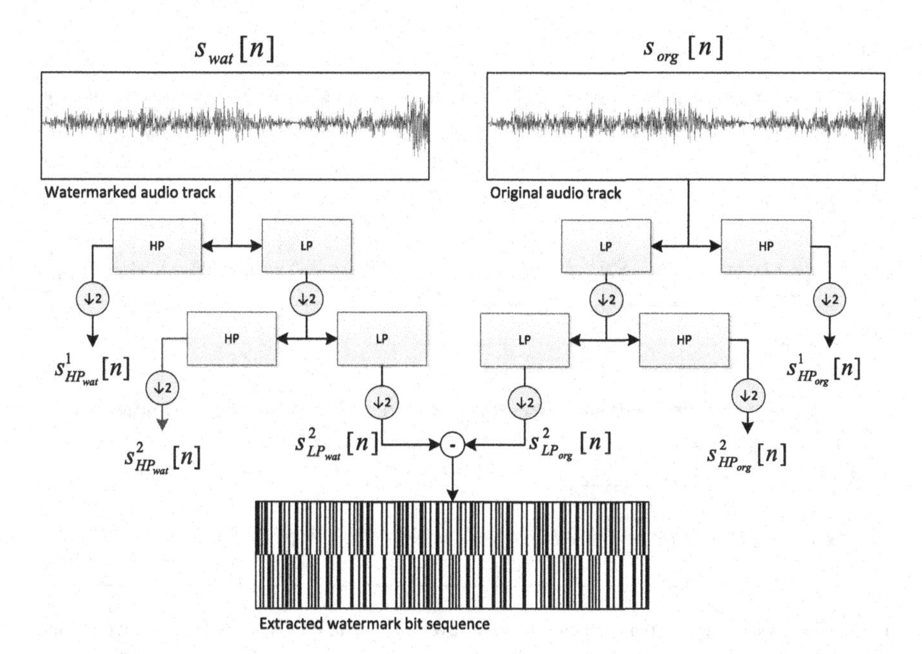

Fig. 7 Watermark extraction process

desirable in audio applications. Moreover, the watermarking strength can be maintained adequate, so it does not introduce any audio quality degradation.

The watermark extraction scheme is shown in Fig. 7 [14]. For the extraction of the watermark two audio tracks are required: the original uncompressed one (S_{org}) and the watermarked (S_{wat}). If the watermarked file is to be converted into a different format then some preprocessing must be performed, first.

The difference between the low frequency parts of the second level of DWT of two files represents the watermark signal. The mathematical formula of it is presented by formula (8). The obtained difference signal is examined in order to find the expected beginning sequence of the watermark and afterwards it is divided into non-overlapping frames containing 256 (quarter of 1024) samples each. From each frame one watermark bit is extracted. The signal differences are summed up within the whole frame. Finally, the binary sequence is translated into text and it can be presented in a human-readable form.

$$bit = \begin{cases} 0, \Rightarrow t \leq \sum_{n=0}^{255} s_{LP_{wat}}^{2}[n] - 0.9 \cdot s_{LP_{org}}^{2}[n] \\ 1, \Rightarrow t > \sum_{n=0}^{255} s_{LP_{wat}}^{2}[n] - 0.9 \cdot s_{LP_{org}}^{2}[n] \end{cases} \tag{8}$$

where t is thresholding level.

Results of thresholding the signal containing differences between the original and watermarked audio streams are presented as a pseudo binary sequence in the

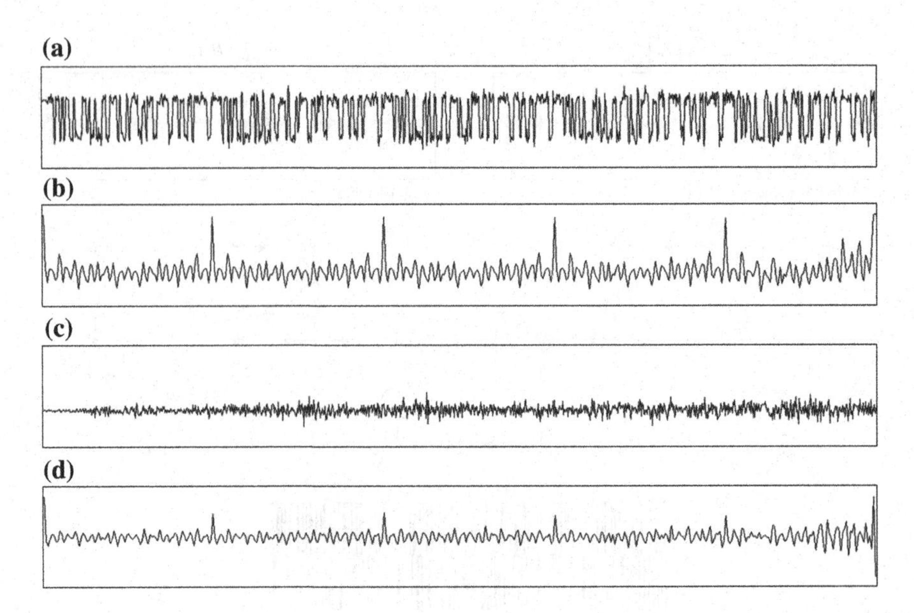

Fig. 8 The watermark signal extracted from: **a** raw audio data, **c** D/A → A/D audio data; **b** and **d** are the corresponding detector responses

bottom part of Fig. 7. The procedure is focused on extracting of the watermark insertion signature, which is embedded before the beginning of each binary sequence of the watermark. Information about the location of the insertion sequence allows for identifying the first bit of the watermark which is required for the proper watermark extraction. The analysis and watermark extraction applied to the unprocessed audio data are errorless, because any additional modifications besides watermarking were not performed. Nevertheless, the variety of possible audio signal modifications encountered in real life, force the necessity of application of the error-prone watermark extraction. Errors at the insertion signature cause the problem with finding the first bit of the watermark sequence. In case of error occurrence in the watermark sequence there is an increased possibility of false positive detection of the beginning of the watermark. The problem with the detection of the first bit of the watermark was solved using the unified correlation function computed using the extracted watermark sequence and the fixed watermark preamble binary representation. The highest similarity between the extracted binary sequence and the preamble binary sequence is always searched for. The preamble is added as binary delimiter between subsequent watermark sequences, it serves also as an indicator of the watermark beginning. Moreover, in case of heavy signal degradation, the correlation function enables detecting watermark inside the audio signal even if the extraction of bit brings bit error rate around 40 %, as it was discovered experimentally by the authors. An example of results of the analysis for the non-attacked watermark sequence is presented in Fig. 8a, the same procedure of extracting the watermark was done for the signal shown in Fig. 8c, after applying the D/A → A/D conversion. The watermark detector responses obtained using the correlation function are presented in Fig. 8b for the original signal and in Fig. 8d for the attacked signal. The audio sample containing classical music was used in the experiment.

The waveforms presented in Fig. 8a, c are significantly different, but both contain the same watermark. The implemented detection algorithm is able to verify the watermark presence inside audio signals even if error rate is not acceptable for the watermarked data recovery. The peaks visible in Fig. 8b, d correspond to the detector positive response. Information about the peak localization and distance between subsequent maxima are sufficient for an accurate watermark validation. There is more than one peak visible, since watermark was embedded repeatedly. The distance between peaks is strongly related with the watermark embedding capacity. The peaks presented in Fig. 8b are significantly greater than extremes observed in Fig. 8d, thus it could be concluded that the D/A → A/D conversion influences negatively the detector reliability. However, the signal degradation does not disable detector completely, since some extremes are still clearly visible.

3.3 Applications of Watermarking in Sound Restoration Service

Each audio watermarking solution creates an additional communication channel employing the host audio signal as a data carrier. The described algorithm was designed as a copyright protection tool for the audio repository. During the development stage many use cases were considered. Several approaches were tested and verified. Brief explanation of the actually developed watermarking applications is given in the following sections.

Digital Rights Management. The created service enables the logged-in user to become acquainted with the practical applications of digital signal processing algorithms—restoration of audio archival recordings. In order to facilitate the restoration procedure, it is necessary to design algorithms for automatic audio content quality assessment and restoration. Such solutions could help to manage efficiently the archive repository. By management one should understand a watermarked service which protects website user's original and restored audio files. The non-blind approach determines that data protection is realized in the form of the detection of an illegal distribution. Audio files downloaded by users are watermarked in the service endpoint. The watermark consists of information necessary to identify the source of multimedia content leakage. The watermark is constructed with fixed binary preamble which contains the web service identifier. The most important data in watermark correspond to: user ID who uploads file to repository, timestamp of uploading, user ID who downloads data from repository and timestamp of downloading. If the suspected file returns back to service with a request of watermark verification, the system recovers users' ID and timestamps. Additional personal data such as: email or IP address may be also read from the database of the service. A comprehensive description of this use case is available in the earlier authors' article [15].

Music Information Retrieval. A more sophisticated services try to provide upgrades that enable users to browse the audio repository using some novel approaches, such as: QBE (Query-by-Example) [17], QBH (Query-by-Humming) [18] or QBM (Query-by-Mood) [19]. MIR (Music Information Retrieval) systems transform the time or frequency domain into fuzzy semantic terms. The repository may also be explored using subjective expressions instead of strictly defined tags (i.e. titles, authors, albums or other). The set of available phrases is usually finite, thus searching is conducted using semantic keywords. An appropriate categorization enables the assignment of audio tracks to adequate membership groups (the groups are linked with some specific sets of phrases). The manual classification of audio tracks is a time consuming and error prone process. In addition, because of the high redundancy of audio signals, parameterization is applied to automate classification procedures. Different types of parameters are extracted from audio files and then crucial features are organized in a vector. Instead of listening to the music and tagging its content (or performing waveform analysis), the feature vector enables the classification of the audio track employing decision systems. Therefore, the main objective of the proposed approach is to embed the feature vector as a watermark inside the audio signal. The extracted feature vector enables the retrieval of audio content—only the encapsulation of a watermark is necessary. In general, any modification of a signal causes a quality degradation. Results of related research and explanation of the proposed use case are more thoroughly described in a previous chapter [14].

Restoration History Container. The principle of the application of watermarking embedding in the digital repository is related to copyright protection. In particular, in such a case the watermarking is applied externally and fingerprinted audio is transmitted outside the service. The novel direction of research is oriented on internal watermarking solutions such as low-level features vectors embedding [14], mentioned previously. On the other hand, the management of restored data requires collecting restoration parameters together with references to the restored audio files. The presented web service enables to perform multiple restorations, what enforces a data acquisition and generating queries to the databases.

The proposed use case assumes that the restoration software extracts watermarks as a part of preprocessing and embeds an updated one in the postprocessing routine. Establishing any additional connection to database is not necessary in this case. Because audio tracks are stored in the same storage space with the restoration software, it is possible to use the same non-blind architecture. If the track is restored for the first time, in the preprocessing step no data are recovered. In case of multiple restorations, firstly watermark is extracted, from which information about the recent restoration is recovered. Eventual restoration iteration may employ previously used parameters to improve the quality of restoration i.e. a different algorithm may be used to improve distortions detection quality.

The proposed watermarking functionality is comparable to the metadata container. An internal usage of this functionality enables to predict a set of factors having a possible negative influence on the watermarked data caused by the restoration algorithms application. There is a high probability that the watermarked audio track will be restored more than once, in such case restoration service may depreciate the embedded watermark. The described restoration algorithms are the threats for the embedded watermark, the restoration procedures may delete a part of watermarked data. The influence of restoration procedure on watermark robustness is presented in the following section.

4 Experiments

In this section experiments devoted to restoration and watermarking algorithms assessment, were described. The efficacy of the parallel restoration procedure was examined in order to prove its usability. Then the quality and robustness of the proposed watermarking method were investigated.

4.1 Parallel Restoration Procedure Assessment

In this experiment the efficacy of the proposed parallel restoration schema was examined. The experiment was carried out on a computer running Windows 7 OS, equipped with the i7-2630QM processor (4 cores/8 threads) and 8 GB RAM. During the evaluation the duration of restoration process was measured. It was done in order to find a relation between the number of parallel tasks (further referred to as threads) and the reduction of the processing time. The proposed schema was tested for two algorithms: clicks and clipping reduction. Additionally two types of distorted signal were analyzed: speech and music. A monophonic speech signal was recorded with the sampling rate equal to 22.05 kS/s and its duration was 44.49 s. The music was recorded in stereo with the sampling rate equal to 44.1 kS/s. The duration of the music signal was 57.27 s. Distortions in both recordings were simulated (clicks/clipping). Clicks were added according to the Eq. (1). Three values of clicks saturation were considered: 0.5, 1 and 5 %. The clipping was simulated for four different values of clipping level: 0.9, 0.7, 0.5, and 0.3.

In Table 2 the time-duration of the clicks restoration process, obtained for different number of threads, was depicted. In brackets, ratios between the length of the recording and the time-durations of the restorations were shown. This measure illustrates the capability of real-time processing of the analyzed algorithms. If its value is equal or higher than one, then the algorithm enables a real-time signal restoration. In the last column the improvement of the efficacy was depicted. It was calculated as the ratio between the restoration time obtained for one thread and the duration of the processing for 8 threads. It can be seen that when the clicks saturation

Table 2 Clicks reduction duration achieved for different numbers of threads

Distortion saturation (%)	No. of threads				
	1	2	4	8	Improvement
Speech					
0.5	251.59 (0.18)	158.98 (0.28)	110.10 (0.4)	76.80 (0.58)	3.28
1	279.67 (0.16)	173.38 (0.26)	118.26 (0.38)	81.69 (0.54)	3.42
5	417.48 (0.11)	243.74 (0.18)	153.73 (0.29)	109.72 (0.41)	3.80
Average	316.25 (0.15)	192.03 (0.24)	127.36 (0.36)	89.40 (0.51)	3.50
Music					
0.5	1087.18 (0.05)	721.70 (0.08)	456.45 (0.13)	322.04 (0.18)	3.38
1	1245.57 (0.05)	794.8 (0.07)	511.89 (0.11)	345.69 (0.17)	3.60
5	2660.17 (0.02)	1531.23 (0.04)	902.32 (0.06)	677.96 (0.08)	3.92
Average	1664.31 (0.04)	1015.91 (0.06)	623.55 (0.1)	448.56 (0.14)	3.71

is low (0.5 %), the processing time is lower than for higher saturations. This phenomenon is related to the principles of the detection and restoration algorithms. Both perform a interpolation of the samples which were recognized as clicks, therefore for a higher saturation of clicks more samples have to be interpolated. The average efficacy improvement achieved for the speech was 3.5 and for music 3.71. It can be expected that this value will be close to 8 (not 4), since the improvement was calculated in the relation to the duration of processing performed using 8 threads. The reason of this discrepancy is the fact that the CPU of the computer could handle 8 threads but it has only 4 cores. Additionally, the improvement is also reduced by the algorithm computation overhead. It is important that the restoration could not be performed in real-time for this hardware configuration. Therefore a greater number of threads is needed in order to restore signal in real-time.

In Table 3 efficacy assessment results obtained for the clipping reduction algorithm are presented. Comparable to the clicks reduction, the processing time increases with the number of distorted samples. But this algorithm performs real-time signal restoration (it was proven possible when the clipping level is lower than 0.5). Naturally, this capability is not only related to the clipping level but also to the content of the recording. Another interesting observation is that for a low distortion levels (0.7 and 0.9), the usage of the parallel processing did not improve the efficacy. This fact is related to the algorithm overhead—when there are a few samples that should be interpolated, the division and concatenation of the signal frames is more computationally demanding than the interpolation process.

4.2 Watermarking Quality Assessment

In this section subjective and objective quality of the proposed watermarking method were investigated. Additionally watermarking robustness against various methods of attacks was examined.

Table 3 Clipping reduction duration achieved for different numbers of threads

Distortion saturation (%)	No. of threads				
	1	2	4	8	Improvement
Speech					
0.3	89.82 (0.5)	49.38 (0.9)	32.35 (1.38)	24.51 (1.82)	3.66
0.5	24.38 (1.82)	15.15 (2.94)	10.13 (4.39)	7.91 (5.62)	3.08
0.7	6.80 (6.54)	4.96 (8.97)	3.63 (12.26)	2.83 (15.72)	2.40
0.9	1.22 (36.47)	1.47 (30.27)	1.14 (39.03)	1.02 (43.62)	1.20
Average	30.56 (1.46)	17.74 (2.51)	11.81 (3.77)	9.07 (4.91)	3.37
Music					
0.3	737.79 (0.08)	479.45 (0.12)	325.20 (0.18)	225.45 (0.25)	3.27
0.5	91.78 (0.62)	77.25 (0.74)	52.93 (1.08)	39.22 (1.46)	2.34
0.7	5.90 (9.71)	12.32 (4.65)	8.63 (6.64)	6.89 (8.31)	0.86
0.9	1.19 (48.13)	8.28 (6.92)	5.76 (9.94)	4.48 (12.78)	0.27
Average	209.17 (0.27)	144.33 (0.4)	98.13 (0.58)	69.01 (0.83)	3.03

Subjective Tests. Signal degradation can be assessed through a variety of evaluation techniques including subjective quality measures. Thus, audio signals with the watermark added were tested subjectively. A group of experts had to choose a better audio recording (in the terms of a better audio quality and an absence of negative artifacts such as noise or clicks) from a pair of audio samples, i.e. the original and the watermarked ones. There were 20 experts employed, and two test series for three values of α formula (8) were performed to enhance a reliability of the experiment. Each series contained two pairs of audio samples corresponding to specific music genres. If the watermarked file was assessed better than the original one, then one point was added to the pool, otherwise one point was subtracted. The final result for a specific genre shows differences between audio tracks that were compared. A positive value means that the watermarked file was assessed better than the original file, while a negative value signifies that the watermark was audible and detected by the expert. A value close to zero means that the watermark was not detected and the audio files were comparable in terms of their quality. The aim of experiments was also to compare the influence of the watermarking strength for three values of α parameter, namely: 0.01, 0.05 and 0.1. The obtained results are presented in Fig. 9.

It can be seen that the embedded watermark is not always perceptually transparent. Moreover, the results differ for various music genres. The degradation of quality caused by the watermarking differs for various cases and depends on both music genre and on the watermarking strength. The smallest value of α parameter does not alter the signal quality. Obtained results are related to the Gaussian normal distribution, results being out of the range of double standard deviation were ignored., since the experts assessed each audio sample similarly. The audible distortions were not reported for the watermarked recordings. The resulting sum of ratings has a positive value for each music genre. Increasing the watermarking strength from 0.01 to 0.05 led to a positive degradation rate for *Classical* genre.

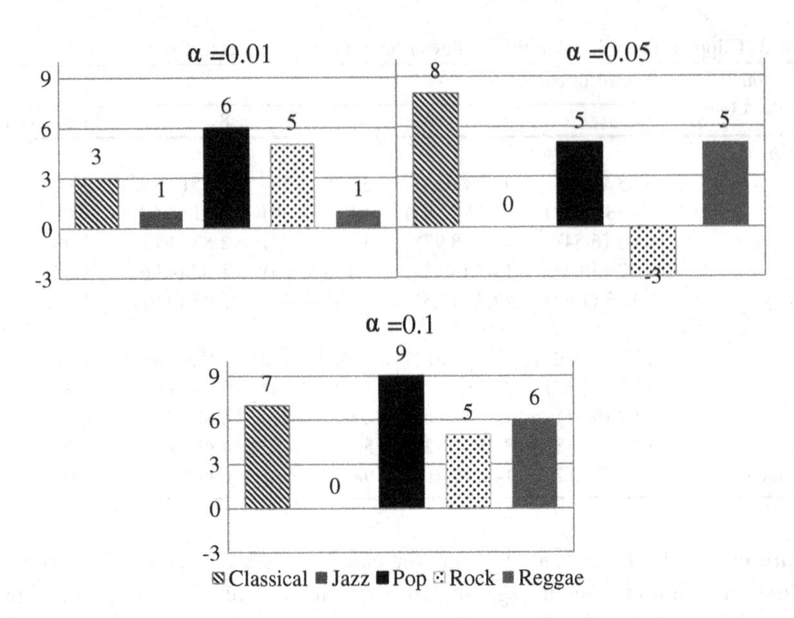

Fig. 9 Results of the subjective tests obtained for equals to 0.01, 0.05 and 0.1

This occurred probably due to the slight audio signal colorization, which may have a positive impact on the audio quality, especially for the *Classical* music category. Other recordings were assessed with the absolute value of notes being less or equal to 6, considering the 120 test series, the obtained results are not conclusive. In the last part of the experiments, some larger values of α (0.05 and 0.1) were examined. Such a modification brought a difference in the experts' judgments. Tracks representing *Classical* and *Pop* genres were assessed higher than other ones. Nevertheless, it is surprising that *Jazz* was assessed equally for different watermarking strength. It may be concluded that a signal quality depends on genre and on the dynamics of the music excerpt. Audible distortions, caused by the watermarking, are not always noted negatively by the listeners. Obviously, the strength of the watermarking embedding procedure is crucial for audio quality, thus a greater value of α means a higher audio quality degradation.

Objective Analysis. To characterize the perceived audio quality as experts would do in listening tests, the PEAQ (Perceptual Evaluation of Audio Quality) [20] analysis that enables a comparison of audio signals based on the human hearing perceptual model was applied. Moreover, the perceptual audio quality degradation caused by the proposed algorithm was compared mutually for different music genres. The result of this comparison is presented in the ODG (Objective Difference Grade) scale. Values of the ODG may range from 0.0 to −4.0, where 0.0 signifies that signals are identical (without differences), and −4.0 means that differences between signals are not acceptable. The measurements were performed employing the advanced mode of PEAQ and the Opera software [21]. The results of the PEAQ analysis are presented in Fig. 10.

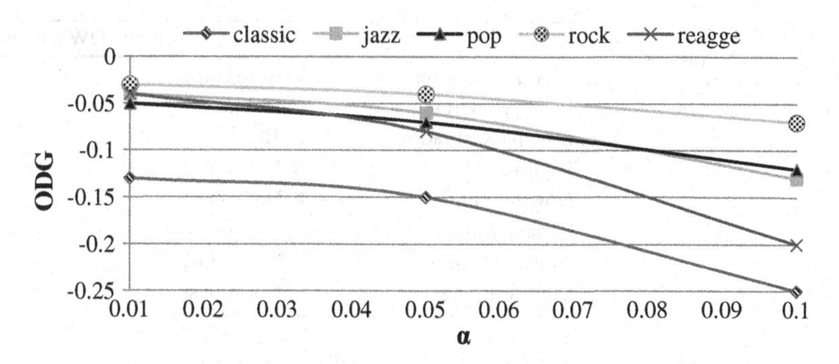

Fig. 10 Results of PEAQ analysis for music genres

The PEAQ analysis confirms the results of the previous subjective assessments only partially. It proves that the quality of music watermarked with the proposed method depends on the music genre. The highest quality degradation occurs for the *Classical* genre. The differences between signals for *Jazz* and *Pop* are comparable. The least degraded are music excerpts belonging to the *Rock* genre. The *Reggae* degradation lies in the middle between the results obtained for *Classical* and *Pop* genres. The quality degradation for music genres caused by the proposed algorithm falls into the range of (−0.03 to −0.25), meaning that the quality was not reduced, practically. It is due to the fact that the proposed watermarking algorithm has a particularly negative impact on the music tracks which have harmonic structure such as *Classical* music.

Watermark Robustness. The tracks obtained in previously described experiments were analyzed accordingly to the errorless watermark extraction after signal modifications. There are several possible attacks that can affect the robustness of the watermark, i.e. resampling, filtration, time warping, cropping, lossy compression, etc. All types of unexpected processing may pose a threat to the extraction of a bit sequence hidden in the embedded watermark. Simple modifications of audio signals are treated as potential watermark attacks described in the literature [22]. A possible processing is not always aimed at the watermark deletion or corruption. However, either intentional or unintentional, the modifications are sufficiently dangerous and can lead to a watermark degradation. The results of the simulations are presented in Table 4.

Simulations of attacks show that the proposed watermarking scheme is not robust to all types of the modifications. The lossy audio compression and low-pass filtration do not affect the watermark harmfully, while the other types of modifications are sufficiently destructive to the blind detector. The occurrence of lossy compression inside multimedia transmission channel is the most probable one, because of the common usage of audio file compression. The designed algorithm and previous research reported in the literature [15] were oriented on an improvement of the watermark robustness for the lossy compression attacks with regards to some commonly known audio formats such as MP3, AAC, VORBIS.

Table 4 Watermark extraction for the proposed algorithm

Attack	Proposed algorithm (DWT domain)
D/A–A/D conversion	✘ Not robust
Lossy compression	✔ Robust
Low-pass filtration	✔ Robust
High-pass filtration	✘ Not robust
Downsampling	✘ Not robust
Upsampling	✘ Not robust
Noise reduction	✘ Not robust
Time warping	✘ Not robust

Employing the watermarks as metadata container internally in the presented web service requires measuring the negative influence of the restoration routines on embedded watermarks. Several restoration algorithms were described briefly in the Sect. 2 and of the previous chapter [1]. Each of the proposed algorithms modifies the audio signal in a specific way. The possible signal processing oriented towards audio quality enhancement from watermark point of view is treated as attacks. The intentionally (artificially) distorted audio signals including music and speech representative samples were engaged in experiments. The extraction results, in meaning of Bit Error Rate (BER), after signal noise reduction are collected in Fig. 11. Three types of noise were investigated i.e. white, brown and pink noise. For each noise type three Signal to Noise Ratios (SNR) (in range from 5 to 15 dB) were considered.

Results presented in Fig. 11 are not particularly promising. Almost each extracted watermark sequence has more than 40 % of erroneously recovered bits. It means that the algorithm will not be able to detect the watermark. Only a slight difference is observed for speech signal affected by the white noise and SNR value equals to 15 dB. Decreasing of BER is observed for the watermarking strength parameter increase. For other tracks the α parameter does not affect the extraction quality. This phenomenon may be explained with fact that the noise elimination algorithm modifies in particular the whole signal, thus it does not changes much single frames. The watermark multiplication has not any positive influence on the extraction capability. A subsequent experiment engaged the spectrum expansion algorithm, bringing results that are shown in Fig. 12.

The spectrum expansion algorithm modifies audio signal in specific way i.e. for tracks with intentionally added white noise watermark detector is still blind. For a few tracks distorted with brown and pink noise it is possible to detect the watermark, especially for speech and music with noise added on the SNR level equals 5 dB. For each type of noise the decreasing BER trend is visible in relation to the increasing watermark strength. Even after watermark detection it is still infeasible to recover the watermarked information. Similarly as for the noise reduction procedure, the spectrum expansion algorithm modifies the whole signal. Another experiment takes into the consideration an influence of clipping reduction algorithm on the errorless watermark extraction. The obtained results are presented in Fig. 13.

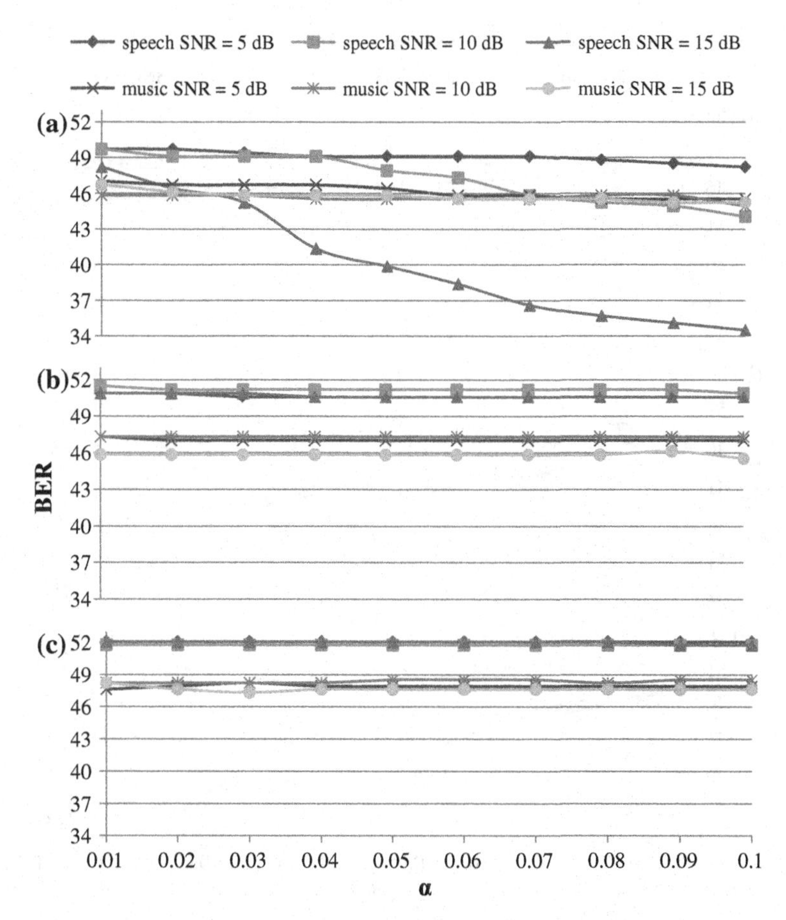

Fig. 11 BER obtained for the watermark extraction algorithm achieved after the noise reduction for the signal affected by **a** white noise, **b** brown noise and **c** pink noise

The restoration of music signals with the clipping reduction algorithm application destroys watermarks and the blind detector action. The speech signal structure enables to reduce clipping in smaller groups of audio frames, therefore the extraction efficiency of it from the restored speech signals is significantly higher than the extraction efficiency from the restored music signals. For each artificially clipped speech and music signal with level equal to 0.3 it is possible to detect the embedded watermark, provided the watermarking strength was greater or equal to 4. The watermark recovery remains still impossible, but the detector response is much more reliable in this case. The last experiment involves a synthetically distorted audio tracks used in order to analyze an influence of clicks reduction routines on the errorless watermark extraction. The results of this analysis are presented in Fig. 14.

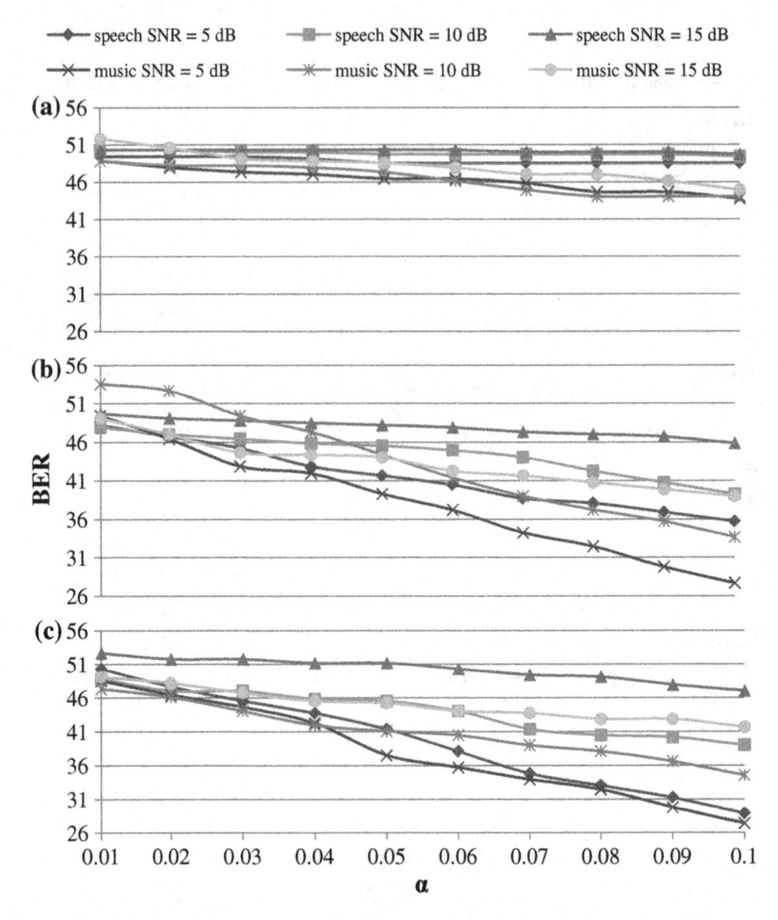

Fig. 12 BER obtained for the watermark extraction algorithm achieved after the spectral expansion for the signal affected by: **a** white noise, **b** brown noise and **c** pink noise

Clicks reduction algorithm affects the watermarks embedded in each type of audio signals. The greater α parameter value, the less number of errors occurs during the watermark extraction. Only for a single example (distorted with clicks using saturation parameter equal to 0.5) it was possible to transmit the watermark with error rate smaller than 25 %. Therefore, for this recording it was possible to extract the watermark information correctly, provided the error correction codes are used. The presented experiment could be concluded with the statement that there is no correlation visible between signal structure and BER.

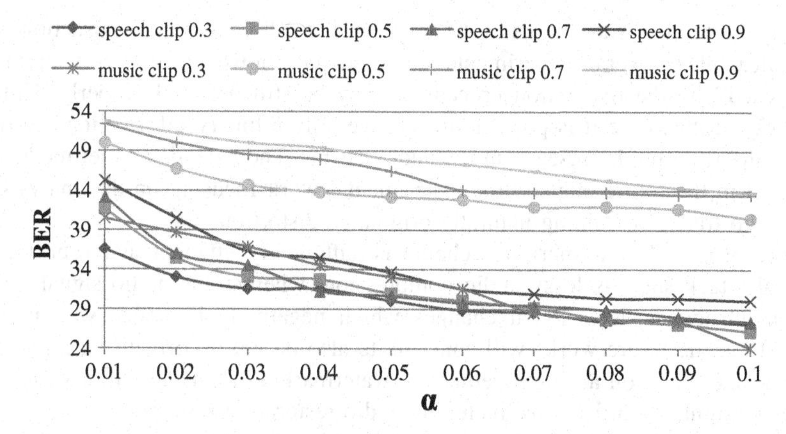

Fig. 13 BER obtained for the watermark extraction algorithm achieved after the clipping reduction

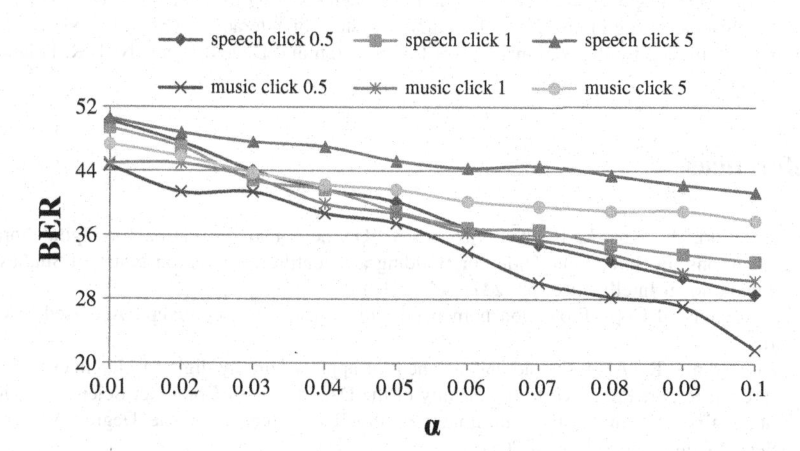

Fig. 14 BER obtained for the watermark extraction algorithm achieved after clicks reduction

5 Summary

It was shown that the proposed parallel restoration schema improves the functionality of the audio restoration web service. The average decrease of restoration duration for clicks reduction was around 3.5 times for the speech and 3.71 for the music (Table 2). For the clipping reduction it was equal to 3.37 (speech) and 3.03 (music) (Table 3).

The proposed application of watermarking does not work properly on artificially distorted audio tracks. The restoration may algorithms improve the audio quality, therefore, watermarks can be also deleted or crushed. Each restoration routine such as noise reduction, spectrum expansion, clipping or clicks reduction affects the hidden watermarks in the way that the subsequent data extraction will be infeasible. Several experiments have proved, however, that a watermark "attacked" by the restoration procedures may be still detected properly. Unfortunately, in this case it is possible to retrieve only a binary information (yes/no) about the watermark presence in the audio carrier. The extended experiments are planned in the future, namely the experimental work focusing on the analysis of the audio tracks containing naturally originated distortions.

The proposed watermarking scheme is sufficiently efficient and robust to a typical attack such as lossy audio compression, whereas the audio signal restoration introduces deep signal changes which negatively influence watermarks. Therefore, the future works will concentrate also on the watermarking algorithm enhancement and on an improvement of watermarking robustness to a variety of types of signal modifications, including audio restoration routines.

Acknowledgments The research is supported within the project No. SP/I/1/77065/10 entitled: "Creation of universal, open, repository platform for hosting and communication of networked resources of knowledge for science, education and open society of knowledge", being a part of Strategic Research Programme "Interdisciplinary system of interactive scientific and technical information" funded by the National Centre for Research and Development (NCBiR, Poland).

References

1. Czyzewski, A., Kostek, B., Kupryjanow, A.: Online sound restoration for digital library applications. In: Intelligent Tools for Building a Scientific Information Platform Studies in Computational Intelligence, pp. 227–242 (2011)
2. Sony Extended Copy Protection homepage, http://cp.sonybmg.com/xcp/. Accessed 2 May 2013
3. Venkataramu, R.: Analysis and enhancement of apple's fairplay digital rights management. A Project Report Presented to the Faculty of the Department of Computer Science San Jose State University. In: Partial Fulfillment of the Requirements for the Degree Master of Science, Computer Science (2007)
4. Microsoft Windows Media Digital Rights Management homepage, http://www.microsoft.com/windows/windowsmedia/licensing/default.mspx. Accessed 27 May 2013
5. Janus Patent U.S. Patent No. 7,010,808
6. Verance homepage, http://www.verance.com/. Accessed 27 May 2013
7. Cinavia homepage, http://www.cinavia.com/. Accessed 27 May 2013
8. AWT (2013) Audio Watermarking Toolkit homepage, http://www.coral-interop.org/. Accessed 27 May 2013
9. Coral Consortium homepage, http://www.coral-interop.org/ Accessed 27 May 2013
10. Godsill, S., Rayner, P.: Digital Audio Restoration-A Statistical Model-based Approach. Springer, London (1998)
11. Esquef, P., Biscainho, L.: A double-threshold-based approach to impulsive noise detection in audio signals. In: Proceedings of X European Signal Processing Conference (EUSIPCO) (2000)
12. Janssen, A., Veldhuis, R., Vries, L.: Adaptive interpolation of discrete-time signals that can be modeled as autoregressive processes. IEEE Trans. Acoust. Speech Signal Process. **34**, 317–330 (1986)

13. Dutoit, T., Marqués, F.: Applied Signal Processing: A Matlab-Based Proof of Concept. Springer, Berlin (2009)
14. Cichowski, J., Czyżyk, P., Kostek, B., Czyzewski, A.: Low-level music feature vectors embedded as watermarks. In: Intelligent Tools for Building a Scientific Information Platform Studies in Computational Intelligence. pp. 453–473 (2013)
15. Czyżyk, P., Janusz, C., Czyzewski, A., Bozena, K.: Analysis of impact of lossy audio compression on the robustness of watermark embedded in the DWT domain for non-blind copyright protection. In: Communications in Computer and Information Science, pp. 36–46. Springer, Berlin (2012)
16. Cichowski J., Czyzewski, A., Bozena, K.: Analysis of impact of audio modifications on the robustness of watermark for non-blind architecture. In: Multimedia Tools and Applications, pp. 1–21. Springer, USA (2013)
17. Helen, M., Lahti, T.: Query by example methods for audio signals. In: 7th Nordic Signal Processing Symposium. pp. 302–305 (2006)
18. Lu, L., Saide, F.: Mobile ringtone search through query by humming. In: IEEE International Conference on Acoustics, Speech and Signal Processing, pp. 2157–2160 (2008)
19. Plewa, M., Kostek, B.: Creating mood dictionary associated with music. In: 132nd Audio Engineering Society Convention, p. 8607. Budapest, Hungary (2012)
20. RECOMMENDATION ITU-R BS. 1387-1, Method for objective measurements of perceived audio quality. (2001)
21. OPERATM Voice and Audio Quality Analyzer, http://www.opticom.de/products/opera.html. Accessed 08 April 2013
22. Lang, A., Dittmann, J., Spring, R., Vielhauer, C.: Audio watermark attacks: from single to profile attacks. In: Proceedings of the 7th Workshop on Multimedia and Security, pp. 39–50 (2005)

Overview of the Virtual Transcription Laboratory Usage Scenarios and Architecture

Adam Dudczak, Michał Dudziński, Cezary Mazurek and Piotr Smoczyk

Abstract This chapter outlines findings from the final stage of development of the Virtual Transcription Laboratory (http://wlt.synat.pcss.pl, VTL) prototype. VTL is a crowdsourcing platform developed to support creation of the searchable representation of historic textual documents from Polish digital libraries. This chapter describes identified usage scenarios and shows how they were implemented in the data model and architecture of the prototype. Last chapter presents current usage of the portal and the results of basic benchmarks conducted in order to assess performance of transcription process in VTL.

Keywords OCR · Digitisation · Post correction · Digital libraries

Presented results have been developed as a part of PSNC activities within the scope of the SYNAT project (http://www.synat.pl) funded by the Polish National Center for Research and Development (grant no SP/I/1/77065/10).

A. Dudczak (✉) · M. Dudziński · C. Mazurek · P. Smoczyk
Poznan Supercomputing and Networking Center, ul Z. Noskowskiego 12/14, 61-704 Poznan, Poland
e-mail: maneo@man.poznan.pl
URL: http://www.man.poznan.pl

M. Dudziński
e-mail: michald@man.poznan.pl

C. Mazurek
e-mail: mazurek@man.poznan.pl

P. Smoczyk
e-mail: smoq@man.poznan.pl

R. Bembenik et al. (eds.), *Intelligent Tools for Building a Scientific Information Platform: From Research to Implementation*, Studies in Computational Intelligence 541,
DOI: 10.1007/978-3-319-04714-0_15,

1 Introduction

Virtual Transcription Laboratory (http://wlt.synat.pcss.pl, VTL) is a crowdsourcing platform developed to support creation of the searchable representation of historic textual documents from Polish digital libraries. Previous work [1, 2] focused on evaluation of the training capabilities of the OCR service, the core part of VTL portal. This chapter presents findings from the final stage of VTL development. Early version of portal was released in September 2012 as a publicly available BETA. Thanks to its wide availability, it was possible to gather feedback from relatively big group of users, adjust the underlying technical architecture and verify the initial assumptions regarding service usage scenarios.

The following chapter explains details of abstract model of transcription process which was developed as a part of initial work on the VTL prototype. This model was one of the initial sources of usage scenarios.

2 Overview of Transcription Model

During initial phases of development of the prototype, several other transcription tools and projects were analysed including T-PEN,[1] TILE,[2] Trove Newspapers [3], Distributed Proofreaders,[3] Transcribe Bentham,[4] WikiSource.[5] These examples deal mainly with image-to-text transformation. In order to have a more abstract view on transcription process the mentioned analysis was extended to cover audio–visual materials. Figure 1 presents an abstract transcription model which can be applied to any kind of materials which needs to be converted to the textual representation for increasing its retrieval capabilities.

For the purpose of this analysis it was assumed that transcription is a textual representation of the document. It can contain both text and metadata necessary to transfer additional features of the original material. The described transcription model can be created manually or via automated conversion. This process might be performed individually or by several transcribers T(1…n). Each of them can prepare their own version of text. In most cases, initial transcription is performed by one entity e.g. instance of a single Optical Character Recognition (OCR) engine or domain expert. Nevertheless in some cases parallel versions may exist (the same image OCRed with different engines). In such a scenario these texts can be used for cross validation.

[1] http://t-pen.org/

[2] http://mith.umd.edu/tile/

[3] http://www.pgdp.net/

[4] http://blogs.ucl.ac.uk/transcribe-bentham/

[5] http://wikisource.org/

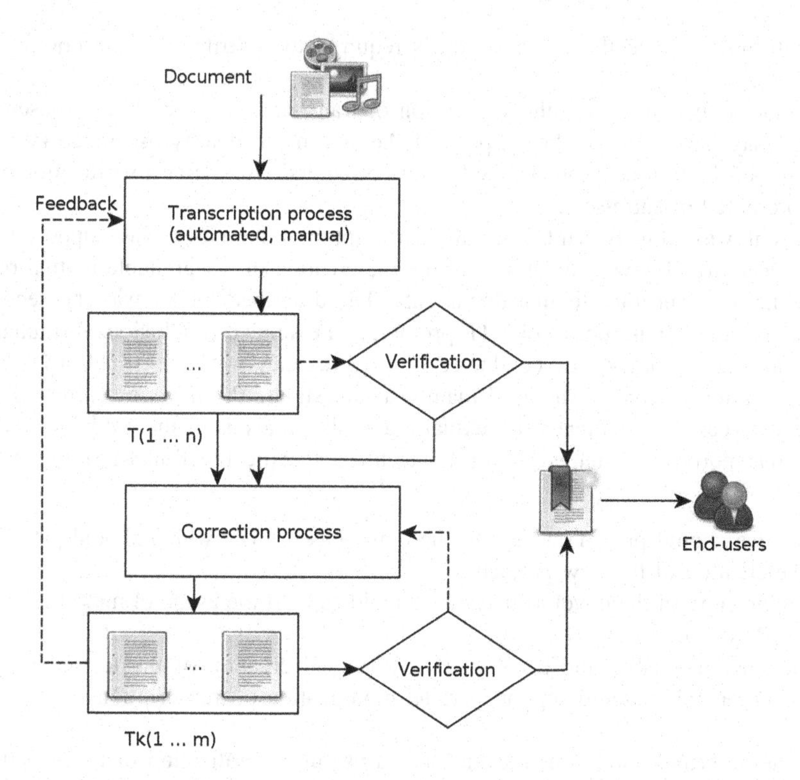

Fig. 1 General overview of the transcription process based on the initial analysis performed in the framework of SYNAT project

The model makes a distinction between verification and correction. Verification does not affect the content of transcription; it is aimed on assessing the quality of the output. As a result of such a verification process supervisor may decide that a given transcription has a sufficient quality and then finish the entire process. This may be the case for most of modern texts which are usually well handled by the OCR software. This scenario was also confirmed in a survey held in 2010 among creators of Polish digital libraries [2]. Because of the lack of human resources and proper tools to perform correction of the OCR output, digital librarians were performing verification just to assess if a so called "dirty OCR" is readable enough. In case of a poor quality of results text was not included in the digital object which was finally published in a digital library.

Verification may also lead to correction, which is to some extent similar to manual creation of the transcription. In the described model, correction is rather a manual task. It was assumed that any automated correction methods can be included as a part of the transcription process. Exactly the same as with the transcription process, correction can be performed by several correctors Tk(1…m). Each of this individuals can work on the parallel versions of the entire text or perform correction on a part of the original object. Proper division of work can significantly reduce the effort which is required to contribute to the correction process.

Exactly the same as with the creation of transcription, alternative representations may be useful for the purpose of the automated quality assurance (verification) of the transcription. The output of correctors' work after a verification can be presented to end users.

As it was already said, Virtual Transcription Laboratory was supposed to deliver a virtual environment supporting the creation of a searchable textual representation of historic textual documents. The described model was the second source of the requirements for VTL prototype, it extended conclusions formulated on top results from survey conducted in 2010 (as they were described in [1, 2]). The mentioned survey was used mainly to investigate the flaws of current digitisation workflows dedicated to textual materials. Analysis of the model served a different purpose, it helped to formulate more specific functional requirements, like:

- service should provide access to the cloud-based OCR which should simplify the creation of the new projects,
- architecture of the developed system should support the usage of more than one OCR engine,
- in order to enable and accelerate crowd-based creation of the textual representation VTL should support parallel modification of transcription.

The described model proved that it is a useful generalization of transcription process. It was developed on the top of several widely recognized projects and tools of this kind. The following chapter describes how it was implemented in a prototype version of the Virtual Transcription Laboratory.

3 Working with the Virtual Transcription Laboratory

The development of the VTL prototype led to its release as a publicly available BETA version in September 2012. Thanks to this it was possible to present the implemented functionality to a large group of future users. This chapter describes all the elements of the implemented transcription workflow as it is depicted in Fig. 2.

3.1 Project Creation

After registering in the VTL portal users can login and create new projects. Projects are a basic unit which organizes the work in VTL. A project contains images,

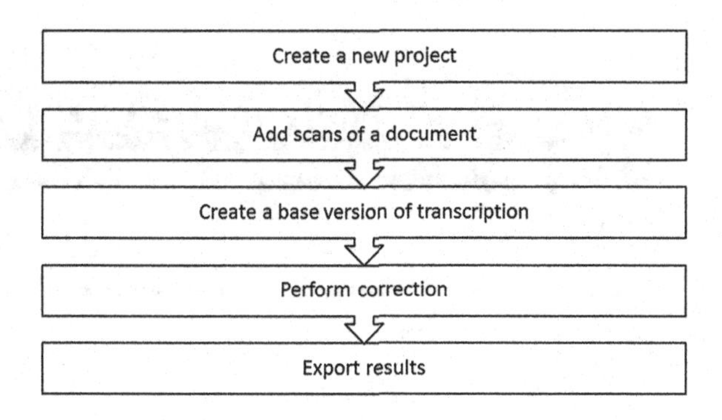

Fig. 2 General overview of the Virtual Transcription Laboratory workflow

some basic metadata and transcription.[6] The creation of the project starts with typing in the information about the transcribed document which include: the name of the project, the title and creator of original document, keywords describing both the project and the document, the type of content (printed, manuscript or other) and finally, the language(s) of the text. This metadata can be keyed manually or imported from the existing description of the object from digital library.

Import of metadata is based on OAI identifiers which are available for all objects in Polish Digital Libraries Federation (DLF, http://fbc.pionier.net.pl) [4]. This identifier has the following syntax:

- Prefix—oai:,
- domain name of the digital repository e.g. www.wbc.poznan.pl,
- locally unique identifier of the document e.g. 1234.

The final identifier is quite self-descriptive: oai: www.wbc.poznan.pl:1234. Users can find this identifier in metadata of a given object in the digital library (see Fig. 3).

DLF keeps track of all changes in physical localisation of the object and gives access to service which transforms OAI ids into concrete URL of given object. In the case of migration of a given digital library to new software which result in changes in the structure of URLs. An OAI identifier does not change and DLF will update its registry. As a result, all references to given documents will still be useful.

The described mechanism can also be used to implement a basic integration between existing digital libraries and VTL. Each digital library can easily create a URL which will cause creation of a new VTL project based on the object available in their repository. The mentioned URL will have the following syntax:

- address of the VTL portal—"http://wlt.synat.pcss.pl/wlt-web"
- address of the new project form—"/new-project.xhtml"
- argument named "oai" with a proper identifier—"?oai = ".
- proper OAI identifier—"oai: www.wbc.poznan.pl:1234".

[6] Details of VTL data model will be outlined in following chapters.

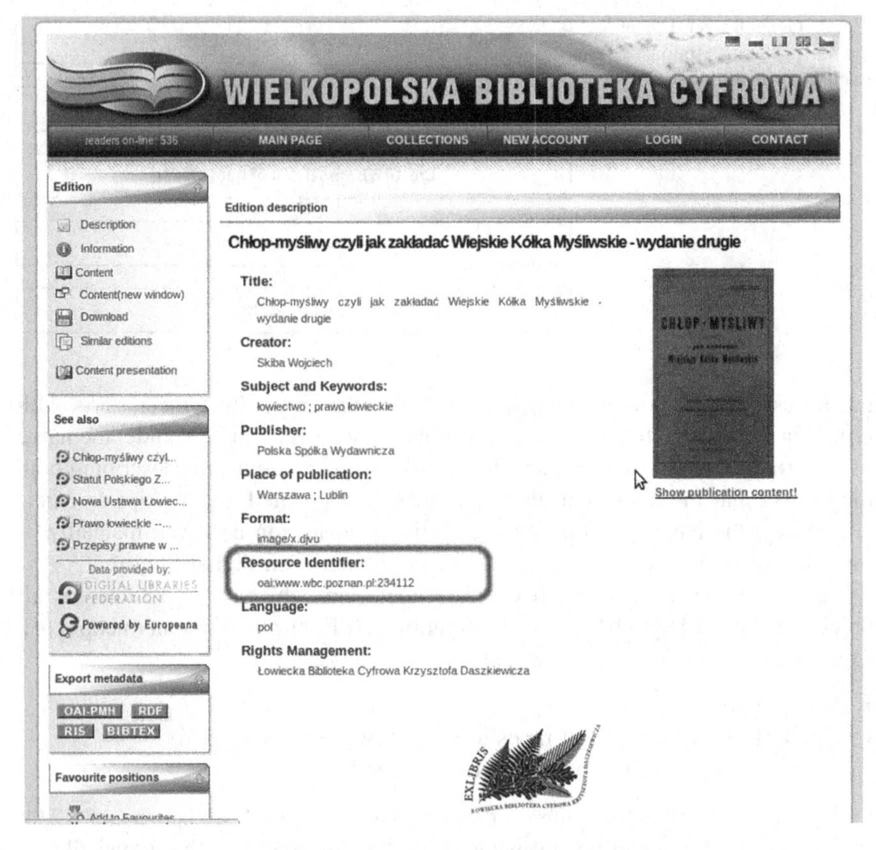

Fig. 3 Exemplary object in a Digital Library of Wielkopolska, *red rectangle marks* OAI identifier, which may be used to import this object into VTL

The provision of such a link on a digital library website next to the document which requires transcription may significantly increase the chance that someone will start the work with this object.

3.2 Import of Scanned Images

The following step includes addition of scanned images. VTL supports upload of files from the hard disk of user's computer or direct import of certain kinds of documents from the digital library.

VTL support upload of images in most widely used graphical formats, including PNG, JPG and TIFF. TIFF files are not well suited for presentation in the web browser, after successful upload they are automatically converted into PNGs. Apart from uploading single files, users can also upload a ZIP archive containing images. VTL will automatically extract the archive and add (or convert if necessary) uploaded files to the project.

The second possibility of adding files is a direct files import from one of the supported digital libraries. The import is based on OAI identifier exactly the same as it was described in the previous chapter.

Operations such as import of files from the remote server or conversion of several dozens of TIFF files is executed as a non–blocking, asynchronous task. The user can close the current web browser window and work on something different. When the asynchronous task is completed VTL will notify user by sending an email message to address associated with the user account. At any time they can also monitor the current execution of a given task by looking at a dedicated page in their profile.

3.3 Base Version of Transcription

Transcription in VTL, apart from text itself contains also its coordinates in the original image and information about annotations/comments. This is a quite rich representation of transcription and as far as authors are aware only T-PEN offers similar level of richness.

There are several ways to create the base version of transcription in VTL. In most cases, the initial version of transcription will be created using VTLcs OCR service. But apart from this, base version of transcription can be created by manual keying of text from images or by import of existing transcription, e.g. results of recognition from standalone OCR engine.

The manual keying is a tedious task but in some cases it might be the only solutions to create searchable text. When performed in VTL, work can be easily distributed among a group of volunteers which will significantly speed up the whole process.

VTL supports import of existing transcription in the hOCR [5] format. It can be created from the output of a standalone OCR engines e.g. Tesseract[7] or OCRopus.[8]

The most comfortable option for creation of the base transcription is the usage of VTL's OCR service. Users can go through the project and invoke OCR processing on each scanned image manually or run a batch OCR on all files in the project. Batch OCR is executed as an asynchronous task, user will be notified via email when the process is finished.

Web interface of the OCR service allows to process the whole page or just a fragment of it. This might be useful while dealing with pages where several languages were used next to each other. User can mark the first half of the page which was printed in one language and run the OCR. When the first part of

---

[7] http://code.google.com/p/tesseract-ocr/

[8] http://code.google.com/p/ocropus/

processing is over, user can go back to the OCR interface and process the rest of the page using different language profile.

As it was already mentioned, OCR service returns not only text but also its coordinates. This might be useful when working with manuscripts. Current OCR service does not offer support for connected scripts but it features a good accuracy at marking boundaries of text lines. In this usage scenario users can use only the coordinates of lines (because the quality of recognised text will probably be very poor). This would work as a template and simplify the process of a manual keying of text.

3.4 Correction of the Transcription

The course of correction process depends mainly from the intentions of project owner. Projects are by default marked as private but their owner can decide to share the work with selected VTL users (workgroup) or make it public. Public projects can be modified by all logged users.

The work of project's editor consists of the following steps:

1. go to the project page,
2. select page from the project or click "Go to the transcription editor",
3. review the text and verify if the content of a scan corresponds to the recognised text,
4. when ready with a given page, editor can go to the next page.

Every text line in the transcription has a status, after verification/correction it goes to a "checked" state. This transition can be done manually (editor marks line as checked) or automatically (if they spend more than 5 s working on a given line). Thanks to this mechanism project owner can track the progress of correction process. VTL can also select pages which have the highest number of unchecked lines and redirect project editors to these pages. The project owner can mark all lines at a given page as "checked" or "unchecked". This feature should help to implement more sophisticated correction workflows.

VTL supports the parallel modification of transcription. If two users work on the same file their changes will be automatically merged. All modifications of the project are tracked, project owner can withdraw changes performed by other users at any time.

3.5 Export of Results

When a quality of transcription is sufficient, the project owner can export final text in one of three output formats:

1. hOCR—can be used to generate alternative document formats e.g. PDFs, DjVu files,
2. plain text—for processing in popular text editors e.g. Microsoft Word,
3. ePUB—format dedicated to mobile readers, may be a subject of further processing in applications like Sigil[9] or Calibre.[10]

Adding new output formats is relatively easy and the list of supported formats will be extended in the future.

4 Relevant Implementation Details

This chapter describes selected issues documenting the internal structure, flexibility and scaling capabilities of VTL.

4.1 Data Model

Thanks to the release of VTL BETA version it was possible to validate its performance and flexibility against actual needs of users. Its current shape (see Fig. 4) is the results of a few major updates which took place in the course of last year.

The data model consists of three layers: first keeps track of all the user informations, second deals with project details (grey rectangles on Fig. 4), the last one models transcription (yellow rectangles). The user can own the projects, which groups the elements related to the original document, includes basic metadata (e.g. title, authors, type and language etc.) and files with scanned images. At the moment the project can have only one transcription. But thanks to the separation between information related to original object and transcription it should be possible to store several transcriptions for one set of files.

Transcription is built out of content areas (corresponding to project's files) which contain text lines. Text lines are in fact essential for user's interface metaphor which is implemented in VTL's transcription editor. Each text line is divided into text segments, these segments can be annotated with several annotations (e.g. emphasize, comments). An Annotation manager tracks which annotations are used in a given transcript.

As it was already mentioned, all changes made in transcription are tracked. Version information is kept on the level of content areas. When the user enters the transcription editor to work with a given file, VTL creates a new working copy from the newest version of the corresponding content area. All changes are saved

[9] http://code.google.com/p/sigil/

[10] http://calibre-ebook.com/

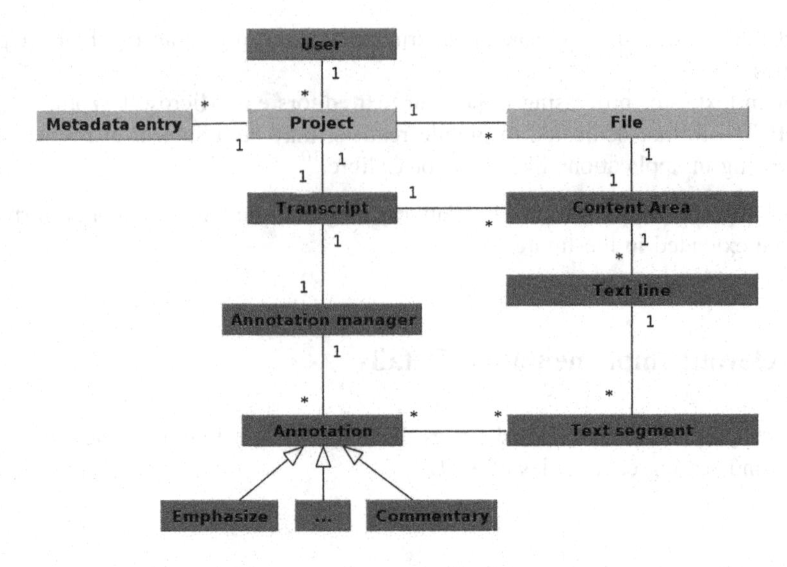

Fig. 4 Data model used as a base of Virtual Transcription Laboratory

automatically in this working copy, at the end of user's session all changes are published and become visible for other users of the portal.

4.2 Structure of the VTL Deployment

Virtual Transcription Laboratory was developed using Java Enterprise Edition 6 technology stack with some additional non-Java dependencies including: Tesseract, ImageMagick, DjVuLibre and Calibre.

At the moment, VTL is using two servers but thanks to its fine grained structure (see Fig. 5) most of the components can be moved to separate machines. There are three main components worth mentioning:

- VTL portal—responsible for serving content to end-users.
- Content services—group of several services implementing the most important VTL backend services e.g. OCR service, import and conversion of files from digital library.
- File storage service—simple, transactional file store, which offers remote access to files (mostly images) used by other components of VTL.

Each of the mentioned components uses its own database, which can be also moved to the remote server machine. In case of performance problems this should allow to scale system up. According to Piwik[11] web analytics engine which is used

[11] http://piwik.org

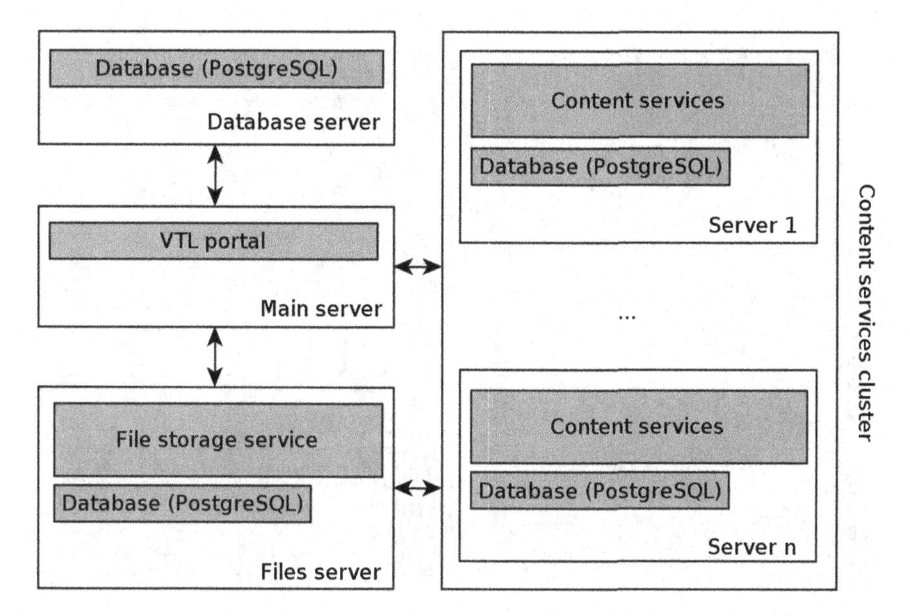

Fig. 5 Virtual Transcription Laboratory scaling capabilities. Diagrams show possible distribution of VTL components among different server machines

to monitor current usage of VTL, the average page generation time calculated is 4 s. This average was calculated on top of data from more than 1500 visits and took into account all VTL pages.

The following chapter describes the most important outcomes of various activities performed as a part of the VTL development since its public release.

5 Evaluation of the Virtual Transcription Laboratory Prototype

In early stages of the Virtual Transcription Laboratory development it was assumed that portal will find its users mainly among scholar, hobbyists (as stated in [3, 6]), teachers and digital librarians. Initial usage scenarios included creation of digital editions of historical books and increasing the retrieval capabilities of documents from digital libraries. In order to validate these assumptions as soon as possible in the course of the SYNAT project, early BETA version of VTL was publicly released. The release was followed by series of presentations and articles. These events were supposed to involve potential future users of VTL and help the development team to gather feedback.

As a sort of side effect of VTL promotion, Poznań, Supercomputing and Networking Center (PSNC) started to organize workshops dedicated to digitisation of historical documents for the purpose of humanities research. These workshops

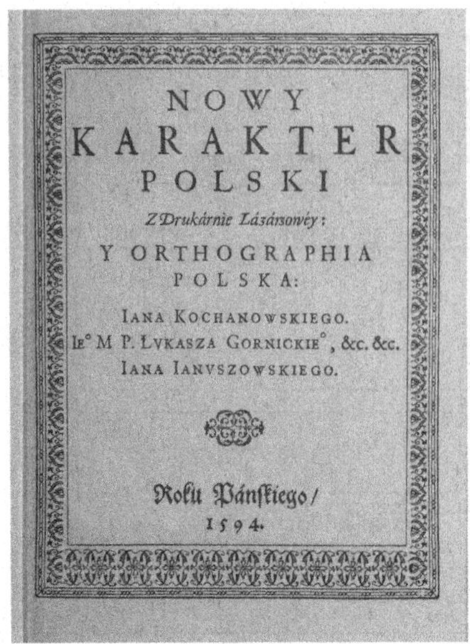

NOWY
KARAKTER
POLSKI
Z Drukárnie Lázárowéy:
Y ORTHOGRAPHIA
POLSKA:
IANA KOCHANOWSKIEGO.
IE° M P. LVKASZA GORNICKIE°, &c. &c.
IANA IANVSZOWSKIEGO.

Roku Páńskiego/
1594.

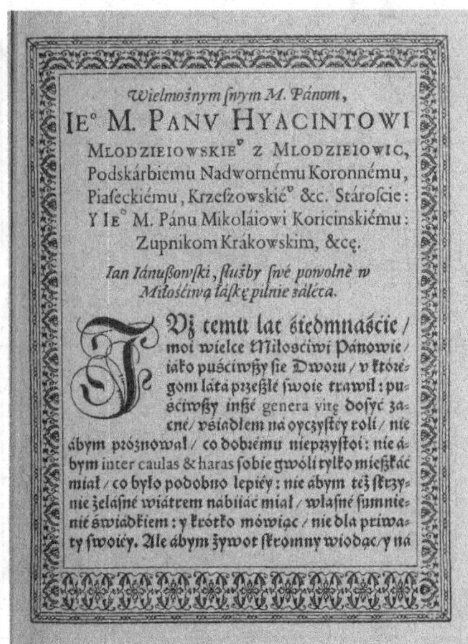

Wielmożnym swym M. Pánom,
IE° M. PANV HYACINTOWI
MLODZIEIOWSKIE° z MLODZIEIOWIC,
Podskárbiemu Nadwornému Koronnému,
Piáseckiému, Krzeszowskié° &c. Stárośćie:
Y IE° M. Pánu Mikoláiowi Koricinskiému:
Zupnikom Krákowskim, &cę.

Ian Iánußowski, służby swé powolnè w Miłośćiwą łáskę pilnie záléca.

TVż temu lat śiedmnaśćie/ moi wielce Miłośćiwi Pánowie/ iáko puśćiwßy sie Dworu/ v którego m láta przeßłe swoie trawił: puśćiwßy inßé genera vitę dosyć zacné/ vsiadłem ná oyczystéy roli/ nie ábym próżnował/ co dobrému nieprzystoi: nie ábym inter caulas & haras sobie gwóli tylko mießkáć miał/ co było podobno lepiéy: nie ábym też skrzynie żelásné wiátrem nabiiáć miał/ własné sumnienié świadkiem: y krótko mówiąc/ nie dla priwaty swoiéy. Ale ábym żywot skromny wiodąc/ y ná

Fig. 6 Pages from of Jan Januszowski's book "Nowy Karakter Polski"

were covering much broader scope than the VTL itself, participants had a chance to learn basic concepts related to digitisation, learn more about OCR, see examples of interesting experiments conducted on top of large corpora of old documents, etc. In period between March and May 2013, development team was able to train and discuss with over than 70 people whose professional interest was in some way convergent to VTL profile. This group included students from faculties of history, archival studies, digital librarians and some widely recognize scholars interested in digital humanities. Development team was able to encourage some of them to try VTL in real projects (Fig. 6).

One examples of the project created in VTL during mentioned period was the transcription of Jan Januszowski's book "Nowy Karakter Polski" from 1594[12]. The transcription was a collaborative effort of students from one of the major Polish universities.

5.1 Current Usage

Since the public release, VTL portal was visited more than 1583 times by 700 unique users. Average time spent on the site, was 8 min 42 s. These statistics show that there is still quite significant number of users who end their session at the main page of the portal.

[12] Project is publicly available here: http://wlt.synat.pcss.pl/wlt-web/project.xhtml?project=40.

Table 1 Basic statistics showing the current amount of data stored in VTL

	April 2013	June 2013
Number of registered users	88	141
Number of projects	141	246
Percent of empty projects (%)	24	22
Number of files	3996	5821

Table 2 Simple benchmark assessing performance of transcription process in VTL

Activity	Time (min)
Scanning and post-processing	14
Creation of a project in VTL and file upload	10
Batch OCR	2
Text correction	45
Export of ePub	2

Overall time spent on transcription of 18 pages—1 h 14 min

Table 1 presents the difference in number of registered users, created projects and the number of files till April and June 2013. It shows that the amount of content in portal is constantly growing. Taking into consideration quite high average time spent on site, it seems that users are able to successfully work with the portal in its current shape.

Chapter 16 summarizes results of some basic experiments aimed at evaluation of transcription performance.

5.2 *Evaluation of the Transcription Performance*

In order to evaluate the performance of transcription process in VTL, a simple experiment was conducted. The experiment was covering the whole transcription process starting from scanning, through creation of a project in VTL, batch OCR, text correction and export of results. A short (18 pages long), relatively modern (published in 1920s) brochure was selected[13] as a subject of the experiment.

Table 2 shows the results of this simple benchmark. Overall time spent on creation of digital edition of 18 pages—1 h 13 min. This results in average of 2 min 30 s necessary to perform the correction of one page. If a given object is available in digital library scanning and post-processing can be skipped. This change can significantly reduce the overall time in the case of longer documents. Usage of existing scans may also simplify the process of text correction because most likely images from digital library offer better quality than home-made scans.

[13] Original object is available in Digital Library of Wielkopolska (oai: www.wbc.poznan.pl: 234112).

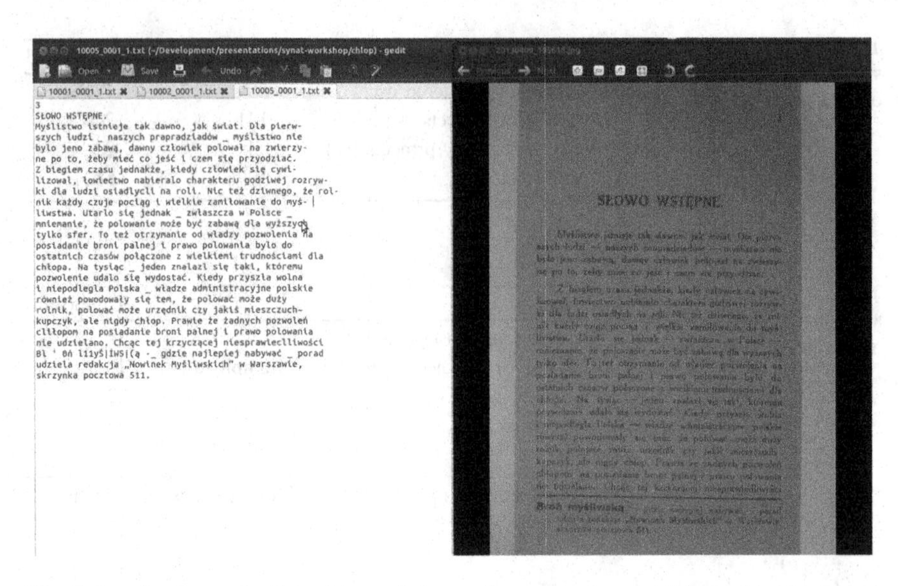

Fig. 7 Baseline for evaluation of transcription performance was established using simple text editor and image viewer

The authors are not aware of any tools which implement workflow similar as VTL. In order to establish a baseline for this benchmark, scope was limited just to text correction. Result of VTL's OCR for the same document was exported to text files. Text correction was performed using a text editor and an image viewer displayed next to each other (see Fig. 7). The overall time spent on this task was 41 min (appr. 2 min 20 s per page). The difference is very small in favour of regular text editor. Nevertheless it needs to be taken into account that VTL gathers much more data than a simple Notepad, like interface e.g. text line coordinates. Document which was used for the purpose of this benchmark had a simple one-column layout, it was relatively easy to confront text on the scan with the corresponding line in text file. This task may become significantly harder when dealing with multicolumn documents like newspapers. Thanks to information of text coordinates VTL users can handle this kind of content easily. This scenario may be subject of further evaluation.

During one of the workshops mentioned an the beginning of this chapter, participants were asked to use VTL to collaborate on text correction. Six of them decided to participate in this experiment, mentioned workshop was a first contact with VTL for 5 of them. By the time experiment started they had appr. 20 min to experiment with VTL transcription editor. After that time all participants were asked to pick up a page from one publicly available project. After 25 min the whole group managed to correct 17 text pages (approx. 1 min 30 s per page). It shows the potential of crowd-based text correction and proves that VTL can be used as crowdsourcing platform.

6 Summary

This chapter summarized the outcomes from the development of Virtual Transcription Laboratory. This chapter starts with the description of the general transcription model which was created in the course of SYNAT project. This abstract model was used as one of the sources of requirements for the development of the Virtual Transcription Laboratory prototype. Initial assumptions regarding usage scenarios of VTL were evaluated after public release of the prototype in September 2012. This chapter outlined most important outcomes from this important part of prototype evaluation. In addition to this real life test, development team conducted series of small benchmarks. Results of these experiments shows that VTL features similar performance in terms of the text correction when comparing to simple, widely-known tools. It ought to be stressed that despite the time necessary to perform correction is similar, VTL gathers much more text-related (e.g. text lines coordinates) data. This data may be very helpful for correction of documents with complex layouts e.g. newspapers. Apart from improved user experience, VTL enables a crowd-based correction, which is hardly achievable with traditional tools.

References

1. Dudczak, A., Kmieciak, M., Werla, M.: Creation of textual versions of historical documents from polish digital libraries. Lecture Notes in Computer Science, vol. 7489, pp. 89–94. Springer, (2012)
2. Dudczak, A., Kmieciak, M., Mazurek, C., Stroiński, M., Werla, M., Weglarz, J.: Improving the workflow for creation of textual versions of polish historical documents. In: Bembenik, R., Skonieczny, Ł., Rybiński, H., Kryszkiewicz, M., Niezgódka, M. (eds.) Intelligent Tools for Building a Scientific Information Platform: Advanced Architectures and Solutions, pp. 187–198. Springer, Berlin Heidelberg (2013)
3. Holley, R.: Many Hands Make Light Work: Public Collaborative OCR Text Correction in Australian Historic Newspapers, pp. 1–28. National Library of Australia Staff Papers, Canberra (2009)
4. Lewandowska, A., Werla, M.: PIONIER network digital libraries federation—interoperability of advanced network services implemented on a country scale. Comput. Methods Sci. Technol. 119–124 (2010)
5. Breuel, T.: The hOCR microformat for OCR workflow and results. In: 9th International Conference on Document Analysis and Recognition ICDAR 2007, vol 2, pp. 1063–1067. (2007)
6. Górny, M., Mazurek, J.: Key users of polish digital libraries. Electron. Libr. **30**(4), (2012)

University Knowledge Base: Two Years of Experience

Jakub Koperwas, Łukasz Skonieczny, Marek Kozłowski, Henryk Rybiński and Wacław Struk

Abstract This chapter is devoted to the 2-years development and exploitation of the repository platform built at Warsaw University of Technology for the purpose of gathering University research knowledge. The platform has been developed under the SYNAT project, aimed at building nation-wide scientific information infrastructure. The implementation of the platform in the form of the advanced information system is discussed. New functionalities of the knowledge base are presented.

Keywords Digital library · Knowledge base · Scientific resources · Repository

1 Introduction

The last decade have shown an increased interest of many universities in the systems concerning research data management and access to publicly funded research data. Various approaches can be observed at universities in Poland—some have been

J. Koperwas (✉) · Ł. Skonieczny (✉) · M. Kozłowski (✉) · H. Rybiński (✉) ·
W. Struk (✉)
Institute of Computer Science, Warsaw University of Technology, Nowowiejska 15/19,
00-665, Warszawa, Poland
e-mail: J.Koperwas@ii.pw.edu.pl

Ł. Skonieczny
e-mail: L.Skonieczny@ii.pw.edu.pl; lskoniec@ii.pw.edu.pl

M. Kozłowski
e-mail: M.Kozlowski@ii.pw.edu.pl

H. Rybiński
e-mail: H.Rybinski@ii.pw.edu.pl; hrb@ii.pw.edu.pl

W. Struk
e-mail: W.Struk@ii.pw.edu.pl; W.Struk@elka.pw.edu.pl

R. Bembenik et al. (eds.), *Intelligent Tools for Building a Scientific Information Platform: From Research to Implementation*, Studies in Computational Intelligence 541,
DOI: 10.1007/978-3-319-04714-0_16,

very enthusiastic in building infrastructure for research repositories storage, some other rather reluctant in supporting academic staff in meeting more demanding requirements for research practice, quite often due to lack of the idea how to provide suitable motivation on one side and assistance on the other, leading to a successful university research knowledge database. It turns out that important success factors are harmonizing solutions concerning both organizational issues, as well as functional features of the software.

In the chapter we summarize our experience of 2-years development and exploitation of the software $\Omega - \Psi^R$, which has been implemented at Warsaw University of Technology (WUT) for the purpose of building the WUT Knowledge Base, focusing on these software functionalities, which we see as the features essential for acceptance of the system by the university research community. The software was developed within the framework of the SYNAT project [1], aimed at building nation-wide scientific information infrastructure in Poland.

In [2] we have presented the main assumptions that were underlying the basic platform features. In particular, we have shown how the functional requirements for a university knowledge base and repository influenced our choices concerning the system architecture, software components, and the system configuration. In [2] we have also presented our experience in converging our imaginations concerning the system with the exploitation needs for running a large and sophisticated information system in a distributed University environment, and covering the needs of quite heterogeneous groups of users. In this chapter we will focus on the new features of the system, which in our opinion are making the system more attractive for various groups of the university community.

The chapter is organized as follows. In the next section we will recall basic assumptions and functionalities of the UKB, however our main objective is to confront them with the university business needs, rather than providing technical details. Section 3 will be devoted to recent extensions of the $\Omega - \Psi^R$ system, which in our opinion have crucial importance in accepting the system by various university actors. The new functionalities concentrate on analytics, providing such new possibilities like looking for experts and/or teams, illustrating the main research areas of the university teams or units. The new functionalities are supposed to play an important role in (1) showing the research "value" of researchers and the teams; and (2) giving means to the university top level management for strategic planning of the research development. Another group of the new system functionalities are related to the data acquisition issues, integrating within the system various important external and internal data. In Sect. 4 we will present the extended system flexibility, showing that with its generic functionality the system is ready to be adapted for the needs of any other university. Finally, in Sect. 5 we will provide our further plans and the conclusions.

2 A Brief Look at the Knowledge Database at WUT

As described in [2] the starting requirements for the university knowledge base were rather typical, focused on the repository functions. The main aim of the repository was to build institutional publication repository services, based on the open access idea to the most possible extent.

At a given point it was decided that in terms of access the system should be a centralized university repository, but distributed as much as possible in terms of data acquisition. Organization-wise, the WUT Central University Library was considered from the very beginning as a repository owner, responsible for organizing all the procedures around the knowledge base, including, inter alia the university staff training, data quality, overall regulations, etc. Hence, the Research Knowledge Database in its early stage was thought as a central entry point for the entire publications repository, however it was also very clear that the database should be created and maintained in a distributed way, by the entire university community, i.e. by the university units, bodies, but also individual staff, in accordance with the principle that the publication and information about the publication is entered to the repository in the place where it was created. Practically, in some cases it means that the data process is performed at the university units (institutes and/or faculties) by a specialized staff, but for some other cases (smaller university units) it also could mean that the process is performed by crowdsourcing, i.e. the data about publication are made by the authors. This approach gives us a guarantee that the information is up-to-date and complete, but not necessarily correct, especially in the cases, where the crowdsourcing is applied for data acquisition.[1]

Since the very beginning it was clear that the system cannot be limited to the repository functions only. The university's research knowledge base should cover a vast and heterogeneous repertory of data concerning various aspects of the research activities, and the knowledge base should be a central entry point for information about researchers and their activity records, including inter alia the projects run at the university, along with various project documents and data, but also presentations, patents, etc., as well as various levels diplomas and theses, starting from B.Sc. through M.Sc. to Ph.D theses. All these data types should be strongly interconnected, building a semantically rich database. With such variety of the content in the university knowledge base the software should be flexible enough, so that demands for new object types and new relationships would be fairly easy to implement. The flexibility should be reflected in providing administrative tools that enable defining new data structures, and then for the new objects

[1] Involving researchers directly into the data acquisition process was presumed as a psychologically important factor for achieving the data completeness. Bearing in mind a possible drop down of data quality, unavoidable for such approach, a variety of new tools guarantying high level of acquisition process have been developed recently—they are mainly based on web mining and will be presented in Sect. 3.

make it easy to define new data entry worksheets and new search screens. The proposed solutions for an upgraded flexibility of the system went further. Actually, the system additionally provides scripting means for

- validation tools supporting data entry process quality
- means for visualizing search results, custom views, defining sorting rules, etc.
- access rights rules, defining access to various objects, as well as synthetic information that can be obtained from the system.

All the tools of this kind have been implemented within the system in such a way that in most cases the system development does not require programmers intervention (for more details see [2]).

Given the software ready for running the whole University Knowledge Base, one of the encountered problems of the highest importance was to reach a satisfactory starting point for filling the knowledge base with the contents. To this end a number of modules have been developed for loading data from various sources. Many sources have been identified within the University (the legacy databases of the Central Library, the local legacy databases run at various faculties, data from the faculty reports, often in the Excel form, etc.), but additionally some procedures have been adopted to acquire data from WEB. The data from the legacy databases have been exported and translated to the new system. In some cases the missing source documents have been found in private archives of the staff (this was mainly the case of the older Ph.D theses). More difficult situation was with many important publications of the University staff. Fortunately, the situation has been recognized at the beginning of the SYNAT project, and concurrently a special research was performed within the project, referring to the problems of data acquisition from WEB. As a result of this research, a specialized data acquisition module has been designed and implemented. It was given the name Ψ^R (for Platform for Scientific Information Retrieval), which is now integrated within the system $\Omega - \Psi^R$. The data acquisition module is described in more details in another chapter of this book, in [3].

Because of the heterogeneity of data formats that were identified for loading to the new system, it became clear that advanced algorithms for merging data from various sources into the knowledge base have to be worked out. We have elaborated such algorithms, some are presented in [4].

As already mentioned above, the functionality of the system $\Omega - \Psi^R$ is not limited to a typical repository. It has been decided that the repository functionality, although important (or even indispensable) is thought as a way towards building a knowledge base for the overall university research activities. It is therefore crucial for the system to gather and integrate various types of information referring to the research activities. The completeness and integration of information referring to the research activities gives the system development team the extraordinary possibilities to implement analytical functionalities, being of interest to the university authorities, academic society, as well as the students and young researchers. The new possible functionalities cover *inter alia*:

1. possibilities to perform search for experts and teams
2. evaluation and characteristics of the activities of particular individuals, teams, and university units, and presenting the map of research, visualizing stronger and weaker fields of research.

In the next section we will sketch how these functionalities are implemented. Bearing in mind that the quality of the analytical computations strongly depends on the completeness of information, we start Sect. 3 with the presentation of the data acquisition process.

3 New Functionalities of the System

3.1 Data Acquisition

This section presents a solution for automatic acquisition of bibliographic entries from the web. This process consists of three steps, depicted in Fig. 1: searching for publications, extracting bibliographic metadata and finally merging entries into university knowledge base. The first step is realized by the Ψ^R module [3]. The module can be seen as a focused web crawler. It delegates user-given queries to various search engines, executes them in a periodical manner and consecutively refines them to improve precision and recall of results. We use it to search the Internet for the publications of WUT authors. Besides publications acquisition the Ψ^R module is being used to retrieve up-to-date information about conferences and journals.

The second step is performed by the Zotero software [5]. Zotero is a free and open-source Firefox extension which helps managing bibliographic data and related research materials. Notable features include automatic detection of bibliographic entries on websites, retrieving these entries, and converting between many formats e.g. BibTex, MODS, RefWorks. Zotero supports bibliography detection in a variety of sources including publishers' databases, like Springer, Scopus or IEEE, as well as, publication search engines like Google Scholar or CrossRef. It is worth mentioning that the system $\Omega - \Psi^R$ itself is supported by Zotero as well.

For the purpose of data acquisition we implemented the Zotero-based web server, named bib2py. We use it to convert websites containing information about WUT publications into BibTex format. As Zotero was developed as a browser extension, it was not straightforward to build the web-server application on the top of it. Several changes in the Zotero source code were made in order to automate the process of bibliographic data extraction and to eliminate the necessity of the user interaction. Moreover, in our web-server we utilized a plugin to Zotero, named "Scaffold", which is a Firefox extension designed for developing and testing the Zotero translators.

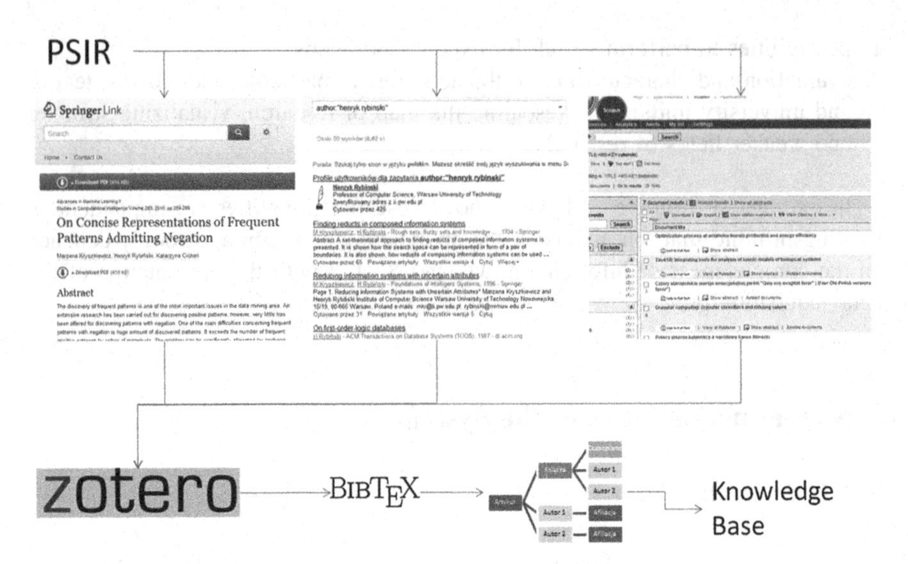

Fig. 1 Data acquisition process

The bib2py module provides the functionality that enables user to manage the process of bibliography extraction from the predefined collection of research resources. The service takes as an input either a single URL or the list of links that point to the websites containing bibliographic data, e.g.:
http://link.springer.com/chapter/10.1007/978-3-642-32826-8_40, or
http://dx.doi.org/10.1016/0306-4379%2884%2990025-5.

For each URL it tries to extract the bibliographic metadata existing on the webpage using an appropriate Zotero translator. When the bibliographic data are successfully extracted from the website, they are exported into the specified citation format (e.g. BibTex). The details of the bib2py architecture are presented on Fig 2. Note that in order to speed up the bibliographic data retrieval we can process different resources in parallel.

The last step, i.e. importing BibTex into the repository, is performed by the $\Omega - \Psi^{R}$ software. It converts BibTex into a native xml format, which represents publications in the form of a tree-like structure (Fig. 3). The tree nodes represent bibliographic elements, which might be shared between many publications, e.g. authors, books, journals, series.

Each tree node might have its own properties (e.g. title, name, surname). As presented in [2], the repository data structure (based on the JCR data model) is composed of objects (or nodes). Therefore, while importing an XML tree, the tree nodes are stored in the repository independently (like objects), so that they can be reused in linking to other publications, as well as, they can be embedded locally in the “document tree”, so that to allow keeping “historical data”. The splitting of “global data” from the local ones makes possible to modify data on either site, e.g. authors affiliation change or surname change can be performed at a global level of

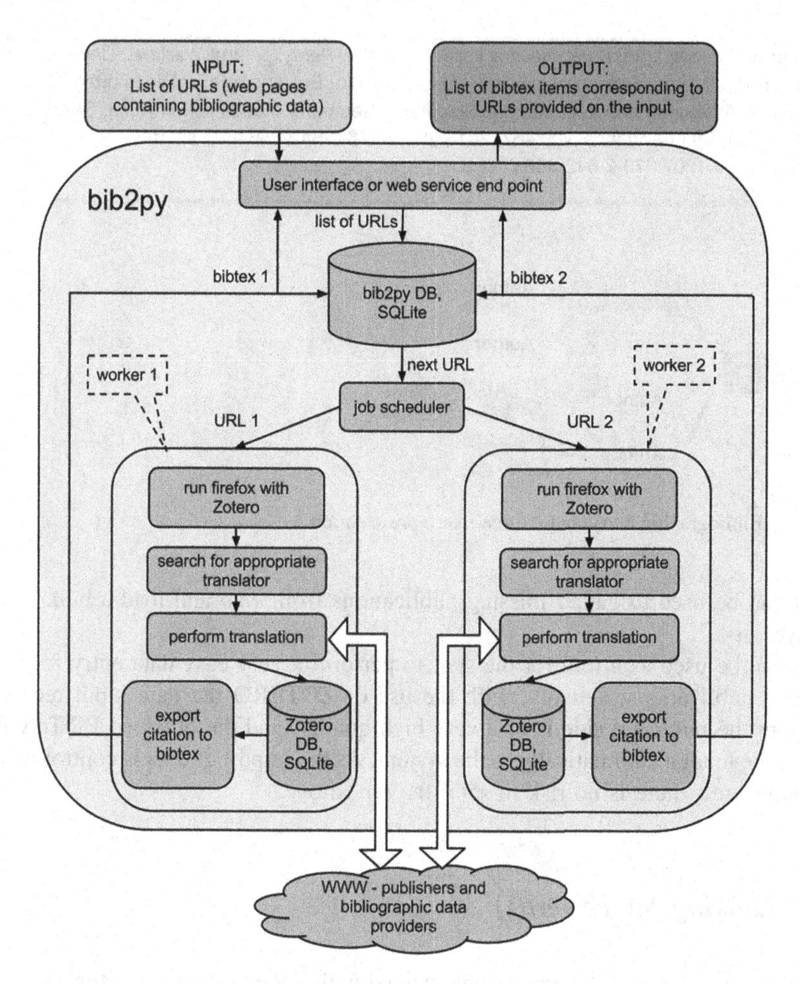

Fig. 2 Zotero as a server—bib2py architecture

the PERSON object, whereas the local author subtree at a given publication tree stores the historic value of the affiliation). During the import, every element of the tree has to be matched against the contents in the repository, then merged and integrated with the existing "objects". We have implemented a number of general methods for the tree matching, as well as specific matching methods for identifying publications and authors. They are presented in more details in [4].

It is worth noting that the presented data acquisition functionality does not cover all the possible functions of the module Ψ^R. This functionality addresses two main needs:

Koperwas Jakub Janusz, Skonieczny Łukasz , Rybiński Henryk , Struk Wacław : Development of a University Knowledge Base, w: Intelligent Tools for Building a Scientific Information Platform: Advanced Architectures and Solutions / Bembenik Robert *[i in.]* (*red.*), Studies in Computational Intelligence, vol. 467, 2013, ISBN 978-3-642-35646-9, ss. 97-110, DOI:10.1007/978-3-642-35647-6_8

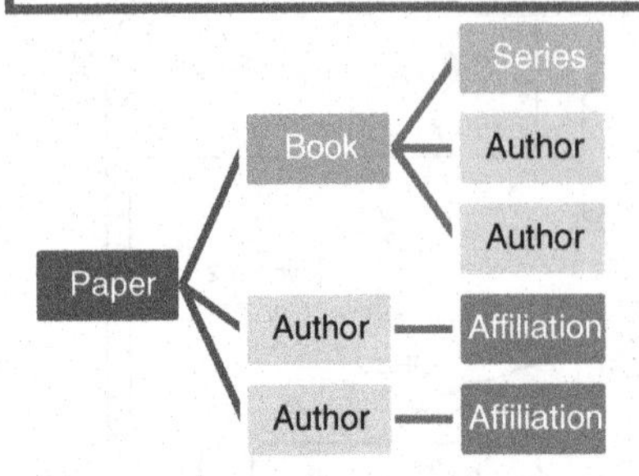

Fig. 3 Bibliographic entry and its tree-like representation

- it can be used to gather missing publications from web and load a bulk to the system
- it can be used as a tool for the staff to perform a very easy data entry for their new publications; actually, with the use of ZOTERO the data input received from the publisher side is of a very high quality, and the obtained BibTeX file can be loaded automatically to the system. As the import of data is controlled for duplicates, there is no risk to spoil the repository.

3.2 Looking for Expert(s)

One of the interesting features, implemented within the system is the functionality of looking for experts, potential candidates for the teams and projects, within the university, as well as for the external users who are looking for the cooperation.

The applied search is not based on what the staff declares or writes to the CV. Instead, it is based on what the knowledge base contains concerning various activities of the researcher. It means that the implementation of the functionality makes sense if the following conditions are satisfied:

1. when the knowledge base represents a kind of "semantic network" by means of variety of interconnected objects, such as publications, patents, projects directed and/or participated, expert's involvement in the conference program committees,[2] etc.

[2] This information is planned for being applied when the acquisition process based on web mining is extended on searching for the involvement into conferences PC.

2. when the knowledge base is as much complete as possible.

The expert search is based on the following components:

- Major fields of the expertise suggestion; there is a module which extracts the main areas of expertise from the publications data (keywords provided by the authors), tags extracted automatically from the papers contents (by semantic processing of the objects, see the next section), the research maps,[3] which are built based on journals subject area, assigned to the papers by the Ontology for Scientific Journals (see [6], and the next section), as well as the researcher's affiliation description, which in turn can also contain area tags, aggregated from the tags of publications, assigned to the "affiliation" unit;
- Search engine: this relies on the $\Omega - \Psi^{R}$ search engine, build on top of the Apache Lucene library; it provides full-text search and a rank on the basis of the well-known TF/IDF measure;
- A ranking module, which is definable by the system administrator with a special scripting language; Such a module can implement a very specific rank algorithm, which can provide specific weights to the particular evaluation elements (e.g. impact factor of the journals, number of citations, special ranks for managing projects, etc.); one can have many ranking algorithms defined, so that the end user can specify the ranking, depending of his/her specific needs. For example, a student looking for his PhD supervisor can provide a search for experts in a given domain but then can sort them by number of supervised theses;
- Result presentation: the results are presented in the form of table with the authors portfolios, containing all the details about the authors activities. The results are enriched with the ranking score bar.

The ranking algorithm can be roughly presented as follows:

- All the knowledge base resources are searched with a specified search phrase (formulated the same way as for searching publications, theses, etc.);
- For all the result items returned, the persons being the publication authors, theses supervisors, project leaders, etc. are extracted and a table for authors is built with all the "achievements" for each person (publications, theses, projects, affiliations, etc.).
- For each "achievement" record for each person in the table the score is calculated according to the selected algorithm;
- For each author the calculated achievements scores are added and the final "authors score" is provided;
- The table is sorted by the final scores and presented as output.

---

[3] For the internal needs, the module presents the tags in the form of a vector, and it visualizes it for the end-users as a word cloud. The word cloud can be "calculated" for the authors, and for the affiliations by aggregating cloud vectors assigned to the papers, supervised theses, run projects etc. This helps the user to pick the most probable area of expertise rather than test the casual phrases.

Fig. 4 A search for experts in a domain and the result page

A simple example for calculating the one "achievement record" score is given below[4]: multiply the following parameters:

- the Lucene relevance score
- the author role weight ("article author", "book author", "book editor", "phd author", "phd supervisor", "bachelor author", "bachelor supervisor", "master author", "master supervisor", "project member", "project leader", "author profile")
- IF of the journal (for the journal papers), IF for the book series (for the chapters or books), a defined value for a book chapter or book if the IF are not known.

In Fig. 4 a screen for searching for an expert and the result screen are provided.

[4] This algorithm causes that publications where the keyword occurred frequently (for example in full text, extracted paper keywords, journal name, journal keyword) are scored higher, moreover the journal impact factor increases the ranking.

3.3 Semantic Processing

Semantic processing in the WUT Knowledge Base is though as a process aiming at enriching the objects by adding semantically meaningful descriptions and/or modifications. The processing is performed on the repository documents (publications, theses, patents, etc.), and it consists in:

1. Scientific domain classification—adding subject tags that classify the objects to given scientific domains;
2. Extraction of semantically meaningful descriptors from the text and assigning them to the "semantic description" field
3. Identifying of synonyms, relevant acronyms, and other close meaning terms based on the semantically meaningful descriptors;
4. looking for translations of the extracted descriptors, and acronyms (for the English texts—to Polish, for the Polish texts—to English).

The target goal of those activities is manifold, the most important reasons are: the processing (1) is mainly used for building maps of research areas for individual researchers, and then, with the use of their research characteristics, for propagating the researchers interest to the affiliation related university units descriptions, and building the research maps for these units (from laboratories to faculties); the processing (2) and (3) are also used for building interest vectors of the researchers (used for building the word clouds), but mainly they are used for improving the search parameters, such as precision and recall; The processing (4) is used to improve multilingual information retrieval.

In the system $\Omega - \Psi^R$ the semantic processing is using two special semantic resources:

1. Ontology for Scientific Journal [6] (in the sequel OSJ), mainly used for performing task (1) above;
2. the Wikipedia resources, mainly used for the tasks (2–4).

Below we describe the usage of the resources in more details.

3.3.1 Publication Classifier

Scientific domain classification is the task consisting in providing a publication with one or more relevant tags, assigning the publication to one or more scientific classes. In $\Omega - \Psi^R$ we have decided to use the OSJ ontology as a classification schema. OSJ is a three level hierarchy, ended with the leaves on the last (fourth) level, which are simply scientific journal titles, so the path in OSJ from the root to a leaf (i.e., a journal title) assigns domain tags to the papers from the journal. The OSJ classification schema, covers 15,000 peer-reviewed scientific journals, and it is translated by more than 22 international experts who volunteered their time and expertise, making the tools available to worldwide scientists. The levels in the OSJ hierarchy are respectively domain, field, and subfield. For example the journal "Artificial organs" in OSJ is classified as follows:

domain–*Applied Sciences*,
field–*Engineering*,
subfield–*Biomedical engineering.*

Clearly, OSJ can be used straightforward for assigning tags to all the papers published in the journals that are contained in the OSJ list. The problem appears for the publications out of the OSJ journal lists, as well as theses, publications being conference papers, chapters in the books, etc. To this end, we have designed and implemented Bayesian classifier, which was trained on the OSJ papers. So, the science domain classifier works as follow for each document:

1. If the document is a paper from the OSJ list, take the tags assigned by OSJ to the journal;
2. Otherwise, use the Bayesian classifier on the available metadata, preferably including title, keywords and abstract, and use the result tags to classify the document.

The classifier provides only the tags from the second level of the ontology.[5] While experimenting, we verified two solutions: one classifier for all the OSJ fields, or a tree of specific classifiers, each node representing a "specialized" classifier. The experiments have shown that the solution with the tree of "specialized" classifiers outperforms one common classifier. The tree of classifiers is a hierarchical structure with the depth 1, where each node represents a specialized classifier. The root is a classifier for the first OSJ level, its children are composed by six classifiers at level 2 (for each OSJ domain there is one fields classifier constructed). An Average accuracy (10-fold cross validation) in a tree mode has reached 85 %, whereas in the case of a single classifier for all the OSJ fields the accuracy was about 60 %.

3.3.2 Semantic Indexing

For implementing semantic indexing, special semantic resources, like domain ontologies, or thesauri, play a crucial role in enhancing the intelligence of Web, mainly by means of enterprise search, and in supporting information integration. Nowadays, most of the semantic resources cover only specific domains. They are created by relatively small groups of knowledge engineers, and are very cost intensive to keep up-to-date with the domain changes. At the same time, Wikipedia has grown into one of the central knowledge sources, maintained by

[5] The first level of OSJ is too general, it has six broad categories: Natural Sciences, Applied Sciences, Health Sciences, Economics and Social Sciences, Arts and Humanities and General, whereas the third level is too detailed, and there is a problem with finding out a training set with a uniform distribution of categories and representative number of examples per category.

thousands of contributors. Bearing in mind that the whole University research domain cannot be covered by one specific domain ontology, we have decided to apply Wikipedia (Polish and English) as a semantic knowledge resource and implement Wikipedia-based semantic indexing of documents in the $\Omega - \Psi^{R}$ system information included in Wikipedia.

Wikipedia contains 30 million articles in 286 languages, including over 4.2 million in the English part. Additionally, it extensively uses ontology called DBpedia, which is a crowd-sourced community effort to extract structured information from Wikipedia and make this information available on the Web. DBpedia allows one to ask for sophisticated queries against Wikipedia, and to link various data sets on the Web to Wikipedia data. Since a few years, both Wikipedia and DBpedia are used in research in many areas involving natural language processing, in particular for information retrieval and information extraction (see e.g. [7–9]). Below we present how we use Wikipedia for the needs of semantic processing in the $\Omega - \Psi^{R}$ system.

Wikipedia articles consist of free text, with structured information embedded, such as *infobox tables* (the pull-out panels that appear in the top right of the default view of some Wikipedia articles, or at the start of the mobile versions), categorization information, images, geo-coordinates and links to external Web pages. This structured information can be extracted and arranged in a uniform dataset, which can be queried. In our project we use Wikipedia in two approaches—*term oriented* and *text oriented.* The first one describes a given term by other words, which come from the Wikipedia content. The module responsible for this task is named WikiThesaurus. The text oriented approach consists on extracting keywords from a text using dictionary of terms (defined as Wikipedia article's titles). The module responsible for extraction is called WikiKeywordsExtractor. Both modules are working on a previously prepared database, built from the Wikipedia content.

WikiThesaurus is a module providing semantic information about an analyzed term. It is inspired by the Milne and Medelyan works [10, 11]. It processes data in two steps. First, given a term extracted from the processed document, it searches for an article with the title equal to, or at least containing this term. Then, the found article is processed in order to extract its labels,[6] senses, translation, or first paragraphs. In order to retrieve related topics to a given term we retrieve all the articles that are siblings in a domain, links-in and links-out. The functionalities provided by WikiThesaurus are as follows:

- Retrieving senses for a term
- Retrieving short description for a term
- Retrieving alternative labels for a term
- Retrieving translations for a term
- Retrieving related topics to a term.

[6] They are manually edited, and assigned to the articles by Wikipedia editors.

WikiKeywordsExtractor is used to extract keywords from a text using knowledge included in Wikipedia. The main idea and some partial solutions come from [12]. The module processes the input text in three steps:

- Text preprocessing
- Terms filtering
- Terms evaluation.

3.3.3 Text Preprocessing

KeywordExtractor processes the input text in such a way that for each term it looks in the Wikipedia preprocessed base for anchors, i.e. links to other articles or proper names. With the found anchors it builds a list of candidate keywords.

3.3.4 Term Filtering

KeywordExtractor filters the candidates in order to remove the irrelevant terms (like articles describing pronunciation), or solve the polysemy problems. The latter one refers to the terms referring to the articles being the disambiguation pages, containing multiple meanings of the given term. For this case, the module has to choose only one meaning. The decision depends on categories matching, and the number of other candidates having link-in/out to a given meaning. The cleaned list of candidate articles is passed to the final phase.

3.3.5 Term Evaluation

KeywordExtractor ranks terms (being candidate titles) using measures based on the common descriptors. Each candidate article is described by a bag of words (bow in the sequel), constructed from titles of related articles (as siblings, links-out, links-in). In the same way we describe the analyzed text—a bag of words is built from terms, which are titles of candidates returned from the filtering phase. Next, each candidate is compared to the bow of the analyzed text. The final measure is a derivative of cardinality of the intersection of the two sets (see Formula 1).

$$Rank\,(candidate,\,text) = \frac{|Bow(candidate) \cap Bow(text)|}{|Bow(text)|} \tag{1}$$

where *Bow(candidate)* is the set of terms describing the candidate article, and *Bow(text)* is the set of terms describing the analyzed text.

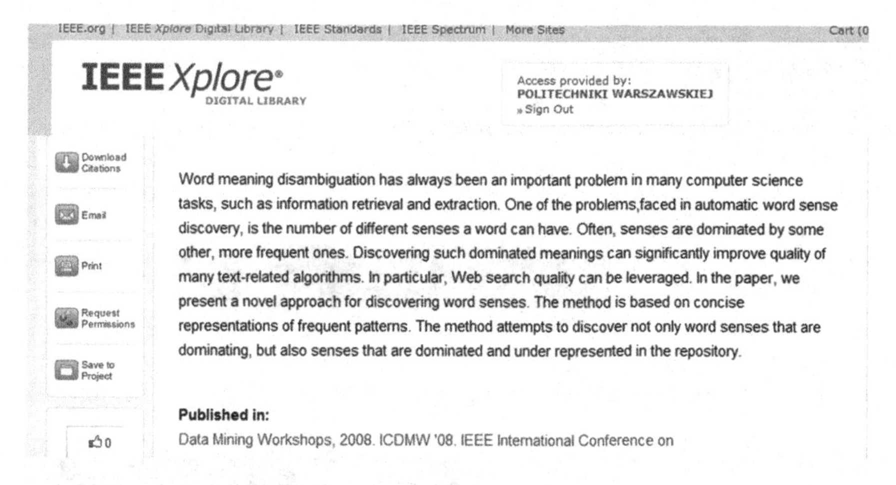

Fig. 5 The example of abstract used in keywords extraction process

Generally, those candidates, which have higher cardinality measure, are ranked higher.[7] Finally, only the candidates ranked above a given threshold are accepted. The titles of those candidates are the final list of keywords.

The example of keywords extraction is demonstrated below. The input text is the abstract from Fig. 5. The extracted keywords are:

`information, information retrieval, web search engine, computer science, word sense, World Wide Web.`

The extracted keywords are used to create research maps which facilitate expert search described in previous section. Publication keywords are aggregated on the authors' level and further on the affiliations' level creating cloud of words which visually depicts research areas of interests for the given person or institution.

4 Flexibility Improvements

As described in [2], the system architecture is strongly oriented for providing high flexibility in defining various aspects of the system applications. A main idea behind this requirement was to provide easiness in defining custom installations and make possible expanding the system in course of its life, without the need of (re)programming the software. In [2] we described how the data structures, views, access rights and reports can be customized. Recently, practically all of these aspects have been improved. Here we will show how the system can be embedded in university and faculty homepages.

---

[7] Other similarity measures are now under tests.

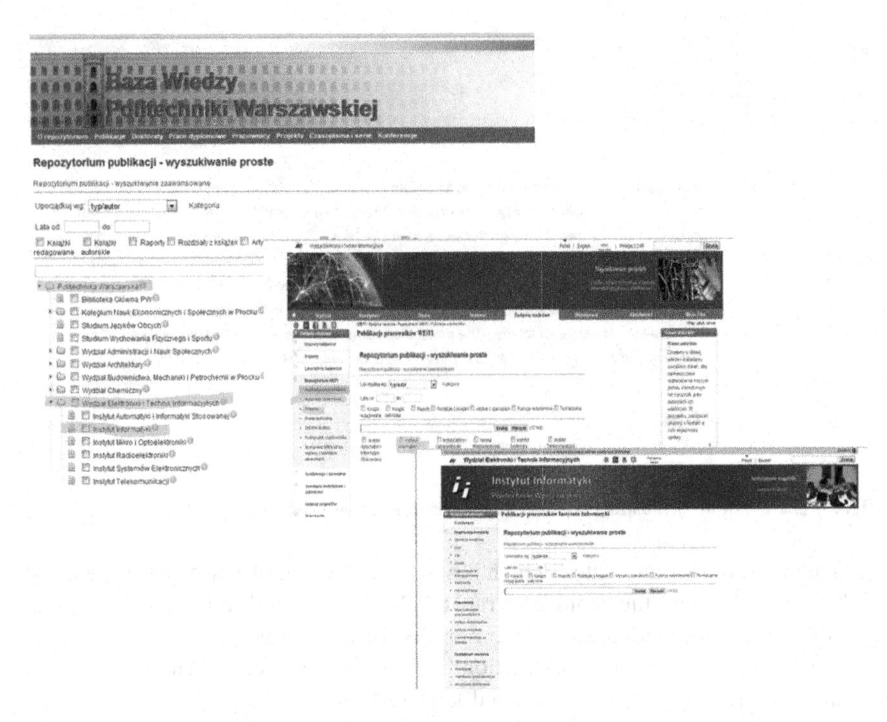

Fig. 6 The main system interface and the embedded screens for local pages

4.1 Embedding of the System

As described earlier in this chapter, the system serves as a centralized repository for the whole university. On the other hand, each faculty or department may want to promote their achievements on their local homepages. In order to make possible view of the knowledge base form the point of view of an university unit, two various solutions have been provided, namely:

- it is possible to embed the system screens within an external system, and by defining a "level" of the university unit all the search screens limits to this unit (and sub-units respectively);
- a number of the system services have been exposed, so that it is possible to "implement" easily an appropriate "window" with all the system functionalities, restricted to the corresponding unit (and sub-units). The functionalities of the system are exposed via both SOAP and REST web services.

The first scenario assumes that the system is included in the faculty home page as an *iframe*. This however does not mean that the "local" system interface will look and behave as the central one. While embedding the system, there is a possibility to declare a value for the "faculty" parameter, which, in turn, provides the system information on:

- how to filter all the data to that specified unit
- which custom skinning (CSS, row formats, etc.) should be used.

This scenario is suitable in most cases identified at the university. It will be further extended to the possibility of defining local custom access control, local reports and other backend user functionalities. The feature is illustrated on Fig. 6. However if the flexibility of embedding mechanism is not sufficient, the REST web services are exposed, and one may define their custom views, GUI, and functionalities, as desired.

5 Conclusions

It turns out that both the WUT Knowledge Base installation and the $\Omega - \Psi^{R}$ system itself are very successful and are attracting many attention. Efforts devoted to the dissemination of scientific achievements start to bring initial results. Usage statistics of WUT Knowledge Base show increasing interests from visitors from all around the world, especially western Europe and North America. It is expected that those effects could be enforced if the authors commitment was increased. Therefore the system will provide in the near future more crowd-sourcing features, in particular making possible uploading full texts by the authors themself, providing social media integration etc.

On the other hand the $\Omega - \Psi^{R}$ system was presented to other Polish universities, who requested the possibility to evaluate the system and are seriously considering using it. Installing the system on more than one university could enable cross-university resources sharing, and thus lead to the development of completely new functionalities increasing synergy in Polish science. For instance, quick access to peers work form the other universities will be possible, looking for a team functionalities can help building project teams consisting of researchers from various universities. Some comparative statistics would also be possible, which would stimulate the competitiveness.

Both the $\Omega - \Psi^{R}$ system, and its implementation as the WUT Knowledge Base seem to significantly contribute to the irreversible global trend of aggregating and sharing the scientific achievements.

References

1. Bembenik R., Skonieczny Ł., Rybiński H., Niezgódka M. (eds.): Intelligent Tools for Building a Scientific Information Platform, Studies in Computational Intelligence, vol. 390, 2012, Springer, ISBN 978-3-642-24808-5, p. 277 doi:10.1007/978-3-642-24809-2

2. Koperwas J., Skonieczny Ł., Rybiński H., Struk W.: Development of a University Knowledge Base. In: Bembenik R. et al (eds.) Studies in Computational Intelligence. In: Intelligent Tools for Building a Scientific Information Platform: Advanced Architectures and Solutions, vol. 467, ISBN 978-3-642-35646-9, (2013), pp. 97–110, doi:10.1007/978-3-642-35647-6_8
3. Adamczyk T., Andruszkiewicz P.: Web Resource Retrieval System for Building a Scientific Information Database. (2013)
4. Andruszkiewicz, P., Gambin, T., Kryszkiewicz, M., Kozlowski, M., Lieber, K., Matusiak, A., Miedzinski, E., Morzy, M., Nachyla, B., Omelczuk, A., Rybinski, H., Skonieczny, L.: Synat/passim-report 4, b11 stage. Technical report (2012)
5. Zotero: http://www.zotero.org/
6. Ontology of Scientific Journal, Classification of Scientific Journals, http://www.science-metrix.com/eng/tools.htm
7. Gabrilovich E., Markovitch S.: Overcoming the brittleness bottleneck using Wikipedia: Enhancing text categorization with encyclopedic knowledge. In: AAAI (2006)
8. Gabrilovich E., Markovitch S., Wikipedia-based semantic interpretation for natural language processing. J. Artif. Intell. Res. **34**, 443–498 (2009)
9. Medelyan O., Milne D., Legg C., Witten I.H.: Mining meaning from Wikipedia. Int. J. Hum. Comput. Stud. **67**(9), 716–754 (2009)
10. Milne D., Medelyan O., Witten I.H.: Mining domain-specific thesauri from Wikipedia: a case study. In: Proceedings of the IEEE/WIC/ACM International Conference on Web Intelligence, Hong Kong, China, 2006, pp. 442–448
11. Milne D., Witten I.H.: An effective, low-cost measure of semantic relatedness obtained from Wikipedia links. In: Wikipedia and Artificial Intelligence: An Evolving Synergy, Chicago, IL, 2008, pp. 25–30
12. Medelyan O.: Human-competitive automatic topic indexing. Ph.D thesis, University of Waikato, Hamilton (2009)

CLEPSYDRA Data Aggregation and Enrichment Framework: Design, Implementation and Deployment in the PIONIER Network Digital Libraries Federation

Cezary Mazurek, Marcin Mielnicki, Aleksandra Nowak, Krzysztof Sielski, Maciej Stroiński, Marcin Werla and Jan Węglarz

Abstract During 3 years of the SYNAT project PSNC was working on the architecture for aggregation, processing and provisioning of data from heterogeneous scientific information services. The implementation of this architecture was named CLEPSYDRA and was published as an open source project. This chapter contains overview of the CLEPSYDRA system design and implementation. It also presents the test deployment of CLEPSYDRA for the purpose of the PIONIER Network Digital Libraries Federation, focusing on aspect such as agent-based data aggregation, data normalization and data enrichment. Finally the chapter includes several scenarios for future use of the system in the national and international context.

C. Mazurek · M. Mielnicki · A. Nowak · K. Sielski ·
M. Stroiński · M. Werla (✉) · J. Węglarz
Poznań Supercomputing and Networking Center, Ul. Z. Noskowskiego
12/14, 61-704 Poznań, Poland
e-mail: mwerla@man.poznan.pl

C. Mazurek
e-mail: mazurek@man.poznan.pl

M. Mielnicki
e-mail: marcinm@man.poznan.pl

A. Nowak
e-mail: anowak@man.poznan.pl

K. Sielski
e-mail: sielski@man.poznan.pl

M. Stroiński
e-mail: stroins@man.poznan.pl

J. Węglarz
e-mail: weglarz@man.poznan.pl

R. Bembenik et al. (eds.), *Intelligent Tools for Building a Scientific Information Platform: From Research to Implementation*, Studies in Computational Intelligence 541,
DOI: 10.1007/978-3-319-04714-0_17,

Keywords Data aggregation · Data enrichment · Metadata processing · REST architecture · Distributed systems

1 Introduction

Since 1999 PSNC is active in the field of digital libraries, working towards creation and development of digital libraries infrastructure dedicated to research and educational use [1]. Cooperation with many cultural and scientific memory institutions lead to creation of tens of digital libraries in Poland, starting with the Digital Library of Wielkopolska (http://www.wbc.poznan.pl/) in 2002 [2].

In 2007, with around 15 digital libraries and 80,000 digital objects PSNC publicly started a new service, metadata aggregator allowing to search descriptions of objects from distributed sources [3]. In 2012 this aggregator, called Digital Libraries Federation and available at http://fbc.pionier.net.pl/, included in its indexes over a million of objects from around 80 different sources.

Increasing number of sources, number of objects and the heterogeneity of data and data access interfaces, was a key motivation to start research and development works focused on a new architecture for aggregation, processing and provisioning of data from heterogeneous, distributed information services. These works were undertaken by PSNC in the frame of the SYNAT project, funded by the National Centre for Research and Development.[1] Beside design of a new architecture, works conducted during the project led to prototype implementation and deployment of this architecture, in the environment of the Polish Digital Libraries Federation.

This chapter in the next section introduces briefly the mentioned architecture. Section 3 describes several details of its implementation—it presents open source CLEPSYDRA software framework. Section 4 is a report from the test deployment of CLEPSYDRA, and Sect. 5 presents the potential usage scenarios of this deployment in the future. The chapter ends with a summary, including several remarks about the development and deployment of the presented system.

2 Overview of the Architecture for Data Aggregation, Processing and Provisioning

The design of new architecture for aggregation, processing and provisioning of data from distributed, heterogeneous information services was preceded by a set of requirements which the new system should fulfill. The following key functional requirements were identified [4]:

[1] See Acknowledgements section at the end of this chapter for more details.

- Store and access large amounts of heterogeneous data records, supporting organization of the data in the context of objects, records (describing objects) and collections (containing objects), and also allowing selective and incremental access to the data.
- Aggregation of data from heterogeneous sources, assuming many access protocols and data formats, active and passive nature of the sources, and their volatility.
- Processing of aggregated data, based on small operations which can be combined in larger chains, resulting in automated capabilities to map data between different formats, clean, normalize and enrich the data in many ways.

The non-functional aspects of the system were based mostly on the experiences in administration of the Digital Libraries Federation, and emphasized aspects like scalability, robustness, availability and interoperability as well as maintainability, extensibility and operability. Also it was assumed that the three functional areas described above (aggregation, processing, access) should be independent one from another, allowing to perform maintenance works in any two of three mentioned areas without influencing the third one.

The general schema of the final architecture is presented on Fig. 1. It consists of three main functional layers (storage, aggregation, processing) and asynchronous messaging bus. Two additional elements visible on the diagram represent external components—data sources which are aggregated and applications or services which re-use the aggregated and processed data.

The data aggregation layer is responsible for periodic communication with aggregated systems and the delivery of new or updated information to the storage layer. This functionality is delegated to a number of agents—small independent services specialized in specific types of data sources. Agents register themselves in a dedicated managing component which assigns data sources to them and monitors whether the agents are alive and working. Agents deliver data directly to the storage layer.

The storage layer is the most critical element of the system, therefore its implementation must be very scalable and have the availability as good as possible (ideally 100 %). This layer is responsible for data storage and data access and it is the part of the system which allows data transfer between all other layers. This layer uses the messaging bus to notify all interested parties (including other layers of the system) about changes in the data (create, update and delete).

The data processing layer reacts on notifications from the storage layer, received via the messaging bus. Such notifications may initiate predefined processing rules, which define how data should be processed e.g. upon new or updated data in schema X from source Y, execute the following chain of processing operations. As mentioned above, the processing is designed as a set of small reusable operations, working on entire records (like mapping of records between particular formats—HTML to XML—or schemas—MARC to PLMET) or on smaller pieces of data (normalization of date, named entity recognition, etc). The next section describes the prototype implementation of the described architecture.

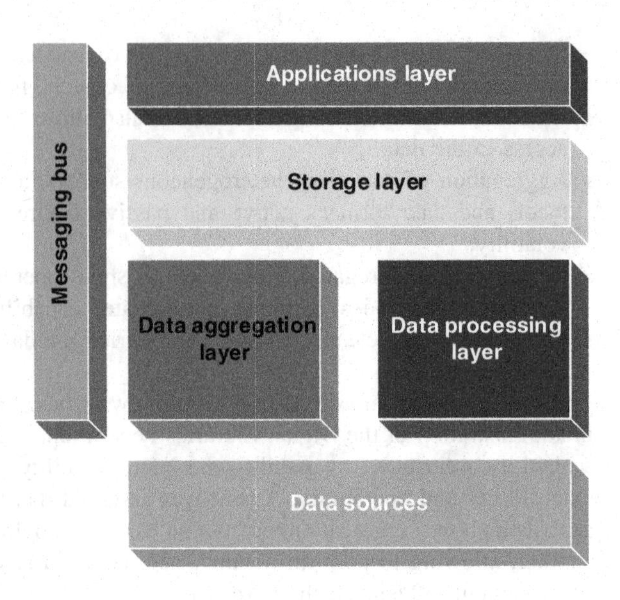

Fig. 1 Layers of the architecture for data aggregation, processing and provisioning

3 CLEPSYDRA Framework

The name of the CLEPSYDRA framework was inspired by the results of recent research presenting the architecture of the internet protocols shaped in the hourglass form, with the IPv4 being in the neck [5]. In this context, the CLEPSYDRA framework as depicted on Fig. 2, enables the flow of the data from heterogeneous data providers to consumers. The top bulb represents the aggregation layer, which extracts data. The data goes through storage and processing layers and falls into lower bulb where consumers can reuse it.

The implementation of CLEPSYDRA consists of the following components and modules:

- CLEPSYDRA Aggregation
  - –Agents—responsible for communication with data sources
  - –Agents Manager—responsible for the agent's management
  - –Sources Manager—holds information about data sources registered in the system
  - –Schemas Manager—holds information about data schemas registered in the system
- CLEPSYDRA Storage—responsible for storage and access to data, based on NoSQL databases.
- CLEPSYDRA Processing
  - –Processors—perform particular data processing information
  - –Processing Manager—manages the processing of new and modified data records.

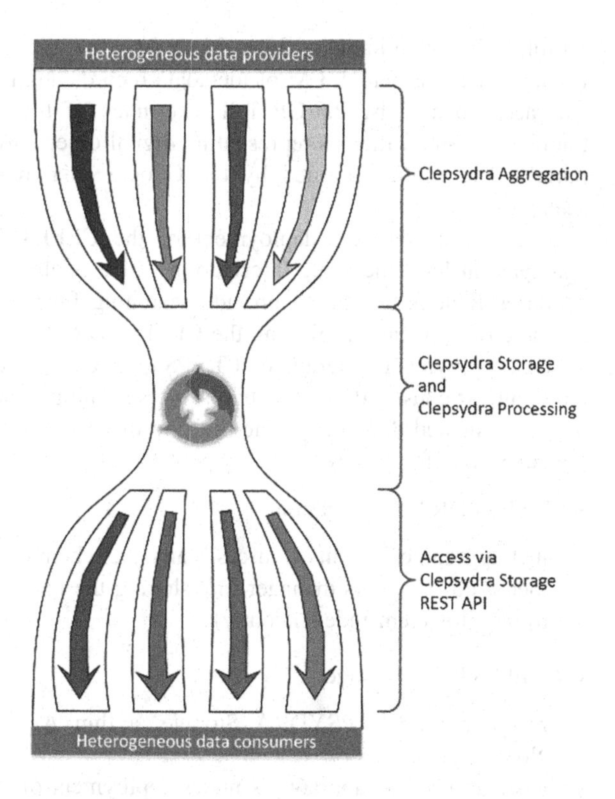

Fig. 2 CLEPSYDRA components presented in the hourglass layout

The implementation of CLEPSYDRA was done in Java EE and therefore the asynchronous messaging layer was based on the Java Message Service (JMS). The CLEPSYDRA storage layer was based on the Apache Cassandra NoSQL database. The entire system was designed and implemented in the REST architectural approach, providing high interoperability and scalability based in the stateless HTTP communication. The system has been released as an open-source software, on the Apache License ver. 2.0 and its available at http://fbc.pionier.net.pl/pro/clepsydra/. The next chapter provides the description of the test deployment of this system in the environment of the PIONIER Network Digital Libraries Federation.

4 Test Deployment of CLEPSYDRA in the PIONIER Network Digital Libraries Federation

Digital Libraries Federation (DLF) portal which is available at http://fbc.pionier.net.pl/ operates on a software which was initially designed in 2006 [6]. Since that time the amount of data processed by this system has increased more than

10 fold.[2] In the middle of June 2013 the core of the old system was Oracle Database with around 1.3 M of metadata records, taking 4.7 GB of storage. This was accompanied by 2.5 GB full text index of the metadata and 120 GB of thumbnails (small images representing digital objects available via the federation portal). The traffic recorded by the Google Analytics service is presented in Table 1.

The aim of the test deployment of the CLEPSYDRA framework was to aggregate at least the same information which is already available in the Digital Libraries Federation and to provide searching functionality on top of it, as an example of application utilizing the CLEPSYDRA Storage layer.

The deployment diagram of CLEPSYDRA is presented on Fig. 3. The test environment consisted of 11 virtual servers running on a high performance cluster with a dedicated disk array. The division of CLEPSYDRA components between machines was the following:

- CLEPSYDRA Aggregation
  - –met-aggregator: agents, sources manager, schemas manager
  - –met-storage: agents manager (not sharing the machine with agents, to be able to monitor them independently)
- CLEPSYDRA Storage
  - –met-storage: CLEPSYDRA Storage, a thin API component wrapping the NoSQL database
  - –cassandra1—cassandra6: six nodes deployment of Apache Cassandra NoSQL database, with replication factor set to 3.
- CLEPSYDRA Processing
  - –met-processor: processors, processing manager.

Beside the CLEPSYDRA components mentioned above, two servers were utilized for the prototype version of the DLF Portal, using the data gathered in CLEPSYDRA deployment. One of these machines is dedicated for building full text search index of the aggregated metadata, using the Apache Solr software. MQ Broker component which is located on the met-aggregator server provides the asynchronous messaging environment, mentioned earlier. This communication approach is used not only for communication between CLEPSYDRA components. A dedicated module, Metadata Indexer, also subscribes for notifications and on this basis feeds the Solr-based indexing module with proper data.

The beta version of DLF Portal consists of the portal itself, two open interfaces for external applications (persistent identifiers resolver—ID Gateway—and search API—Open Search), replica of the Solr index, strictly for searching purposes, and

[2] See http://fbc.pionier.net.pl/stats/owoc/owoc_all_n_en.png for up-to-date details.

Table 1 Digital Libraries Federation portal traffic statistics according to Google Analytics

	2011	2012	2013
Visits/year	749,065	1,014,288	(est.) 1,250,000
Pageviews/year (sum)	3,119,635	4,157,335	(est.) 4,327,129
Pageviews/day (avg)	8,547	11,390	11,855
Pageviews/day (max)	14,926	17,175	17,841
Pageviews/hour (avg)	357	475	494
Pageviews/hour (max)	1,292	1,894	1,728

the Cassandra Storage component which helps to access binary data stored in the Cassandra cluster.

After few months of activity, in the middle of June 2013, the test environment aggregated 17.7 M of metadata records. The six Cassandra nodes occupied together over 1 TB of storage space, distributed as follows:

- Node1: 123.05 GB
- Node2: 204.15 GB
- Node3: 152.34 GB
- Node4: 220.55 GB
- Node5: 191.81 GB
- Node6: 148.4 GB.

Out of these 17.7 M of objects representations, 2.1 M were thumbnails and 15.6 M were metadata records (mostly in MARC or Dublin Core-based schemas). There were no major technical problems with aggregating and processing of the data, also the implementation of test application was relatively simple because of well-designed CLEPSYDRA API and the availability of asynchronous notifications. The prototype version of DLF portal is one of the five CLEPSYDRA Usage scenarios which are described in the next section.

5 Performance Tests

A suite of tests was performed in order to verify whether the developed system will be able to handle the traffic generated by users of Digital Libraries Federation portal and to process new data records and data updates made daily by the aggregated data source. These tests covered data storage and data processing layers.

5.1 Data Storage Layer Tests

The following functional aspects of the Data Storage Layer were covered by performed tests:

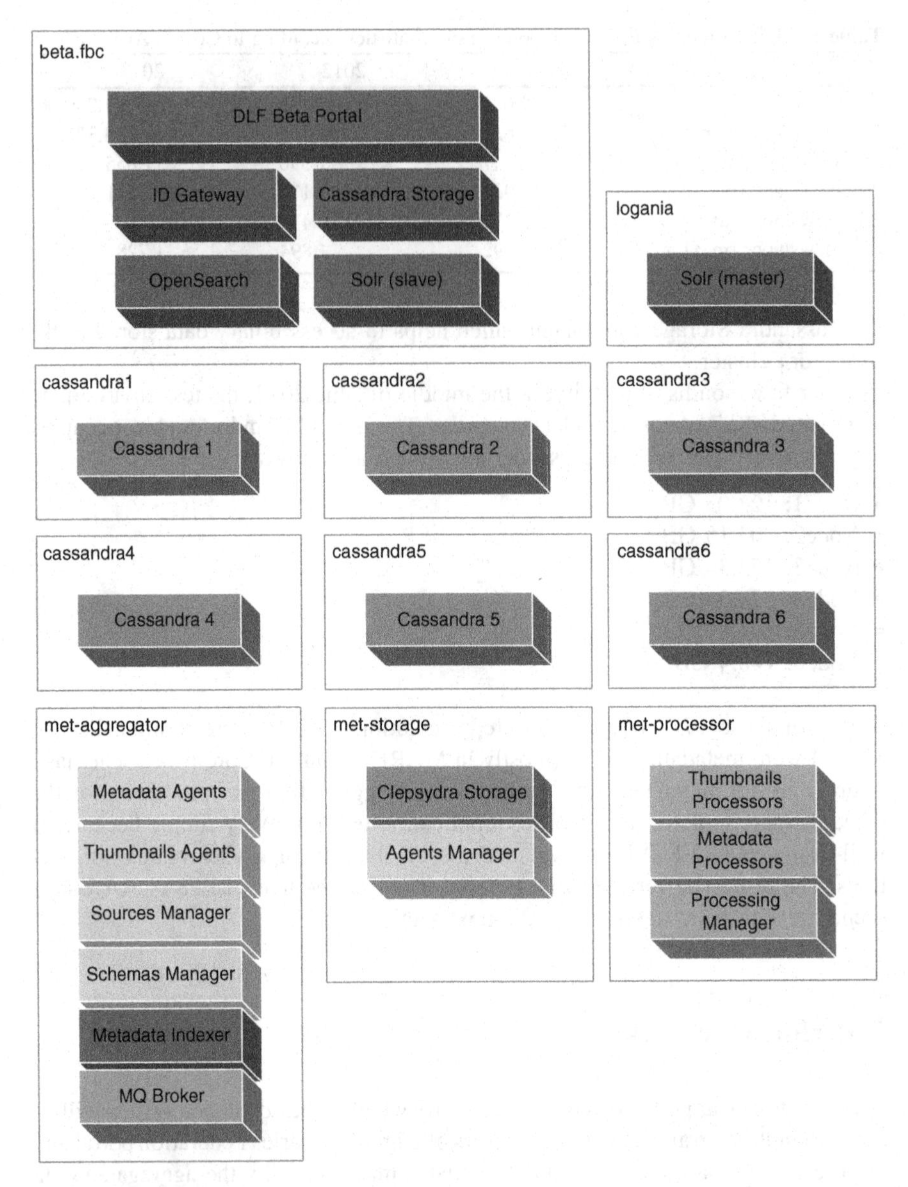

Fig. 3 Deployment diagram of the CLEPSYDRA framework in the environment of the PIONIER Network Digital Libraries Federation

- **Data access**: 10 parallel clients were set up to access random data records with the highest possible speed (there was no programmatic delay between consecutive requests). Each client was accessing the same single instance of Clepsydra Storage component (see Fig. 3).

- **Data deletion**: single client requested deletion of a large set of data records from particular source.
- **Data insertion**: 10 parallel clients were set up to insert different data records with the highest possible speed (there was no programmatic delay between consecutive requests). Each client was accessing the same single instance of Clepsydra Storage component (see Fig. 3).

The performance of the developed system measured during the above operations is presented in Table 2.

5.2 Data Processing Layer Tests

In order to test the performance of data processing the system was configured to do a bulk transformation of data records, consisting of three consecutive data conversions. Table below depicts the overall performance of this operation. It also includes the time necessary to access the data records which have to be processed and the time of storing of the output data record.

Operation	Avg. throughput (records per hour)	Avg. throughput (records per minute)	Avg. throughput (records per second)
Data transformation	232,883	3,881	65

5.3 Hardware Load

During these tests the CPU load on the server which was hosting the Metadata Storage component was almost all the time at the level of 80 %. The server which was hosting all components of the Data Processing Layer had also high load, with maximum going up to 90 %.

The average outgoing network data transfer was around 4 Mbps (with maximum of around 7 Mbps). The average incoming network data transfer was also around 4 Mbps (with the maximum of around 5 Mbps). On the Cassandra database nodes which were serving the Metadata Storage component the following maximum values were measured:

- Maximum outgoing data transfer on a node: 870 kbps
- Maximum incoming data transfer on a node: 660 kbps
- Maximum disk usage: 40 % disk busy time.

Table 2 Average performance of data storage layer operations

Operation	Avg. throughput (records per hour)	Avg. throughput (records per minute)	Avg. throughput (records per second)
Data access	654,060	10,901	182
Data deletion	1,217,927	20,299	338
Data insertion	302,997	5,050	84

5.4 Tests Results Interpretation

The general conclusions which can be drawn from these tests are following:

- The new system in the tested deployment is fast enough to serve data for the purpose of Digital Libraries Federation portal. The portal had peaks of up to 2,000 pageviews per hour and the system is ready to serve over 600,000 data records per hour.
- The new system in the tested deployment is fast enough to process data aggregated from Digital Libraries Federation data sources on an ongoing basis. The usual increase of new records is around 3,000 per day and the data insertion and processing rates are much higher.
- The new system in the tested deployment is fast enough to process data in cases when new large data source will be connected to it. Largest Polish digital libraries reached in the middle of 2013 the size of 200,000 digital objects, and test results show that the system should be able to store and process such amount of data in few hours. Of course in such case the limitation will be the speed of data acquisition from the new data source.
- The new system in the tested deployment is fast enough to expose data for Europeana, which requires access to data dumps once per month. Current data dump, including 1.1 M records from institutions which are cooperating with Europeana, should be obtained from the system in around 2 h.
- The comparison of performance of data access and data insertion versus data processing shows that the overall processing time (which includes data access and data insertion) is faster than could be expected, basing on access and insertion times. The reason for that is the fact that the processing is done internally, within the cluster of servers on which CLEPSYDRA system is installed. This eliminates the delay caused by data transfer to and from the client, which obviously influenced access and insertion tests.
- In case of very high data access traffic the potential bottleneck will be the single instance of Clepsydra Storage component, as during the tests higher load was observed at this component than on Cassandra database nodes. This is not very surprising, and in such case the solution will be to deploy additional servers with new instances of Clepsydra Storage.

Tests described above do not measure precisely all performance aspects of the system. Such complex tests are planned for a later stage when the system will be prepared to replace old Digital Libraries Federation infrastructure, which is planned for the end of 2014. Because of the complexity and uniqueness of the system it was also impossible to make any kind of comparative tests of CLEPSYDRA performance.

6 CLEPSYDRA Usage Scenarios

A natural step following the design and implementation of the new architecture for aggregation, processing and provisioning of data was to define possible usage scenarios of the developed system. In the context of the SYNAT project and the Digital Libraries Federation five scenarios were defined. They are listed below:

1. Integration of CLEPSYDRA with the SYNAT Platform developed by ICM (University of Warsaw), with the main aim of providing the aggregated data to the INFONA portal (http://www.infona.pl/) developed as the main interface to the platform resources. This scenario will be implemented as in collaboration between ICM and PSNC, during continuation of the SYNAT project.
2. Construction of semantic Knowledge Base and use of semantic web technologies to provide new ways to explore and use the aggregated data. This scenario was deeply elaborated in the A10 stage of the SYNAT project, also led by PSNC.
3. Deployment of CLEPSYDRA in the environment of PIONIER Network Digital Libraries Federation to provide more data with higher quality, and create new generation of the Digital Libraries Federation Portal. This scenario was already initially explored, as described in the previous chapter, and will be developed further, to finally replace the old DLF system.
4. Reuse of the metadata processing modules to support the creation of high-quality metadata on the level of data sources. This scenario assumes that digital library software providers will be interested in utilizing CLEPSYDRA capabilities to provide features like date or language normalization, or even named entity detection on the level of particular digital libraries. PSNC as a provider of the dLibra Digital Library Framework (http://dlibra.psnc.pl/) is interested in implementing this scenario in the future, when the CLEPSYDRA deployment in the Digital Libraries Federation environment will be available and maintained as a production-quality service.
5. Creation of thematic cultural heritage portals based on the selection of content aggregated in CLEPSYDRA e.g. portal about collections of Polish museums. This scenario is now investigated in the cooperation with selected Digitisation Competence Centers of the Polish Ministry of Culture and Cultural Heritage.

The above scenarios show high potential of the CLEPSYDRA framework and its deployment focused on aggregation of data from Polish cultural and scientific information services. As mentioned above, some of the scenarios are already in progress, other will be investigated in the future. Another interesting possibility is to reuse CLEPSYDRA as a basis or at least component of the European cultural heritage data cloud which is developed in the frame of the Europeana Cloud project,[3] in which PSNC is one of the main technical partners.

7 Conclusions

This chapter introduced CLEPSYDRA Data Aggregation and Enrichment Framework. The motivation for its creation comes from the infrastructure of digital libraries developed in the PIONIER Network since 2002. Increasing amount of data available in Polish digital libraries and its arising heterogeneity and distribution caused that the initial design of the aggregation service made in 2006 became outdated. Works undertaken in the SYNAT project allowed to design and implement new architecture and to deploy it in the test environment.

This test start of the new system empirically proved that the distributed, modular, service-based architecture compliant with the REST approach was good choice. The prototype implementation of Digital Libraries Portal and its metadata indexing service confirmed that strong separation of aggregation, processing and storage layers enables applications based on CLEPSYDRA to be highly available during high load of aggregation or processing activities.

Problems which were observed during this test phase were mostly associated with the nature of aggregated data. The most burdensome aspect of administration of the CLEPSYDRA deployment was related to the need of updating (or at least verification) of metadata processing rules upon each change of metadata schema in particular data source. Initial assumption that the metadata schemas used by data sources should be relatively stable came out to be not so safe. Automated mechanism for detection of changes in sources metadata schemas was developed, but the verification/update of metadata processing rules still remains manual procedure requiring human participation. Minimization of this effort can be one of the aspects of future works.

Normalization of metadata came out to be another tedious activity—especially in case where normalization of particular metadata element (like "language" or "type") was very useful for the end user interface, but there was no standardized list of values. For example mapping of "language" was relatively easy, as the output was limited to the set of values listed in the ISO 639-3 standard. The mapping of "type" element appeared to be much more problematic. It was decided to have two types of mapping in this case. One was producing four unique values:

[3] Project website: http://pro.europeana.eu/web/europeana-cloud.

TEXT, IMAGE, AUDIO and VIDEO (as defined in ese:type element of the Europeana Semantic Elements metadata schema [7]). The second mapping was supposed to generate more descriptive and precise types, but turned out to be impossible to implement, as the input data contains several typologies mixed in one metadata element. For example one data source uses simple terms like "book", "magazine" and another is more precise and uses terms like "novel", "story", "short story", "monthly", "biweekly", "daily". Mapping to the most general terms is the only possible solution here, but even that is not always simple, as for example word "journal" in Polish can be understood as a daily newspaper, but also as a diary. The proper solution requires close cooperation with experts coming from data sources domains (libraries, museums, archives, ...) and was also partially undertaken in the SYNAT project, especially in the cooperation with the National Museum in Warsaw. But the complete solution is still a work in progress.

Finally some practical obstacle and remark: from the beginning it was known that the prototype implementation of CLEPSYDRA will be released as an open source software, but the decision about the license was not made in the beginning, but left to the stage where the software will be mature enough to be publicly released. And at this stage it came out that not all external dependencies were legally compliant with the open source license we wanted to use. As a result it was necessary to carefully review licenses of tens of dependencies and make sure that we are not violating them. This experience clearly showed that licensing decision should be made at the very beginning, and then respected (or verified) each time when there is a need to add a new external dependency.

The future works related to CLEPSYDRA framework will be connected to scenarios listed in Sect. 5 of this chapter. Beside extensive stress tests are planned to give more information about the performance limits of the test deployment and to detect potential efficiency bottlenecks of the designed architecture.

Acknowledgements Works described in this chapter are financed as a part of the SYNAT project, which is a national research project aimed at the creation of universal open repository platform for hosting and communication of networked resources of knowledge for science, education, and open society of knowledge. It is funded by the Polish National Center for Research and Development (grant no SP/I/1/77065/10).

References

1. Mazurek, C., Parkoła, T., Werla, M.: Distributed digital libraries platform in the PIONIER network. In: Red. Gonzalo, J., Thanos, C., Verdejo, M.F., Carrasco, R.C. (eds.) Lecture Notes in Computer Science, LNCS vol. 4172. Springer, Berlin (2006). ISBN 3-540-44636-2. ECDL 2006: Research and Advanced Technology for Digital Libraries, Alicante, Spain, 17–22 Wrzesień, pp. 488–491 (2006)
2. Górny, M., Gruszczyński, P., Mazurek, C., Nikisch, J.A., Stroiński, M., Swędrzyński, A.: Zastosowanie oprogramowania dLibra do budowy Wielkopolskiej Biblioteki Cyfrowej. SBP KWE, Warszawa (2003). ISBN 83-915689-5-4. Internet w bibliotekach II, Wrocław, 23–26 Wrzesień (2003)

3. Lewandowska, A., Mazurek, C., Werla, M.: Enrichment of European digital resources by federating regional digital libraries in Poland. In: Research and Advanced Technology for Digital Libraries, Proceedings of the 12th European Conference (ECDL 2008), LNCS vol. 5173, pp. 256–259. Aarhus, Denmark, 14–19 Września (2008)
4. Mazurek, C., Mielnicki, M., Nowak, A., Stroiński, M., Werla, M., Węglarz, J.: Architecture for aggregation, processing and provisioning of data from heterogeneous scientific information services. In: Bembenik, R., Skonieczny, Ł., Rybiński, H., Kryszkiewicz, M., Niezgódka, M. (eds.) Intelligent Tools for Building a Scientific Information Platform: Advanced Architectures and Solutions. Springer, Berlin, pp. 529–546. ISBN 978-3-642-35646-9 (2013)
5. Akhshabi, Saamer, Dovrolis, Constantine: The evolution of layered protocol stacks leads to an hourglass-shaped architecture. SIGCOMM-Comput. Commun. Rev. **41**(4), 206 (2011)
6. Mazurek, C., Stroiński, M., Werla, M., Węglarz, J.: Distributed services and metadata flow in the polish federation of digital. In: International Conference on Information Society (i-Society), pp. 39–46. ISBN 978-0-9564263-8-3 (2011)
7. Europeana Semantic Elements Specification Version 3.4.1. http://pro.europeana.eu/documents/900548/dc80802e-6efb-4127-a98e-c27c95396d57

Author Index

R. Bembenik et al. (eds.), *Intelligent Tools for Building a Scientific Information Platform: From Research to Implementation*, Studies in Computational Intelligence 541,
DOI: 10.1007/978-3-319-04714-0,

Printed in the United States
By Bookmasters